AF356830

Liposomal and Ethosomal Gels: From Design to Application

Liposomal and Ethosomal Gels: From Design to Application

Editors

Maddalena Sguizzato
Rita Cortesi

Basel • Beijing • Wuhan • Barcelona • Belgrade • Novi Sad • Cluj • Manchester

Editors
Maddalena Sguizzato
Department of Chemical,
Pharmaceutical and
Agricultural Sciences
University of Ferrara
Ferrara
Italy

Rita Cortesi
Department of Chemical,
Pharmaceutical and
Agricultural Sciences
University of Ferrara
Ferrara
Italy

Editorial Office
MDPI
St. Alban-Anlage 66
4052 Basel, Switzerland

This is a reprint of articles from the Special Issue published online in the open access journal *Gels* (ISSN 2310-2861) (available at: www.mdpi.com/journal/gels/special_issues/Liposomal_Gels).

For citation purposes, cite each article independently as indicated on the article page online and as indicated below:

Lastname, A.A.; Lastname, B.B. Article Title. *Journal Name* **Year**, *Volume Number*, Page Range.

ISBN 978-3-7258-1436-7 (Hbk)
ISBN 978-3-7258-1435-0 (PDF)
doi.org/10.3390/books978-3-7258-1435-0

Contents

Editorial

Liposomal and Ethosomal Gels: From Design to Application

Maddalena Sguizzato * and Rita Cortesi

Department of Chemical, Pharmaceutical and Agricultural Sciences (DoCPAS), University of Ferrara, 44100 Ferrara, Italy; crt@unife.it
* Correspondence: sgzmdl@unife.it

The use of lipid-based nanosystems for topical administration represents an innovative "green" approach, being composed of materials, defined as GRAS (generally recognized as safe), characterized by low toxicity, biocompatibility, and biodegradability [1]. For instance, vesicular lipid-based delivery systems, such as liposomes [2], ethosomes [3], and niosomes [4], are good candidates for topical applications of active compounds for their dimensions, as well as their similarities, with the epidermal lipids becoming useful in the treatment of different pathologies, mostly in cutaneous diseases [5,6]. The main obstacle for the use of lipid-based nanosystems is their fluidity, which can hamper their in situ permanence. Hence, the possibility to incorporate them into polymeric gelled matrices can enhance the spreadability and adhesiveness onto the skin, mucosae, or other topical districts for biomedical, pharmaceutical, or cosmetic purposes [7–11].

This Special Issue of the Gels journal, entitled "Liposomal and Ethosomal Gels: From Design to Application", includes twelve articles describing the production, characterization, and application of different nanocarriers included in gel systems, providing an interesting overview on gelled nanosystems from their qualitiy by design studies to their in vivo activity evaluation. Different administration routes have been considered, such as the cutaneous, buccal, ocular, and auricular routes, offering a wide spectrum of applications.

Notably, in the first research paper, the optimization of a PEGylated liposomal formulation by the quality by design (QbD) technique has been assessed to investigate the influence of lipid concentration on particle size, encapsulation efficiency, and in vitro release. The selected PEGylated Brucine liposomal emulgel based on jojoba oil showed an improved skin permeation, reflected in a significant anti-inflammatory effect (contribution 1).

An alternative lipid-based formulation, described by Khan et al. for the treatment of atopic dermatitis, revealed that the gel formulation embedding lipid nanoparticles by means of glycerol and sodium alginate maintained the same drug permeation level of the un-thickened nanoparticles, increasing the drug retention thanks to the gel's bioadhesive properties (contribution 2).

Vesicular nanosystems, namely, niosomes, containing ginger extract have been incorporated within the emulgel to obtain a transdermal delivery (contribution 3). The optimization of both niosomes and emulgel productions, obtained by applying QbD technique, demonstrated satisfactory physico-chemical characteristics and formulation enhancements of both skin permeability performance and anti-inflammatory effects due to the synergistic interaction between sesame oil and ginger extract from niosomes–sesame oil-based emulgel. In addition, Shabery et al. (contribution 4) incorporated a niosomal formulation for skin application within the patented palm oil base Hamin-C®. In vitro drug permeability was assessed by a Strat-M™ membrane, revealing higher permeability of both lidocaine and prilocaine when formulated with a cold process.

Niosomes have been also considered for the encapsulation of natural phenolic compounds by using alternative xanthan gum or the thermoreversible polymer Poloxamer 407 as gelling agents to favor cutaneous applications (contribution 5). The preformulative study investigated the influence of the non-ionic surfactant and the hydration medium on the morphology, encapsulation efficiency, and in vitro release. The results showed that both the

Citation: Sguizzato, M.; Cortesi, R. Liposomal and Ethosomal Gels: From Design to Application. *Gels* **2023**, *9*, 887. https://doi.org/10.3390/gels9110887

Received: 6 November 2023
Accepted: 7 November 2023
Published: 9 November 2023

hydration phase and the type of thickening agent were able to control the drug diffusion. Niosomal gels based on xanthan gum revealed higher retention on the application site and no irritative reactions during in vivo patch tests in 95% of cases.

Among vesicular nanosystems, ethosomes represent an intriguing tool for topical applications. They have been considered for the encapsulation of rosehip extract to improve the stability of bioactive compounds, and their inclusion into hyaluronic acid gel allows for obtaining a suitable formulation for cosmetic use (contribution 6). The produced extract-loaded ethosomes showed small sizes, low polydispersity, and good entrapment efficiency and great stability during the time. The permeation study was performed through the artificial biomimetic barrier Permapad®, confirming the increasing of extract permeability when delivered by ethosomes. Lastly, the in vitro release studies, conducted on ethosomes and ethosomal–gel formulations, showed that encapsulation delayed the release of the extract.

In addition to the common use of gel formulations for skin applications, buccal administration has been considered as an innovative application site for local or systemic effects.

Another effective emulgel, optimized by Central Composite Design (CCD) with the QbD method, was described by Iyer and colleagues. (contribution 7). The characterization of clove/cinnamon extracts-loaded emulgels showed globule sizes around 321 nm, contents of each extract around 96%, and good viscosity, spreadability, and extrusion properties. Additionally, the total release of loaded drugs demonstrated efficient anti-fungal potential in counteracting Candida-associated denture stomatitis. Clinical trials confirmed the effectiveness of the treatment with better taste acceptability and no side effects.

The above-mentioned advantages of gel formulations (e.g., adhesiveness, spreadability, and sustained release) can be considered also to achieve the increasing of drug bioavailability. A carboxymethyl cellulose/hydroxypropyl cellulose (CMC/HPC) composite mixture was selected to produce an innovative formula for buccal applications (contribution 8). Carvedilol-loaded bilosomes, with spherical shapes and being 217 nm in diameter, showed a sustained drug release and high buccal permeability across sheep buccal mucosa. Their incorporation into a CMC/HPC nanosponge allowed for increasing the mucoadhesion of the formulation and to control the drug release thanks to the swelling ratio. The overall result was the management of hypertension with superior cardio-protective effects.

An oral in situ gel for the sustained delivery of Buspirone hydrochloride (BH) was developed to achieve a reduced daily dose frequency for the treatment of pediatric anxiety (contribution 9). Mucoadhesive gel was produced, utilizing alginates exhibiting sol-to-gel phase transitions due to pH changes. The formulation, optimized by a QbD study, resulted in an increased bioavailability of the drug as compared to the solution.

Furthermore, Zafar et al. described the use of gelled–lipid nanoparticles for the treatment of bacterial conjunctivitis after ocular application (contribution 10). Nanostructured lipid carriers (NLC) containing erythromycin were loaded into an in situ gel composed of carbopol and chitosan combination. The nanometric size and the high entrapment efficiency allowed to obtain high release and permeation of drugs, evaluated both in vitro and ex vivo on goat corneas, ensuring better antimicrobial activities. The selection of gelling polymers showed an improvement in precorneal residence time and tolerability, in terms of hydration, irritation, and isotonicity.

Likewise, the treatment of ocular diseases was achieved with the development of a sol–gel system composed of Azithromycin-loaded lipid nanocarriers embedded in a thermosensitive gelling agent (contribution 11). The suitability of the formulation as ocular delivery system was assessed in vitro and ex vivo, demonstrating good corneal permeation, ocular tolerance as isotonic and non-irritant, and increased antimicrobial activity.

Lastly, the use of gel formulations has been investigated also for auricular delivery. A biopolymer lipid hybrid microcarrier was investigated for enhanced local Dexamethasone delivery and sustained release at the round window membrane level of the middle ear for the treatment of sensorineural hearing loss (SNHL) (contribution 12). In particular, polysaccharide–protein and pectin–bovine serum albumin were combined with lipids, such as Lipoid S100 and dimethyl-dioctadecyl-ammonium bromide, to obtain a hybrid biopolymer–liposome system. The presence of pectin hydrogel in the shell of the microparticles allowed to increase the microparticles' stability profiles and their swelling behavior in aqueous environments. The sustained release provided by the formulation has been assessed by in vitro release studies that represent a fundamental condition for the prospective in vivo experiments.

In conclusion, the application of gel formulations can reach different districts, offering controlled therapeutic effects. The optimization of the described drug delivery systems as a result of the QbD study led to shedding light on their advantages and drawbacks.

Hence, the importance of gel formulations to increase residence time and adhesiveness of the investigated nanosized delivery systems has been largely demonstrated.

Conflicts of Interest: The authors declare no conflict of interest.

List of Contributions

1. Abdallah, M.H.; Elsewedy, H.S.; AbuLila, A.S.; Almansour, K.; Unissa, R.; Elghamry, H.A.; Soliman, M.S. Quality by Design for Optimizing a Novel Liposomal Jojoba Oil-Based Emulgel to Ameliorate the Anti-Inflammatory Effect of Brucine. *Gels* **2021**, *7*, 219. https://doi.org/10.3390/gels7040219.
2. Khan, A.S.; Shah, K.U.; Mohaini, M.A.; Alsalman, A.J.; Hawaj, M.A.A.; Alhashem, Y.N.; Ghazanfar, S.; Khan, K.A.; Niazi, Z.R.; Farid, A. Tacrolimus-Loaded Solid Lipid Nanoparticle Gel: Formulation Development and In Vitro Assessment for Topical Applications. *Gels* **2022**, *8*, 129. https://doi.org/10.3390/gels8020129.
3. Abdallah, M.H.; Elghamry, H.A.; Khalifa, N.E.; Khojali, W.M.A.; Khafagy, E.-S.; Lila, A.S.A.; El-Horany, H.E.-S.; El-Housiny, S. Ginger Extract-Loaded Sesame Oil-Based Niosomal Emulgel: Quality by Design to Ameliorate Anti-Inflammatory Activity. *Gels* **2022**, *8*, 737. https://doi.org/10.3390/gels8110737.
4. Shabery, A.M.; Widodo, R.T.; Chik, Z. Formulation and In Vivo Pain Assessment of a Novel Niosomal Lidocaine and Prilocaine in an Emulsion Gel (Emulgel) of Semisolid Palm Oil Base for Topical Drug Delivery. *Gels* **2023**, *9*, 96. https://doi.org/10.3390/gels9020096.
5. Sguizzato, M.; Pepe, A.; Baldisserotto, A.; Barbari, R.; Montesi, L.; Drechsler, M.; Mariani, P.; Cortesi, R. Niosomes for Topical Application of Antioxidant Molecules: Design and In Vitro Behavior. *Gels* **2023**, *9*, 107. https://doi.org/10.3390/gels9020107.
6. Sallustio, V.; Farruggia, G.; Di Cagno, M.P.; Tzanova, M.M.; Marto, J.; Ribeiro, H.; Goncalves, L.M.; Mandrone, M.; Chiocchio, I.; Cerchiara, T.; et al. Design and Characterization of an Ethosomal Gel Encapsulating Rosehip Extract. *Gels* **2023**, *9*, 362. https://doi.org/10.3390/gels9050362.
7. Iyer, M.S.; Gujjari, A.K.; Paranthaman, S.; Abu Lila, A.S.; Almansour, K.; Alshammari, F.; Khafagy, E.-S.; Arab, H.H.; Gowda, D.V. Development and Evaluation of Clove and Cinnamon Supercritical Fluid Extracts-Loaded Emulgel for Antifungal Activity in Denture Stomatitis. *Gels* **2022**, *8*, 33. https://doi.org/10.3390/gels8010033.
8. Khafagy, E.-S.; Abu Lila, A.S.; Sallam, N.M.; Sanad, R.A.-B.; Ahmed, M.M.; Ghorab, M.M.; Alotaibi, H.F.; Alalaiwe, A.; Aldawsari, M.F.; Alshahrani, S.M.; et al. Preparation and Characterization of a Novel Mucoadhesive Carvedilol Nanosponge: A Promising Platform for Buccal Anti-Hypertensive Delivery. *Gels* **2022**, *8*, 235. https://doi.org/10.3390/gels8040235.
9. Abdallah, M.H.; Abdelnabi, D.M.; Elghamry, H.A. Response Surface Methodology for Optimization of Buspirone Hydrochloride-Loaded In Situ Gel for Pediatric Anxiety. *Gels* **2022**, *8*, 395. https://doi.org/10.3390/gels8070395.
10. Zafar, A.; Imam, S.S.; Yasir, M.; Alruwaili, N.K.; Alsaidan, O.A.; Warsi, M.H.; Mir Najib Ullah, S.N.; Alshehri, S.; Ghoneim, M.M. Preparation of NLCs-Based Topical Erythromycin Gel: In Vitro Characterization and Antibacterial Assessment. *Gels* **2022**, *8*, 116. https://doi.org/10.3390/gels8020116.
11. Gilani, S.J.; Jumah, M.N.B.; Zafar, A.; Imam, S.S.; Yasir, M.; Khalid, M.; Alshehri, S.; Ghuneim, M.M.; Albohairy, F.M. Formulation and Evaluation of Nano Lipid Carrier-Based Ocular Gel System: Optimization to Antibacterial Activity. *Gels* **2022**, *8*, 255. https://doi.org/10.3390/gels8050255.
12. Dindelegan, M.G.; Pașcalău, V.; Suciu, M.; Neamțu, B.; Perde-Schrepler, M.; Blebea, C.M.; Maniu, A.A.; Necula, V.; Buzoianu, A.D.; Filip, M.; et al. Biopolymer Lipid Hybrid Microcarrier for Transmembrane Inner Ear Delivery of Dexamethasone. *Gels* **2022**, *8*, 483. https://doi.org/10.3390/gels8080483.

References

1. Nyström, A.M.; Fadeel, B. Safety Assessment of Nanomaterials: Implications for Nanomedicine. *J. Control. Release* **2012**, *161*, 403–408. [CrossRef] [PubMed]

2.   Kirjavainen, M.; Urtti, A.; Valjakka-Koskela, R.; Kiesvaara, J.; Mönkkönen, J. Liposome-Skin Interactions and Their Effects on the Skin Permeation of Drugs. *Eur. J. Pharm. Sci.* **1999**, *7*, 279–286. [CrossRef]

3.   Touitou, E.; Dayan, N.; Bergelson, L.; Godin, B.; Eliaz, M. Ethosomes—Novel Vesicular Carriers for Enhanced Delivery: Characterization and Skin Penetration Properties. *J. Control. Release* **2000**, *65*, 403–418. [CrossRef] [PubMed]

4.   Chen, S.; Hanning, S.; Falconer, J.; Locke, M.; Wen, J. Recent Advances in Non-Ionic Surfactant Vesicles (Niosomes): Fabrication, Characterization, Pharmaceutical and Cosmetic Applications. *Eur. J. Pharm. Biopharm.* **2019**, *144*, 18–39. [CrossRef] [PubMed]

5.   Kapoor, B.; Gupta, R.; Gulati, M.; Singh, S.K.; Khursheed, R.; Gupta, M. The Why, Where, Who, How, and What of the Vesicular Delivery Systems. *Adv. Colloid Interface Sci.* **2019**, *271*, 101985. [CrossRef] [PubMed]

6.   Lai, F.; Caddeo, C.; Manca, M.L.; Manconi, M.; Sinico, C.; Fadda, A.M. What's New in the Field of Phospholipid Vesicular Nanocarriers for Skin Drug Delivery. *Int. J. Pharm.* **2020**, *583*, 119398. [CrossRef] [PubMed]

7.   Chai, Q.; Jiao, Y.; Yu, X. Hydrogels for Biomedical Applications: Their Characteristics and the Mechanisms behind Them. *Gels* **2017**, *3*, 6. [CrossRef] [PubMed]

8.   Baloglu, E.; Karavana, S.Y.; Senyigit, Z.A.; Guneri, T. Rheological and Mechanical Properties of Poloxamer Mixtures as a Mucoadhesive Gel Base. *Pharm. Dev. Technol.* **2011**, *16*, 627–636. [CrossRef] [PubMed]

9.   Nsengiyumva, E.M.; Alexandridis, P. Xanthan Gum in Aqueous Solutions: Fundamentals and Applications. *Int. J. Biol. Macromol.* **2022**, *216*, 583–604. [CrossRef] [PubMed]

10.   Zhang, Y.; Ng, W.; Hu, J.; Mussa, S.S.; Ge, Y.; Xu, H. Formulation and in Vitro Stability Evaluation of Ethosomal Carbomer Hydrogel for Transdermal Vaccine Delivery. *Colloids Surf. B Biointerfaces* **2018**, *163*, 184–191. [CrossRef] [PubMed]

11.   Breitsamer, M.; Winter, G. Vesicular Phospholipid Gels as Drug Delivery Systems for Small Molecular Weight Drugs, Peptides and Proteins: State of the Art Review. *Int. J. Pharm.* **2019**, *557*, 1–8. [CrossRef] [PubMed]

Article

Design and Characterization of an Ethosomal Gel Encapsulating Rosehip Extract

Valentina Sallustio [1], Giovanna Farruggia [2], Massimiliano Pio di Cagno [3], Martina M. Tzanova [3], Joana Marto [4], Helena Ribeiro [4], Lidia Maria Goncalves [4], Manuela Mandrone [5], Ilaria Chiocchio [5], Teresa Cerchiara [1,*], Angela Abruzzo [1], Federica Bigucci [1] and Barbara Luppi [1]

[1] Drug Delivery Research Laboratory, Department of Pharmacy and Biotechnology, Alma Mater Studiorum, University of Bologna, Via San Donato 19/2, 40127 Bologna, Italy; valentina.sallustio2@unibo.it (V.S.); angela.abruzzo2@unibo.it (A.A.); federica.bigucci@unibo.it (F.B.); barbara.luppi@unibo.it (B.L.);
[2] Pharmaceutical Biochemistry Laboratory, Department of Pharmacy and Biotechnology, Alma Mater Studiorum, University of Bologna, Via San Donato 19/2, 40127 Bologna, Italy; giovanna.farruggia@unibo.it
[3] Department of Pharmacy, Faculty of Mathematics and Natural Sciences, University of Oslo, Sem Saelands vei 3, 0371 Oslo, Norway; m.p.d.cagno@farmasi.uio.no (M.P.d.C.); m.m.tzanova@farmasi.uio.no (M.M.T.)
[4] Research Institute for Medicines (iMed.ULisboa), Faculty of Pharmacy, Universidade de Lisboa, Avenida Professor Gama Pinto, 1649-038 Lisboa, Portugal; jmmarto@ff.ulisboa.pt (J.M.); hribeiro@campus.ul.pt (H.R.); lgoncalves@ff.ulisboa.pt (L.M.G.)
[5] Pharmaceutical Botany Laboratory, Department of Pharmacy and Biotechnology, Alma Mater Studiorum, University of Bologna, Via Irnerio 42, 40127 Bologna, Italy; manuela.mandrone2@unibo.it (M.M.); ilaria.chiocchio2@unibo.it (I.C.)
* Correspondence: teresa.cerchiara2@unibo.it; Tel.: +39-0512095615

Citation: Sallustio, V.; Farruggia, G.; di Cagno, M.P.; Tzanova, M.M.; Marto, J.; Ribeiro, H.; Goncalves, L.M.; Mandrone, M.; Chiocchio, I.; Cerchiara, T.; et al. Design and Characterization of an Ethosomal Gel Encapsulating Rosehip Extract. *Gels* **2023**, *9*, 362. https://doi.org/10.3390/gels9050362

Academic Editor: Zihao Wei

Received: 31 March 2023
Revised: 22 April 2023
Accepted: 24 April 2023
Published: 25 April 2023

Abstract: Rising environmental awareness drives green consumers to purchase sustainable cosmetics based on natural bioactive compounds. The aim of this study was to deliver *Rosa canina* L. extract as a botanical ingredient in an anti-aging gel using an eco-friendly approach. Rosehip extract was first characterized in terms of its antioxidant activity through a DPPH assay and ROS reduction test and then encapsulated in ethosomal vesicles with different percentages of ethanol. All formulations were characterized in terms of size, polydispersity, zeta potential, and entrapment efficiency. Release and skin penetration/permeation data were obtained through *in vitro* studies, and cell viability was assessed using an MTT assay on WS1 fibroblasts. Finally, ethosomes were incorporated in hyaluronic gels (1% or 2% *w/v*) to facilitate skin application, and rheological properties were studied. Rosehip extract (1 mg/mL) revealed a high antioxidant activity and was successfully encapsulated in ethosomes containing 30% ethanol, having small sizes (225.4 ± 7.0 nm), low polydispersity (0.26 ± 0.02), and good entrapment efficiency (93.41 ± 5.30%). This formulation incorporated in a hyaluronic gel 1% *w/v* showed an optimal pH for skin application (5.6 ± 0.2), good spreadability, and stability over 60 days at 4 °C. Considering sustainable ingredients and eco-friendly manufacturing technology, the ethosomal gel of rosehip extract could be an innovative and green anti-aging skincare product.

Keywords: *Rosa canina* L. extract; polyphenols; antioxidant activity; ethosomal gel; sodium hyaluronate; anti-aging ingredients; cosmetic products

1. Introduction

Recently, many changes in consumer behaviors have occurred due to the coronavirus pandemic and climate changes [1], and one of them is a rising awareness about the cumulative effects of individual consumption behavior on environmental sustainability [2]. This attitude has been encouraged by the principles of sustainable development adopted by the United Nations, suggesting that all people should strive to meet their needs in a way that preserves, protects, and restores the health and integrity of the Earth's ecosystem [3]. This rising attention to promoting a more sustainable lifestyle drives green consumers to purchase green products [4]. Hence, consumer attitude towards sustainability is remarkable even in the cosmetic market trends, with rising requests for brands actively involved in ethical or environmentally sustainable practices [5]. Even though the three dimensions of sustainability (social, economic, and environmental) concern the entire cycle of cosmetic production, the choice of raw materials and the product design are the phases with the highest impact on the sustainability of a cosmetic product [6]. For this reason, the selection of botanical ingredients has gained the cosmetics industry's attention in order to satisfy green consumers' requests [7,8]. Among botanical ingredients, rosehips could represent a valid green and sustainable raw material for the cosmetics industry.

Rosehips are the wild edible pseudo-fruits of *Rosa canina* L., a spontaneous shrub resistant to adverse soil and climatic condition, widespread in Asia and Europe due to its high environmental adaptability. Rosehips are a natural source of bioactive compounds known from ancient times for many health benefits, as they have anti-inflammatory, anti-arthritic, analgesic, anti-diabetic, and antimicrobial properties [9]. Rosehips contain many antioxidant compounds such as ascorbic acid and polyphenols. As reported in the literature, among all polyphenols, many phenolic acids have been identified, such as gallic, ellagic, caftaric, chlorogenic, caffeic, coumaric, and ferrulic acids [10,11]. Flavonoid compounds have also been determined, in particular catechins, epicatechin, and quercetin [12,13], suitable for anti-aging cosmetic action [9,14]. Noticeably, the "anti-aging" segment of the cosmetic market is growing parallel to the "natural" segment [7], and rosehip extract could conjugate the preference for a natural source of ingredients with the issue of the rejuvenation of skin.

Polyphenols and ascorbic acid can scavenge reactive oxygen species to reduce oxidative stress and skin damage. These compounds offer a synergistic anti-inflammatory and antioxidant effect. Moreover, ascorbic acid promotes collagen synthesis, resulting in a tightening effect [15]. However, these antioxidant compounds are not stable to the temperature, light, and oxygen action, and their antioxidant activity could be easily lost [16,17]. Based on this limitation, the encapsulation of natural extracts in nanocarriers (liposomes, phytosomes, ethosomes, niosomes, transferosomes) has been demonstrated to be a valid strategy to preserve the antioxidant properties of bioactive compounds and improve their bioavailability [18,19].

Ethosomes are vesicular systems suitable for botanical extract encapsulation and well-studied for skin applications [20,21]. Discovered for the first time by Touitou et al. [22], they are composed of phospholipids, ethanol, and water. Among all the phospholipids, lecithin is a suitable natural option to prepare ethosomes [6]. Ethanol percentage influences the physicochemical characteristics of lipid vesicles, namely size, polydispersion index, zeta potential, and entrapment efficiency. Moreover, ethanol makes ethosomes more flexible, deformable, and elastic than conventional liposomes and, consequently, increases the ability to penetrate through the different skin layers [23]. Ethosomal suspensions could be incorporated in semisolid formulations to improve topical use; hyaluronic acid (HA) was selected as a gelifying natural polymer to obtain an ethosomal gel.

HA is widely used in cosmetic products to improve skin hydration, elastin, and collagen stimulation. Owing to its tissue regeneration potential, it exhibits anti-wrinkle, anti-aging, and skin rejuvenation properties. Moreover, regarding technological properties, HA is an excellent gelling agent due to its ability to bind water easily [24]. Different concentrations of HA can be used to obtain gels with tunable rheological properties and drug release profiles [25]. Currently, gel formulations are consumers' most appreciated semisolid dosage forms due to their fast hydration, low stickiness, and light consistency.

This work aimed to encapsulate rosehip extract in lecithin-based ethosomes to improve the stability of bioactive compounds. The ethosomes were characterized through size, polydispersion index, zeta potential, entrapment efficiency, *in vitro* release and skin permeation studies. In addition, a biocompatibility assay was performed on WS1 fibroblasts. Finally, we developed a gel based on hyaluronic acid containing rosehip extract ethosomes; the rheological properties and stability over time were assessed to obtain a green anti-aging cosmetic formulation.

2. Results and Discussion

2.1. In Vitro Evaluation of Rosehip Extract Anti-Aging Properties

The determination of antioxidant activity is an effective approach to ensure the quality of an anti-aging cosmetic formulation. In fact, intracellular and extracellular oxidative stress initiated by reactive oxygen species (ROS) is involved in skin aging.

Figure 1 shows the antioxidant activity measured as the percentage of DPPH reduction for different concentrations of rosehip extract and ascorbic acid (AA) solution, a reference substance for the antioxidant action. Rosehip extract at a low concentration (0.05 mg/mL) has a lower antioxidant activity (77.52%) compared to the same concentration of AA (92.40%). The antioxidant activity of rosehip extract increases at higher concentrations, and at 1 mg/mL, the sample and the control are similar (94.08 and 94.24% for rosehip extract and AA, respectively).

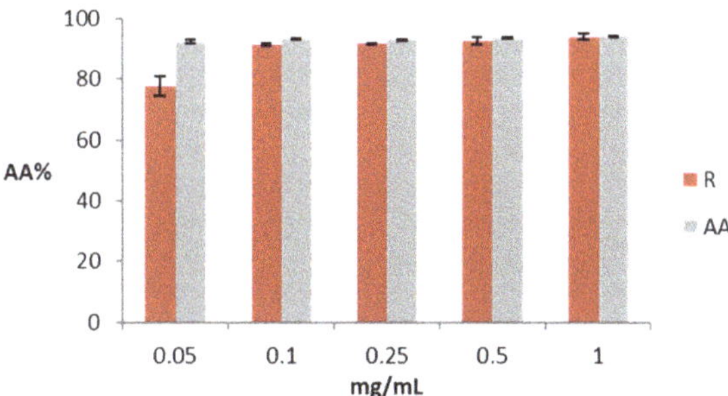

Figure 1. Antioxidant activity of rosehip extract (R) and ascorbic acid (AA). The data are expressed as the mean of three replicate experiments ±SD. Significance: (*) $p < 0.05$.

In addition, the antioxidant activity was assessed as the capacity to reduce the intracellular generation of ROS induced either by a chemical compound (H_2O_2) or by UV light *in vitro* in HaCaT (human keratinocyte cell line). The results are shown in Figure 2a,b.

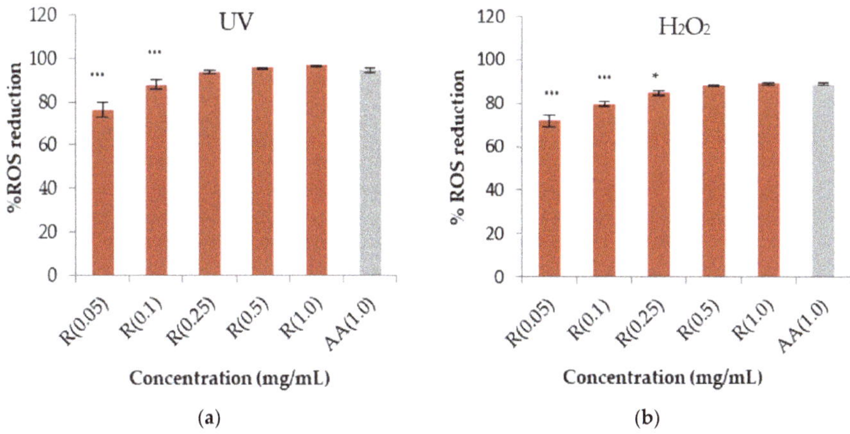

Figure 2. (a,b) Reactive oxygen species reduction by rosehip extract solution (R) at different concentrations after exposure to UV and H_2O_2 radiation of HaCaT cells. The data are expressed as the mean of at least nine replicate experiments ±SD. Significance: (*) $p < 0.05$ versus positive control cells (ascorbic acid, AA), (***) $p < 0.001$.

According to the DPPH assay, a similar trend was found for rosehip extract in the ROS reduction assay, confirming that the highest concentration of rosehip extract is comparable with the control. The IC_{50} was also determined after exposure to H_2O_2 and UV radiation, resulting in 37.2 ± 1.1 and 19.4 ± 1.2 µg/mL, respectively. The antioxidant activity of rosehip extract could be due to the extract's polyphenol and AA content. Based on these results, we selected the concentration of 1 mg/mL of rosehip extract for the continuation of this study as it could be effective in an anti-aging cosmetic product [7,19].

2.2. Preparation and Characterization of Ethosomes: Selection of Raw Materials, Influence of Ethanol, and Preparation Technique

The encapsulation of rosehip extract was performed to protect the antioxidant activity. Specifically, the extract was encapsulated in ethosomes (ET) by using the ethanol injection method. This method was preferred to the thin layer method, the first method used for liposome production, as it is a more green technique. In fact, it avoids pollutant solvents and some evaporation steps, reducing energy consumption, and simple equipment is required to enhance the scalability [26].

Concerning the phospholipid for vesicle preparation, lecithin was preferred as a suitable natural option for cosmetic formulations [27].

Finally, different percentages of ethanol ranging from 10 to 40% were tested to select the most promising nanocarrier based on their final physicochemical properties.

These properties can influence the *in vitro* behavior and, consequently, their cosmetic application; for this reason, all types of prepared phospholipid vesicles were characterized in terms of vesicle size (VS), polydispersity index (PDI), ζ potential (mV), pH, and entrapment efficiency (EE%), as reported in Table 1.

Table 1. VS (nm), PDI, ζ potential (mV), EE (%), and pH of ethosomes unloaded and loaded with rosehip extract.

Samples *	Size (nm)	PDI	ζ (mV)	EE%	pH
ET1	312.4 ± 8.5	0.30 ± 0.02	-52.12 ± 1.33	-	6.41 ± 0.10
ET2	223.8 ± 5.4	0.25 ± 0.02	-53.40 ± 1.70	-	6.30 ± 0.01
ET3	178.8 ± 1.5	0.25 ± 0.01	-53.12 ± 1.26	-	6.22 ± 0.04
ET4	213.6 ± 3.4	0.25 ± 0.01	-54.13 ± 0.87	-	6.15 ± 0.05
ET1R	363.6 ± 7.3	0.35 ± 0.03	-49.57 ± 1.48	68.78 ± 5.08	5.44 ± 0.08
ET2R	288.1 ± 8.6	0.30 ± 0.10	-51.42 ± 0.95	89.42 ± 2.21	5.47 ± 0.09
ET3R	225.4 ± 7.0	0.26 ± 0.02	-52.99 ± 1.40	93.41 ± 5.30	5.53 ± 0.05
ET4R	240.9 ± 1.9	0.29 ± 0.05	-53.19 ± 1.08	87.10 ± 3.94	5.61 ± 0.06

* ET1, ET2, ET3, and ET4: ethosomes containing an ethanol percentage of 10, 20, 30, and 40%, respectively. ET1R, ET2R, ET3R, and ET4R: ethosomes containing rosehip extract and an ethanol percentage of 10, 20, 30, and 40%, respectively. Data expressed as mean $\pm$ SD (n = 3).

The size of unloaded ethosomes containing 10% of ethanol (ET1) is 312.4 ± 8.5 nm, and this decreases to 178.8 ± 1.5 when the ethanol percentage is 30% (ET3). An increasing percentage of ethanol determines a decrease in the ethosome size due to the interposition of ethanol between the hydrocarbon chains, leading to a reduction in vesicular membrane thickness and, hence, in vesicular size [23,28]. When the percentage of ethanol is 40% (ET4), the size increases to 213.6 ± 3.4 nm, as a high percentage of ethanol leads to more leaky vesicles [29]. According to the literature, intermediate percentages of ethanol lead to ethosomes of minimum sizes; instead, the highest ethanol percentages can lead to the disruption of lipid vesicles [30].

In the case of loaded ethosomes, a similar trend related to the ethanol percentage can be observed. However, the vesicle size of loaded samples increases for all formulations (363.6 ± 7.3 nm, 288.1 ± 8.6 nm, 225.4 ± 7.0 nm, and 240.9 ± 1.9 nm for ET1R, ET2R, ET3R, and ET4R, respectively) compared to unloaded vesicles. This enlargement could be attributed to the interposition of the phytochemicals in vesicle structures. In particular, the more hydrophobic compounds are arranged between the phospholipid bilayer, while the more hydrophilic substances are included in the inner aqueous nucleus, determining the vesicle enlargement [31]. The increase in the size of vesicles determined by polyphenol encapsulation is reported for different natural extracts by other authors [32,33].

The polydispersity index is a parameter to measure the homogeneity of the systems. A value of PDI less than 0.3 indicates that the vesicles have similar dimensions, and the system could be considered homogeneous. Among all prepared formulations, those containing 30% ethanol have the lower value of PDI, resulting in 0.25 ± 0.01 for ET3 and 0.26 ± 0.02 for ET3R, indicating good homogeneity.

Concerning the zeta potential (ζ), this value helps predict the system's stability. A value of zeta potential of ± 30 mV indicates that the vesicles tend to repulse each other and maintain the system stability over time. According to the literature, ethanol acts as a negative charge provider for the surface of ethosomes. All the formulations show negative values of zeta potential less than -40 mV that tend to increase with the ethanol percentage [30,34].

Entrapment efficiency (EE) is a parameter related to the delivery capacity of vesicular systems. Among all types of vesicular systems, ethosomes are characterized by a high value of EE. As reported in Table 1, the EE increases with the increase in ethanol up to 30%. ET3R has the highest value of EE ($93.41 \pm 5.30\%$) due to the ethanol effect, meaning that ethanol improves the solubility of hydrophobic and hydrophilic compounds and, consequently, the loading in the vesicles. However, when the ethanol increase is higher, the EE tends to decrease as it could dissolve the membranes [28,30].

Finally, it was observed that the pH value ranges from 6.41 ± 0.10 for unloaded ethosomes to 5.44 ± 0.08 for loaded ethosomes. The lower value of loaded ethosomes could be explained by some organic acid in the extract (malic, citric, ascorbic acid). This lowering of pH is favorable for skin application [35].

Based on these results, the ethosomal suspension of ET3R was selected for the study as it had a small size (<300 nm), low PDI (<0.3), and the highest EE (>90%).

2.3. Vesicles' Physical Stability

The stability of nanocarriers as ingredients for cosmetic formulations needs to be investigated to guarantee the effectiveness and safety of the skincare product. Figure 3a,b shows the variation in the sizes of unloaded (ET3) and loaded ethosomes (ET3R) over a storage period of 20 weeks at 4.0 and 25 ± 2.0 °C. Unloaded ethosome sizes range from a minimum of 178 nm to a maximum of 225 nm at the lower temperature and a maximum of 217 nm at the higher temperature; these results are in agreement with the literature, confirming that among all the nanocarriers, ethosomes have good stability [21]. Concerning the loaded ethosomes, stability studies show that the enlargement in the dimensions due to the incorporation of extract compounds is maintained over the considered period. Comparing Figure 3a,b, a slight increase in loaded ethosome sizes can be noticed at 25 °C due to the increased fluidity of phospholipid vesicles [30]. However, over the considered period, the size of loaded ethosomes remains less than 300 nm, a size value that is favorable for skin penetration [36,37].

2.4. In Vitro Release Studies

The release of polyphenols from ET3R and the rosehip extract solution (1 mg/mL) was performed in a mixture of PBS: EtOH 7:3 v/v at a temperature of 32 ± 1.0 °C with constant stirring at 100 rpm. The Folin–Ciocalteu test was used to determine the TPC released over a period of 24 h.

The *in vitro* release profiles of TPC from the extract solution and from loaded ethosomes ET3R are shown in Figure 4. Rosehip extract exhibited a fast release profile, with $81.67 \pm 6.11\%$ of TPC released after 180 min, and reached the maximum release after 24 h. In contrast, the extract encapsulated in phospholipid vesicles showed a slower release over time, with $73.84 \pm 4.59\%$ of TPC released after 24 h. This release behavior can favor the delivery of bioactive compounds for cosmetic application [21].

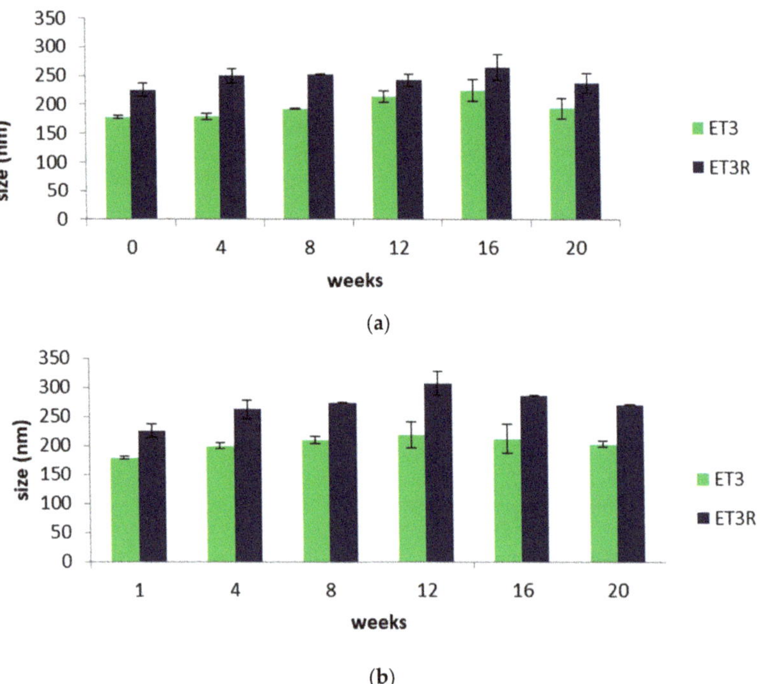

(a)

(b)

Figure 3. (**a**) Size (nm) of unloaded (ET3) and loaded ethosomes (ET3R) during 20 weeks of storage at $4.0 \pm 2.0\ ^{\circ}$C. Data expressed as mean $\pm$ SD ($n = 3$). (**b**) Size of unloaded (ET3) and loaded ethosomes (ET3R) during 20 weeks of storage at $25 \pm 2.0\ ^{\circ}$C. Data expressed as mean $\pm$ SD ($n = 3$).

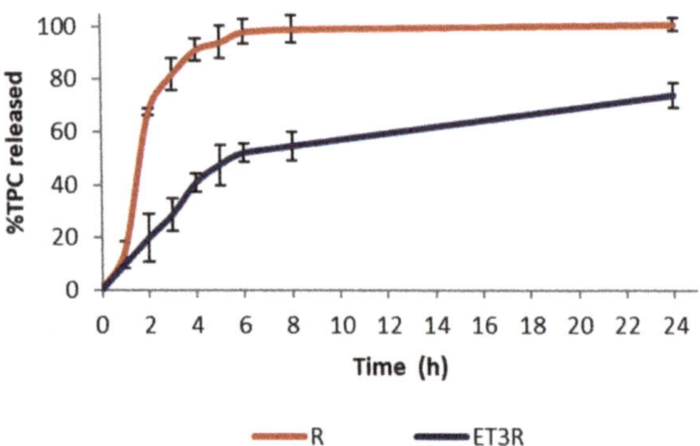

Figure 4. *In vitro* release of TPC in PBS:EtOH (7:3 v/v) from ET3R and R solution. Data expressed as mean $\pm$ SD ($n = 3$).

2.5. In Vitro Permeation Studies Using an Artificial Biomimetic Barrier

The evaluation of skin absorption is essential for cosmetic efficacy and cosmetic safety. Skin absorption includes penetration (the mass of the test substance that enters the skin) and permeation (the mass transferred from the skin to the reservoir compartment fluid) [38]. An *in vitro* permeation test using a 96-well Permeapad® plate was performed to clarify the different aspects of the absorption of bioactive compounds encapsulated in nanocarriers through the different layers of the skin. The permeability studies using this membrane compared the rosehip extract solution and the rosehip extract loaded into the ethosomes.

After 6 h, the amount of TPC was almost undetectable through the Permeapad®
96-well plate (Figure 5). After 24 h, a small amount of TPC (<2%) for the encapsulated
extract was detectable in the receptor compartment, confirming that the encapsulation in
ethosomes contributes to the increase in the permeation of bioactive compounds. Further
investigations are required to determine the type of polyphenols that pass through the
Permeapad® in the receptor compartment after 24 h. A standard Permeapad® has been
developed for quantifying the GI and mucosal drug absorption *in vitro*. The Permeapad®
technology has been used as a starting point for manufacturing a new skin-mimicking
barrier (SMB) suitable for transdermal formulation [38]. However, as this is not a commercial product yet, we focused on the classical 96-well plate PermeaPad®. We regarded
this approach as acceptable for the scope of this work as the goal of this permeation study
was to verify if polyphenols could be potentially transported through a biological barrier.
The prepared formulation seems to be suitable for cosmetic use as its systemic absorption
should be quite limited, according to the *in vitro* permeability testing results within the
application time.

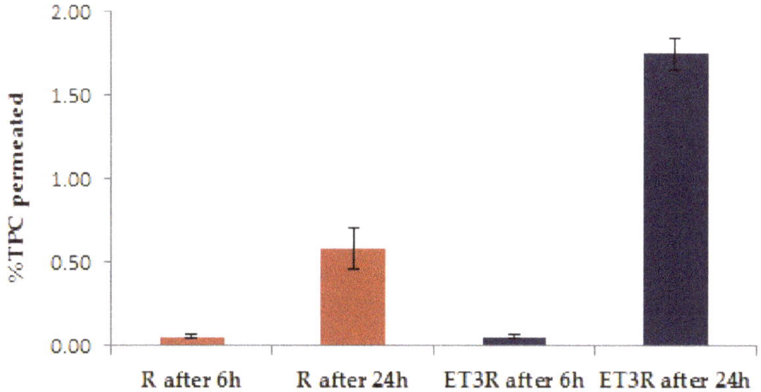

Figure 5. *In vitro* permeation of polyphenols from R solution (1 mg/mL) and ET3R after 6 and 24 h.
Data expressed as mean ± SD (*n* = 3).

2.6. In Vitro Skin Permeation/Retention Studies Using Pig Ear Skin

The retention of polyphenols inside the skin layers was assessed using porcine ear
skin due to its similarity to human skin. In Figure 6, we compare the percentage of TPC
from loaded ethosomes or rosehip extract solution detectable in the porcine skin after 6
and 24 h of permeation.

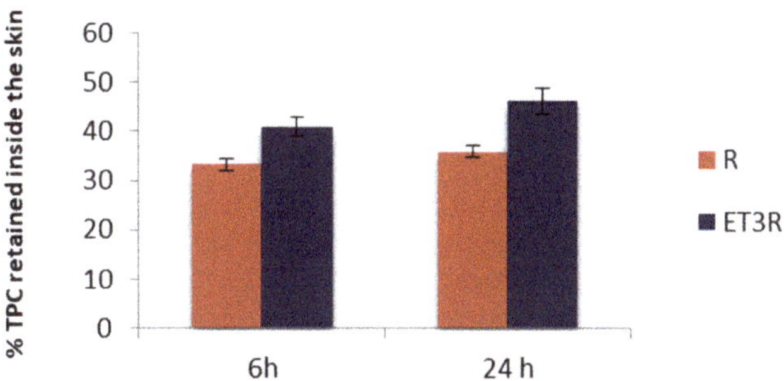

Figure 6. *In vitro* skin retention of TPC from R solution and ET3R after 6 and 24 h. Data expressed as
mean ± SD (*n* = 6).

The TPC percentage amount from ET3R retained inside the skin after 6 and 24 h was 40.99 ± 1.95% and 46.21 ± 2.75%, respectively. For the rosehip extract solution used as control, the TPC retained inside the skin after 6 and 24 h was 33.18 ± 1.21% and 40.92 ± 1.17%, respectively. These results suggested that after 6 h, ET3R increased the retention of TPC in comparison to the control ($p < 0.05$).

The higher amount of TPC found in the skin for loaded ethosomes could be due to the ethosomes' properties as penetration enhancers. As reported in the literature, this is the result of two combined effects: (i) the "ethanol effect", which increases the fluidity of phospholipids in the stratum corneum and (ii) the "ethosomes effect", which promotes the fusion of vesicles with skin lipids, favoring the penetration into the deeper layers of the skin [30]. Moreover, ethosomes' lecithin assures higher hydration to the skin compared to the free extract solution, which can promote skin penetration [39]. TPC was not detected in the receptor compartment after 6 h for all samples, confirming the data obtained with the Permeapad® membrane. These data confirm that the prepared formulations are suitable for cosmetic use as the bioactive compounds are intended to have a predominantly local effect [40]. In particular, the concentration of polyphenols in the epidermis and dermis is favorable for the cosmetic action of a skincare product [41].

2.7. In Vitro Cytotoxicity Assay

The biocompatibility of the ethosome suspensions was tested through an MTT assay on WS1 fibroblasts, the primary cells of connective tissue, producing collagen and elastin, and the target of the topical formulations.

WS1 fibroblasts were treated for 24 h with the rosehip extract solution, or unloaded or loaded ethosomes, in a range of concentrations between 5 and 100 µg/mL. The data showed that none of the treatments had a detrimental effect on WS1 cells (Figure 7). However, at the highest concentration, only the rosehip extract encapsulated in the ethosomes produced a significative increment in cell viability, as indicated in the increased absorbance of metabolized formazan salt in these samples. In similar studies, other authors reported that ethosomes are safe and present excellent skin tolerability [21].

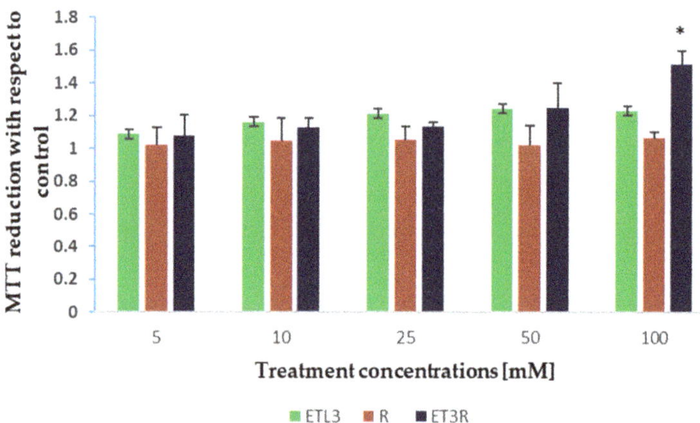

Figure 7. MTT assays on WS1 cells treated for 24 h with increasing concentrations of ET3, R solution, and ET3R. Effects on cell viability were normalized in comparison with control cells and analyzed using one-way ANOVA (* $p < 0.05$).

2.8. Development of Rosehip Ethosomal Gel

2.8.1. Physicochemical Characterization of Rosehip Ethosomal Gels

Gels are currently the preferred formulations for anti-aging products and the most suitable semisolid formulation to incorporate phospholipid vesicles [42]. Compared to liquid suspensions, hydrogel formulations increase consistency and adhesiveness and facilitate skin application. For this reason, ethosomal suspensions were integrated into low-molecular weight hyaluronic gels. Among natural gelling polymers, hyaluronic acid was selected for its moisturizing effect and anti-aging properties. HA percentages of 1 and 2% were selected to increase the viscosity and facilitate skin application, considering that the usual percentage of HA in cosmetic products ranges from 0.2 to 2% [43]. The viscosity (cP), pH value, spreadability (mm^2), and particle size of ethosomal gels were measured at room temperature 48 h after their preparation.

In Table 2, the range of viscosity values is between 305.0 ± 7.0 cP and 403.3 ± 11.5 cP for formulations based on HA1% and between 3166.7 ± 57.7 cP and 3233.3 ± 155.5 cP for those at HA2%. Concerning the effects of the different compounds in the formulations, a higher concentration of the polymer HA increases the viscosity. In the HA1% formulations, the presence of rosehip extract leads to a decrease in viscosity. In contrast, lecithin ethosomes lead to an increase in viscosity in all formulations. A similar result was reported by Ramadon et al. [34]. Concerning the HA2% formulations, viscosity values increase owing to the higher polymer concentration.

Table 2. Viscosity (cP), pH value, spreadability, and particle size of ethosomal gels.

Sample	Viscosity (cP) *	pH *	Spreadability (mm^2) *	Size (nm) *
HA1%	305.0 ± 7.0	6.3 ± 0.1	4435.58 ± 90.28	
R HA1%	250.0 ± 1.0	5.6 ± 0.2	5025.57 ± 218.91	-
ET3 HA1%	403.3 ± 11.5	6.1 ± 0.1	3995.13 ± 128.71	281.5 ± 2.1
ET3R HA1%	397.5 ± 17.1	5.6 ± 0.2	3998.79 ± 319.52	363.5 ± 1.8
HA 2%	3233.3 ± 55.5	6.8 ± 0.2	2404.72 ± 133.28	-
R HA2%	3166.7 ± 57.7	5.6 ± 0.2	2178.11 ± 96.08	-
ET3 HA2%	3233.3 ± 57.7	7.0 ± 0.2	1556.20 ± 126.94	219.4 ± 10.4
ET3R HA2%	3233.3 ± 57.5	5.5 ± 0.2	1566.33 ± 40.34	289.7 ± 6.4

* Values are expressed as mean $\pm$ SD, ($n = 3$).

Regarding pH, formulations containing rosehip extract have a lower pH value, between 5.5 and 5.6, due to some organic acids in the extract (i.e., malic, citric, ascorbic acid). This lowering of pH makes the formulations suitable for skin application [35].

Spreadability is one of the most important characteristics of a semisolid formulation. It is essential to ensure that bioactive substances are distributed well and quickly for their topical delivery at the moment of skin application, increasing consumer compliance [44]. The spreadability of the prepared gels is reported in Table 2. A decrease in spreadability was observed with the increase in viscosity, and the ethosomal gel containing HA1% was more spreadable than the HA2% gel.

Finally, regarding the size of vesicles, an increase in this value can be observed compared to the ethosomal suspensions, which could be due to the wrapping of HA on the surface of vesicles [45].

2.8.2. Rheological Properties of Ethosomal Gels

Rheological measurements are essential to developing a final formulation acceptable for consumers as they give preliminary information about spreading, adhesiveness, or lubrication in skin application.

Figure 8 shows the profile of all prepared gels in terms of the shear rates. The results suggested that gel viscosity increases with the percentage of HA. However, each gel can be classified as a shear-thinning fluid (non-Newtonian system) because the viscosity decreases as the shear rate increases. An oscillation frequency sweep test was performed to assess the viscoelastic behavior of the prepared formulations. This test could provide more information about the effect of colloidal forces and interactions among particles [46]. A preliminary amplitude sweep test defined the fluid's LVER, and the results showed that this region was at 2% shear strain. Figure 9 shows the frequency behavior of HA1% and 2% loaded gels. The higher concentration of HA increases the value of the elastic modulus G'

and viscous modulus G''. At all the measured frequencies, the G'' modulus is higher than G' modulus, indicating that both formulations exhibited fluid-like behavior.

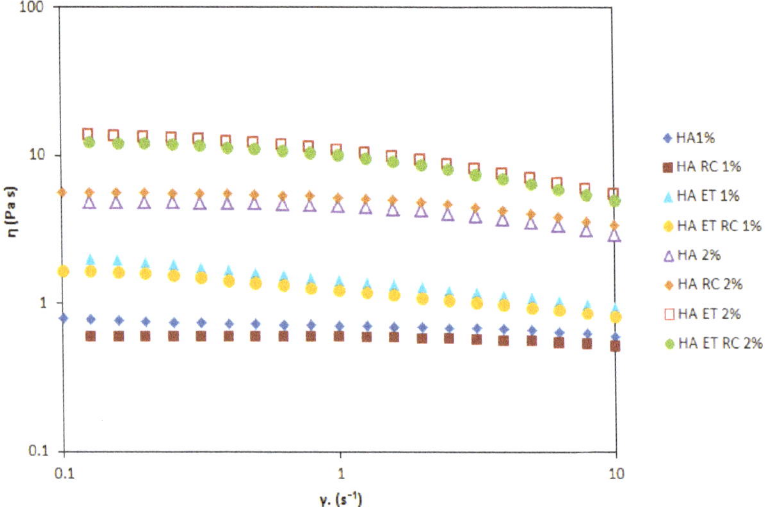

Figure 8. Viscosity as a function of shear rate for ethosomal gels. These curves describe the flow behavior of the sheared and time-independent fluid.

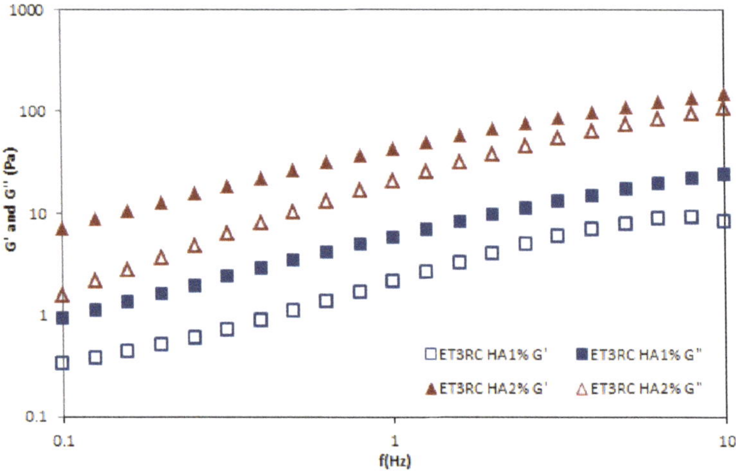

Figure 9. Frequency sweep with shear moduli as a function of the frequency of loaded hyaluronic gels ET3R HA1% and ET3R HA2% at room temperature.

The adhesiveness test of gel formulations measures the force required to separate the product from the test surface. The higher the force to separate these surfaces, the more adhesive the cosmetic product is. Table 3 shows the results of the adhesion test for all prepared gels.

Table 3. Normal force and area under force–time curve results for ethosomal gels.

Formulation	Peak Normal Force–Normal Force (N) *	Area Under Force–Time Curve (N·s) *
HA1%	-0.244 ± 0.020	0.514 ± 0.047
R HA1%	-0.257 ± 0.012	0.502 ± 0.049
ET3 HA1%	-0.399 ± 0.026	0.721 ± 0.056
ET3R HA1%	-0.329 ± 0.015	0.743 ± 0.093
HA 2%	-1.031 ± 0.039	1.294 ± 0.129
R HA2%	-1.039 ± 0.021	1.161 ± 0.143
ET HA2%	-1.436 ± 0.056	1.567 ± 0.192
ET3R HA2%	-1.333 ± 0.036	1.385 ± 0.043

* Values are expressed as mean $\pm$ SD, ($n = 6$).

The data showed that the formulation ET HA2% has the highest adhesiveness, with a peak normal force of -1.436 ± 0.056 N and an area under force–time curve of 1.567 ± 0.192 N·s. Similarly to the viscosity data, the increase in hyaluronic acid concentration and lecithin's presence in ethosomes increase the adhesiveness. In contrast, the presence of the extract decreases the adhesiveness when it is non-encapsulated [47].

2.9. In Vitro Release from Gels

The *in vitro* release studies were performed for gel formulations following the same method applied for ethosomal suspensions to investigate the polyphenols' release.

Figure 10 shows the release profiles of the prepared gel formulations. The TPC released after 6 h is $50.19 \pm 6.60\%$ for HA1% and $30.10 \pm 6.72\%$ for HA2%.

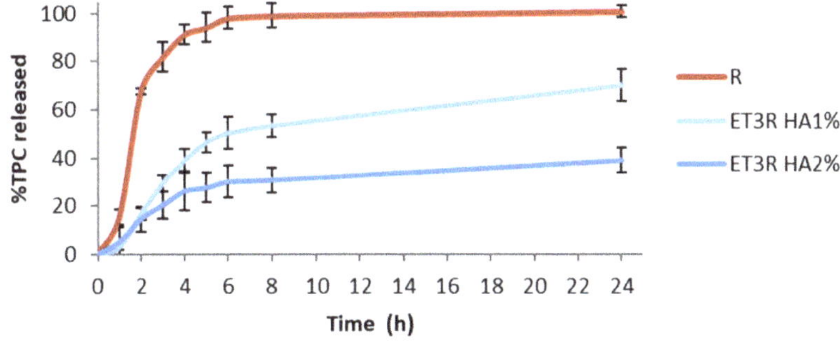

Figure 10. *In vitro* release of TPC in PBS:EtOH (7:3 v/v) from R solution (1 mg/mL) and loaded ethosomal gels (HA1% and HA2%). Data expressed as mean $\pm$ SD ($n = 3$).

The HA2% gel formulation showed a slower release than HA1%, which could be due to the higher viscosity slowing the diffusion of bioactive compounds. Other studies reported that the inclusion of ethosomes in the gel can give an additional modulation in the retention time [20,21]. These results confirmed that HA as the gelling agent could modulate the release of polyphenols, depending on concentration [24].

On the other hand, the release profile from the ET3R HA1% formulation seems more suitable for a skincare anti-aging product, and we selected this formulation to assess its stability over time.

2.10. Stability Studies of Gels

The stability test is fundamental to establishing the shelf life of a cosmetic product and checking that the effects of the environmental conditions do not change its quality and, consequently, its skin benefits. The stability studies for the most suitable gel formulation were carried out for 60 days to assess any change in organoleptic characteristics, viscosity, pH, and vesicle size at 4° and 25° $\pm$ 2.0 °C at specific time intervals. The results are shown in Table 4.

Table 4. Stability of ethosomal gels at different storage conditions.

Gel Formulation *	Temperature (°C)	Time (Day)	Viscosity (cP)	pH	Vesicle Size (nm)
ET3 HA1%	4 ± 2 °C	1	403.3 ± 11.5	6.1 ± 0.1	281.5 ± 2.1
		7	405.5 ± 21.2	6.1 ± 0.1	259.5 ± 12.0
		15	415.0 ± 7.1	6.0 ± 0.1	284.5 ± 1.1
		30	426.7 ± 20.8	6.0 ± 0.1	304.0 ± 8.1
		60	415.0 ± 21.2	5.9 ± 0.1	283.4 ± 5.8
ET3 HA1%	25 ± 2 °C	1	403.3 ± 11.5	6.1 ± 0.1	281.5 ± 8.1
		7	350.5 ± 10.2	6.1 ± 0.1	276.0 ± 18.4
		15	375.1 ± 21.2	6.1 ± 0.1	282.2 ± 12.5
		30	410.5 ± 14.4	5.9 ± 0.1	363.7 ± 18.6
		60	420.0 ± 56.6	5.9 ± 0.1	355.2 ± 8.5
ET3R HA1%	4 ± 2 °C	1	397.5 ± 17.1	5.6 ± 0.2	363.5 ± 1.8
		7	440.3 ± 14.1	5.4 ± 0.2	352.5 ± 19.2
		15	445.5 ± 7.1	5.4 ± 0.2	371.0 ± 19.8
		30	435.0 ± 21.2	5.7 ± 0.2	405.9 ± 17.8
		60	440.0 ± 28.3	5.5 ± 0.2	382.7 ± 9.8
ET3R HA1%	25 ± 2 °C	1	397.5 ± 17.1	5.6 ± 0.2	363.5 ± 1.8
		7	375.5 ± 7.1	5.4 ± 0.2	380.5 ± 10.6
		15	410.0 ± 28.3	5.2 ± 0.2	391.0 ± 46.7
		30	425.0 ± 21.2	5.4 ± 0.2	418.1 ± 7.7
		60	400.0 ± 28.3	5.4 ± 0.2	457.4 ± 8.9

* ET3 HA1% (unloaded ethosomes in hyaluronic acid gel 1%), ET3R HA1% (loaded ethosomes in hyaluronic acid gel 1%). Values are expressed as mean ± SD, (n = 3).

Regarding the organoleptic characteristics, the aspect, color, and odor of gels stored at 4.0 ± 2.0 °C remained similar over the period considered. However, for the gel stored in standard conditions, some changes were observed, namely an increase in the vesicle size of the loaded gel and a color change from cream gold (Pantone 13-0739TCX) to old gold (Pantone 15-0955TCX), likely due to the oxidation process at this temperature; similar color changes in ethosomal gels were reported by Ramadon et al. [34]. Based on the viscosity measurement, an increase over time in refrigerated conditions and an initial decrease in standard conditions were observed, suggesting a direct correlation with the temperature. This increase is higher for gels containing ethosomes, likely due to the more significant effect of temperature on lecithin's physical state. For the loaded gels, the viscosity ranged from a minimum of 397.5 ± 17.1 cP to a maximum of 440.0 ± 28.3 cP at 4.0 ± 2.0 °C. This minimal increase indicates the substantial physical stability of the gel formulation concerning the viscosity up to 60 days at refrigerated temperatures. The pH of loaded formulations remains in the range of 5.1–5.7, indicating that the prepared gel remains suitable for skin application over time.

3. Conclusions

An ethosomal gel of rosehip extract was successfully prepared using hyaluronic acid as a gelling agent. *In vitro* anti-aging properties of rosehip extract were tested through a DPPH assay and ROS reduction test; rosehip extract exhibited antioxidant activity similar to a solution of ascorbic acid. Rosehip extract was encapsulated in lecithin ethosomes and the ethosomal suspensions, and ET3R containing 30% ethanol was selected for its optimal characteristics: small size, low polydispersity, and good entrapment efficiency. The *in vitro* release studies showed that encapsulation leads to a delay in the release of the extract. The permeation and retention studies through several membranes reveal that the amount of TPC in the skin from the encapsulated extract was higher than that from the un-encapsulated extract. Moreover, TPC was not detectable in the receptor compartment after 6 h, confirming the appropriate cosmetic use of the formulation. Finally, hyaluronic acid 1 or 2% (*w/v*) was added to the ethosomal suspension to improve the anti-aging action and facilitate skin application. Ethosomal gel of 1% (*w/v*) hyaluronic acid is a promising cosmetic formulation with a suitable pH for skin use, easy spreadability, and stability regarding viscosity and pH, and it could be considered a green, sustainable, and effective anti-aging cosmetic as it is made with sustainable ingredients, an easily scalable production process, and well-known anti-aging bioactive compounds.

4. Materials and Methods

4.1. Materials

The hydroalcoholic extract was obtained from lyophilized rosehips [48]. Methanol and ethanol were purchased from Sigma-Aldrich (Milan, Italy). Sodium hyaluronate (molecular weight = 800–1200 kDa) was sourced from Farmalabor (Canosa di Puglia, Italy). Folin–Ciocalteu reagent was sourced from Titolchimica (Pontecchio Polesine, Italy). Soy lecithin and other chemicals were purchased from Carlo Erba (Milan, Italy). Sodium phosphate dibasic anhydrous, sodium phosphate monobasic anhydrous, and Dulbecco's modified Eagle medium supplemented with D-glucose were purchased from Sigma-Aldrich Co. (St. Louis, MO, USA). Fetal bovine serum (FBS), L-glutamine, penicillin (1000 U/mL), and streptomycin were purchased from Euroclone S.p.A. (Milan, Italy).

4.2. In Vitro Evaluation of Rosehip Extract Anti-Aging Properties

4.2.1. Antioxidant Activity of Rosehip Extract according to DPPH Method

The antioxidant activity of rosehip extract was determined using the 2,2-diphenyl-1-picrylhydrazyl radical (DPPH, Sigma Aldrich, Milan, Italy) reduction assay as described by Brand-Williams et al. [49] with minor modification. Briefly, 1 mL of rosehip extract solution at different concentrations ranging from 0.05 to 1 mg/mL was added to 1 mL of DPPH methanol solution (0.1 mM). The mixture was stored in the dark for 30 min, then the absorbance was spectrophotometrically measured at 517 nm. Methanol:water 50:50 v/v was used as a blank solution and the DPPH solution as the control. The test was carried out in triplicate. The results are expressed as a percentage of inhibition of the DPPH radical according to the following equation: inhibition % = $[(A_0 - A)/A_0]$, where A_0 is the absorbance of the DPPH control and A is the absorbance of the sample.

4.2.2. Antioxidant Activity of Rosehip Extract according to Reactive Oxygen Species (ROS) Production Measurement

The intracellular production of ROS within cells was assessed with a fluorimetric technique using 2′,7′-dichlorodihydrofluorescein diacetate (H2-DCFDA, Life Technologies, Paisley, UK) [50]. HaCaT (human keratinocyte cell line, CLS, Eppelheim, Germany) sub-confluent cells grown in 96-well plates were incubated for 30 min with 20 µM of H2-DCFDA in the dark at 37 °C. The media was removed and fresh medium was added to the cells before they were exposed to different concentrations of samples (0.05–1 mg/mL) and ascorbic acid (AA, 1 mg/mL) for 1.5 h. Hydrogen peroxide (H_2O_2) solution (500 µM, 1 h) or exposure to UVB light (emission wavelength 312 nm) for 15 min were used for the induction of ROS in cells. After exposure, ROS levels were determined at excitation 485 nm and emission 520 nm wavelengths using a fluorescence microplate reader (FLU-Ostar BMGLabtech, Offenburg, Germany). Data from 9 replicates are reported as the percentage of ROS reduction determined as 100-(fluorescence of exposed cells/fluorescence of unexposed control from the same experiment) × 100. The concentration of the samples that reduced 50% of ROS (IC_{50}) was determined through non-linear regression analysis with GraphPad PRISM® 5 software (San Diego, CA, USA, www.graphpad.com, (accessed on 16 November 2022).

4.3. Preparation of Ethosomes: Selection of Raw Materials, Influence of Ethanol, and Preparation Technique

Ethosomes were prepared through the ethanol injection–sonication method reported by Ma et al. [51] with minor modifications. Briefly, soy lecithin (100 mg) was dissolved in different volumes of ethanol, ranging from 1 to 4 mL, in a covered beaker to avoid ethanol evaporation. The rosehip extract (10 mg) was dissolved in double-distilled water and mixed uniformly with a magnetic stirrer. The ethanolic solution of phospholipids was slowly added to the aqueous solution of rosehip extract with a syringe at 1 mL/min under constant stirring at 700 rpm to obtain four different formulations, named ET1R, ET2R, ET3R, and ET4R, containing an ethanol percentage of 10, 20, 30, and 40%, respectively. The resulting vesicle suspensions were homogenized through ultrasonication for 15 min (Elma Transonic T310, Singen, Germany) to convert large multilamellar vesicles into smaller vesicles [28]. Unloaded ethosomes (ET1, ET2, ET3, and ET4) were prepared as a control.

4.4. Vesicle Characterization

4.4.1. Size and Zeta Potential Measurements

The prepared ethosomes were characterized for their vesicle size (VS) and their polydispersity index (PDI). The ethosome suspensions were diluted (1:800 v/v) in ultrapure water (18.2 MΩ cm, MilliQ apparatus by Millipore, Milford, MA, USA). Size and PDI were measured through PCS (photon-correlation spectroscopy) using a Brookhaven 90-PLUS instrument (Brookhaven Instruments Corp., Holtsville, NY, USA) with a He-Ne laser beam at a wavelength of 532 nm (scattering angle of 90°). The measurements were performed at room temperature with five runs for each determination. The zeta potential measurements were carried out at 25 °C on a Malvern Zetasizer 3000 HS instrument (Malvern Panalytical Ltd., Malvern, UK) after the same dilution.

4.4.2. Entrapment Efficiency

The entrapment efficiency (EE) of the ethosomes was determined through the dialysis method. To evaluate the amount of bioactive compounds encapsulated into the vesicles, samples (1 mL) were purified from the non-incorporated components through dialysis (Spectra/Por® membranes: 12–14 kDa MW cut-off) in water (0.5 L) for 2 h at room temperature, with distilled water refreshed after 30 min (2 L in total amount). At the end of the purification process, the antioxidant activity (inhibition %) of the samples before and after dialysis was measured using the DPPH assay, and the EE was calculated as a percentage of the antioxidant activity after dialysis versus that before dialysis as in the following formula: EE% = (inhibition % dialyzed sample/inhibition% non-dialyzed sample) × 100.

4.4.3. pH Measurement

The pH of various ethosomal suspensions was determined by using a digital pHmeter (Crison Instruments, S.A. Barcelona, Spain). The glass electrode was calibrated with the solutions determined for the equipment (pH of 4.00 and 7.00).

4.4.4. Physical Stability

The physical stability of the unloaded and loaded ethosomes with 30% ethanol (ET3 and ET3R) was assessed by monitoring the size and the PDI over 20 weeks of storage at 4.0 and 25.0 ± 2.0 °C in the dark. For this study, at predetermined periods (0, 4, 8, 12, 16, and 20 weeks) aliquots of vesicle suspensions were diluted in ultrapure water (1:800; v/v), and the change in ethosome size and PDI were measured through DLS as reported in Section 4.4.1.

4.5. In Vitro Release Studies

The polyphenol release profiles from ethosomes were investigated using a Franz-type static glass diffusion cell (15 mm jacketed cell with a flat-ground joint, and clear glass with a 12 mL receptor volume; diffusion surface area = 1.77 cm^2) equipped with a V6A Stirrer (PermeGearInc., Hellertown, PA, USA). A cellulose membrane (MF-Millipore cellulose nitrate 0.22 μm, Sartorius Stedim, Biotech GmbH, Germany) was placed between the receptor and the donor compartments, and 12 mL of a mixture of 3:7 (v/v) ethanol/pH 7.4 buffer was used as the receptor medium. The donor compartment was filled with 1 mL of vesicle suspension. The systems were kept at 32.0 ± 1.0 °C under magnetic stirring (100 rpm/min). Aliquots (0.2 mL) were withdrawn at predetermined intervals, and the release medium was refilled with the same volume. The polyphenols were determined through UV-Vis spectrophotometry using the Folin–Ciocalteu method, and the amount released from ethosomes was compared with the simple dissolution of the extract in the same medium. The polyphenol release profile test was performed in triplicate for each formulation.

4.6. In Vitro Permeation Studies Using an Artificial Biomimetic Barrier

Rosehip polyphenol permeation through an artificial biomimetic barrier was investigated using a high-through put 96-well Permeapad® plate (InnoMe GmbH, Espelkamp, Germany) [52]. The Permeapad® represents the first new approach to investigating drug permeability using a biomimetic artificial barrier. It consists of two regenerated cellulose membranes enclosing a layer of dry phospholipids between them, having a thickness of 0.10 mm. Once hydrated, this barrier forms a liposomal gel that, in structure and composition, reassembles a cellular monolayer and accounts for paracellular drug transport [53]. A mixture of PBS pH 7.4/Ethanol 7:3 (v/v) was used as the acceptor solution (400 μL). The donor (200 μL) compartments were filled using rosehip extract solution (1 mg/mL) and loaded and unloaded ethosomal suspensions. After filling both the acceptor and donor compartments, the plates were incubated in an orbital shaker–incubator (ES-20, Biosan, Riga, Latvia, LV) at 25 °C and 200 rpm for 6 and 24 h, respectively. Samples (100 μL) were taken from the acceptor every 60 min and the withdrawn volume was replaced with fresh release medium. For polyphenol quantification, the samples were directly transferred upon withdrawal to a UV-transparent 96-well microliter plate (Corning Inc., Kennebunk, ME, USA), and the absorbance was measured at 280 nm on a microplate spectrophotometer (SpectraMax 190, Molecular Devices Inc., Sunnyvale, CA, USA). Standard solutions (concentration range: 0.01–0.5 mg/mL) were measured on the same plate, and blank absorption (PBS/Ethanol 7:3 (v/v) or unloaded ethosomes) was deducted from all measurements' UV-Vis spectra. Three replicates were assessed for each sample. The percentage of polyphenols permeated was determined using the calibration curve of rosehip extract (R = 0.9966).

4.7. In Vitro Skin Permeation/Retention Studies Using Pig Ear Skin

In vitro skin permeation and retention studies were performed on Franz diffusion cells using porcine ear skin as a membrane model. Porcine ears were provided by a local slaughterhouse (CLAI, Faenza, Italy), and full-thickness skin was isolated using a method previously reported by Makuch et al. [54]. The excised circular pig skins with 1.60 mm thickness were placed between the donor and receptor compartments, with the inner part of the skin facing the upper inside portion of the cell. Ethosomal suspensions and extract solution (1 mL) were loaded on the skin in the donor compartment. The receptor medium consisted of a 3:7 (v/v) ethanol/buffer pH 7.4 solution and it was maintained under magnetic stirring at a temperature of 32.0 ± 1.0 °C throughout the experiment. For the permeation studies, aliquots (200 mL) were withdrawn after 1, 2, 4, 6, and 24 h from the receptor medium and immediately replaced with an equal volume of fresh medium. The Folin–Ciocalteu test was used to detect the presence of polyphenols in withdrawn aliquots. At the end of the 6 or 24 h permeation study, the skin was removed from the Franz cells and the excess formulation was gently removed using a cotton swab. Then, the skin was cut into tiny pieces, placed into separated beakers, and extracted with 5 mL of a mixture of ethanol/water 50:50 (v/v). Beakers were covered to avoid the evaporation of the solvent and stirred at 300 rpm for 24 h at room temperature. The samples were centrifuged and spectrophotometrically analyzed by using the Folin–Ciocalteu method to determine the amount of TPC retained in the skin. Six replicates were performed for each experiment.

4.8. Cell Viability Studies

WS1 cells (American Type Culture Collection, ATCC, Manassas, VA, USA) were used, and were grown routinely in 5% CO_2/humidified air at 37 °C with Dulbecco's modified Eagle medium supplemented with D-glucose (4.5 g/L), FBS (10% v/v), L-glutamine (2 mM), penicillin (1000 U/mL), and streptomycin (1 mg/mL). Cells were seeded at 5.000 cells/well (15×10^3 cell/cm^2) in a 96-well plate (Corning®, Corning, NY, USA) and incubated for 24 h to allow cell adhesion. After adhesion, the medium was removed and replaced with fresh medium containing the desired concentration of the compounds to be tested. To test the cell viability, an MTT assay was performed according to Calonghi et al. [55]. Briefly, after 24 h of treatment, 0.01 mL of MTT dissolved in DPBS at a concentration of 5 mg/mL was added to each well, and they were incubated for 4 h at 37 °C in the dark. Then, the medium was removed, and the reduced formazan crystals were dissolved in 0.1 mL pure isopropanol and the absorbance at 570 nm was measured using a multiwell plate reader (Tecan, Männedorf, CH). Data were analyzed using Prism GraphPad software (San Diego, CA, USA, www.graphpad.com, (accessed on 16 November 2022)).

4.9. Preparation of Ethosomal Gels

To prepare the ethosomal gels, sodium hyaluronate (HA 1 or HA 2% w/v) was added to 20 mL of vesicle suspension as a gelling agent, and the suspensions were stirred at 200 rpm for 24 h at room temperature to obtain a homogenous gel.

4.10. Physicochemical Characterization of Ethosomal Gels

4.10.1. Measurement of pH, Viscosity, and Spreadability

All gels were characterized in terms of pH, viscosity, and spreadability.

The pH of the gels was determined by using a digital pHmeter (Crison Instruments, S.A. Barcelona, Spain).

The viscosity was measured using a rotational viscometer Multi-Visc Rheometer (Fungilab, Barcelona, Spain) at 4 and 25 ± 2 °C. After thermal equilibration, the hydrogels were sheared using standard spindles (TR8 and TR10 for gels containing HA 1% and HA 2%, respectively) at a maximum rotational speed of 100 rpm. Viscosity values were expressed in cP.

A parallel-plate method was adopted to measure the spreadability [44]. Gel samples (300 µL) were spread between two Sil-Tec medical-grade solid silicone sheets of rubber 304 mm × 304 mm × 0.127 mm (Technical Products Inc., Lawrenceville, GA, USA). A circle of 20 cm diameter and 200 g weight was applied to them for 1 min. The change in diameter was measured. Measurements were performed at room temperature (25 ± 2 °C). Spreadability in mm^2 was calculated by the formula: $S = d^2 \times \pi/4$, where S is the spreading area (mm^2) and d is the spreading area diameter (mm).

Particle size measurements were performed after dilution in ultrapure water following the procedure described in Section 4.4.1.

All measurements were performed in triplicate.

4.10.2. Rheological Measurements

The rheological studies were conducted on HA gels that had been stored at room temperature for 2 days after their manufacture. The investigations were performed with a controlled stress Malvern Kinexus Rheometer Lab+ (Malvern Instruments, Worcestershire, UK) using a plate and plate geometry.

Viscosity measurements were carried out between 0.1 and 10 s^{-1}. All measurements were performed at 25.0 ± 0.2 °C. The viscosity values (Pa·s) were reported as a function of the shear rate.

The viscoelastic properties of the produced fluid gels were measured through small amplitude oscillatory experiments at a fixed strain of 2%, which was within the linear region, using a frequency range between 0.1 and 10 Hz. Measurements were performed at 25.0 ± 0.2 °C.

The tackiness and adhesiveness of gels were measured using the same equipment with a plate and plate geometry (pull-away assay or tackiness testing). A toolkit with the conditions of 0.1 mm/s, 5 mm, and 0.15 gap was selected. In this test, the peak of negative normal force required to separate the two parallel plates holding the gel can be attributed to tack. The area under the force–time curve represents the adhesive strength. Six measurements were performed for each sample.

4.11. In Vitro Release Studies from Ethosomal Gels

The polyphenol release profiles from HA gels were investigated using a Franz-type static glass diffusion cell. Briefly, 0.5 mL of HA gels, unloaded and loaded with rosehip extract, was placed in the donor compartment, and the release studies were performed as reported in Section 4.5.

4.12. Stability Studies of Gels

Stability studies were conducted by keeping the control and the tests of the selected formulations in refrigerated conditions (4 ± 2 °C) and standard conditions (25 ± 2 °C) for 60 days. The formulations were evaluated on the initial day and after 7, 15, 30, and 60 days according to their pH, viscosity, and organoleptic properties checked by visual inspection. The stability studies were performed in triplicate for each type of gel formulation.

4.13. Statistical Analysis

The experiments were performed in triplicate and the results are expressed as mean ± standard deviation (SD). ANOVA was used to determine statistical significance. Differences were deemed significant for $p < 0.05$.

Author Contributions: Conceptualization, T.C. and V.S.; methodology, V.S., G.F., M.M.T., M.P.d.C., J.M., L.M.G., M.M. and I.C.; validation, T.C., V.S., G.F., M.P.d.C., J.M., H.R. and L.M.G.; investigation, V.S., G.F., M.P.d.C., M.M.T., J.M., H.R. and L.M.G.; data curation, T.C. and V.S.; writing—original draft preparation, T.C., V.S., G.F., M.P.d.C., J.M. and L.M.G.; writing—review and editing, T.C, V.S., A.A., B.L. and F.B.; supervision and project administration, T.C. All authors have read and agreed to the published version of the manuscript.

Funding: Fundação para a Ciência e Tecnologia, Portugal (UIDB/04138/2020 and UIDP/04138/2020 for iMed.ULisboa, CEECINST/00145/2018 for J. Marto, and CEEC-IND/03143/2017 for L. M. Gonçalves).

Institutional Review Board Statement: Not applicable.

Informed Consent Statement: Not applicable.

Data Availability Statement: The data are contained within the article.

Acknowledgments: The authors are thankful to Sofia Melo Rocha for her contribution to the work. The authors thank the Fundação para a Ciência e Tecnologia, Portugal (UIDB/04138/2020 and UIDP/04138/2020 for iMed.ULisboa, CEECINST/00145/2018 for J. Marto, and CEEC-IND/03143/2017 for L. M. Gonçalves).

Conflicts of Interest: The authors declare no conflict of interest.

References

1. Jaciow, M.; Rudawska, E.; Sagan, A.; Tkaczyk, J.; Wolny, R. The Influence of Environmental Awareness on Responsible Energy Consumption—The Case of Households in Poland. *Energies* **2022**, *15*, 5339. [CrossRef]
2. Magano, J.; Au-Yong-Oliveira, M.; Ferreira, B.; Leite, Â. A Cross-Sectional Study on Ethical Buyer Behavior towards Cruelty-Free Cosmetics: What Consequences for Female Leadership Practices? *Sustainability* **2022**, *14*, 7786. [CrossRef]
3. Maciejewski, G.; Lesznik, D. Consumers towards the Goals of Sustainable Development: Attitudes and Typology. *Sustainability* **2022**, *14*, 10558. [CrossRef]
4. Kamalanon, P.; Chen, J.-S.; Le, T.-T.-Y. "Why Do We Buy Green Products?" An Extended Theory of the Planned Behavior Model for Green Product Purchase Behavior. *Sustainability* **2022**, *14*, 689. [CrossRef]
5. Fortunati, S.; Martiniello, L.; Morea, D. The Strategic Role of the Corporate Social Responsibility and Circular Economy in the Cosmetic Industry. *Sustainability* **2020**, *12*, 5120. [CrossRef]

6. Bom, S.; Ribeiro, H.M.; Marto, J. Sustainability Calculator: A Tool to Assess Sustainability in Cosmetic Products. *Sustainability* **2020**, *12*, 1437. [CrossRef]
7. Ferreira, M.; Magalhães, M.; Oliveira, R.; Sousa-Lobo, J.; Almeida, I. Trends in the Use of Botanicals in Anti-Aging Cosmetics. *Molecules* **2021**, *26*, 3584. [CrossRef]
8. Tengli, A.; Srinivasan, S.H. An Exploratory Study to Identify the Gender-Based Purchase Behavior of Consumers of Natural Cosmetics. *Cosmetics* **2022**, *9*, 101. [CrossRef]
9. Mármol, I.; Sánchez-de-Diego, C.; Jiménez-Moreno, N.; Ancín-Azpilicueta, C.; Rodríguez-Yoldi, M. Therapeutic Applications of Rose Hips from Different Rosa Species. *Int. J. Mol. Sci.* **2017**, *18*, 1137. [CrossRef]
10. Demir, N.; Yildiz, O.; Alpaslan, M.; Hayaloglu, A.A. Evaluation of Volatiles, Phenolic Compounds and Antioxidant Activities of Rose Hip (*Rosa L.*) Fruits in Turkey. *LWT-Food Sci. Technol.* **2014**, *57*, 126–133. [CrossRef]
11. Elmastaş, M.; Demir, A.; Genç, N.; Dölek, Ü.; Güneş, M. Changes in Flavonoid and Phenolic Acid Contents in Some Rosa Species during Ripening. *Food Chem.* **2017**, *235*, 154–159. [CrossRef]
12. Olsson, M.E.; Gustavsson, K.-E.; Andersson, S.; Nilsson, Å.; Duan, R.-D. Inhibition of Cancer Cell Proliferation in Vitro by Fruit and Berry Extracts and Correlations with Antioxidant Levels. *J. Agric. Food Chem.* **2004**, *52*, 7264–7271. [CrossRef]
13. Nađpal, J.D.; Lesjak, M.M.; Šibul, F.S.; Anačkov, G.T.; Četojević-Simin, D.D.; Mimica-Dukić, N.M.; Beara, I.N. Comparative Study of Biological Activities and Phytochemical Composition of Two Rose Hips and Their Preserves: *Rosa Canina* L. and *Rosa Arvensis* Huds. *Food Chem.* **2016**, *192*, 907–914. [CrossRef]
14. Winther, K.; Wongsuphasawat, K.; Phetcharat, L. The Effectiveness of a Standardized Rose Hip Powder, Containing Seeds and Shells of *Rosa Canina*, on Cell Longevity, Skin Wrinkles, Moisture, and Elasticity. *CIA* **2015**, *10*, 1849. [CrossRef] [PubMed]
15. Michalak, M.; Pierzak, M.; Kręcisz, B.; Suliga, E. Bioactive Compounds for Skin Health: A Review. *Nutrients* **2021**, *13*, 203. [CrossRef] [PubMed]
16. Duru, N.; Karadeniz, F.; Erge, H.S. Changes in Bioactive Compounds, Antioxidant Activity and HMF Formation in Rosehip Nectars During Storage. *Food Bioprocess Technol.* **2012**, *5*, 2899–2907. [CrossRef]
17. Angelov, G.; Boyadzhieva, S.; Georgieva, S. Rosehip Extraction: Process Optimization and Antioxidant Capacity of Extracts. *Open Chem.* **2014**, *12*, 502–508. [CrossRef]
18. Ruiz Canizales, J.; Velderrain Rodríguez, G.R.; Domínguez Avila, J.A.; Preciado Saldaña, A.M.; Alvarez Parrilla, E.; Villegas Ochoa, M.A.; González Aguilar, G.A. Encapsulation to Protect Different Bioactives to Be Used as Nutraceuticals and Food Ingredients. In *Bioactive Molecules in Food*; Mérillon, J.-M., Ramawat, K.G., Eds.; Reference Series in Phytochemistry; Springer International Publishing: Cham, Switzerland, 2019; pp. 2163–2182. ISBN 978-3-319-78029-0.
19. Grgić, J.; Šelo, G.; Planinić, M.; Tišma, M.; Bucić-Kojić, A. Role of the Encapsulation in Bioavailability of Phenolic Compounds. *Antioxidants* **2020**, *9*, 923. [CrossRef]
20. Andleeb, M.; Shoaib Khan, H.M.; Daniyal, M. Development, Characterization and Stability Evaluation of Topical Gel Loaded with Ethosomes Containing *Achillea Millefolium* L. Extract. *Front. Pharmacol.* **2021**, *12*, 603227. [CrossRef]
21. Mota, A.H.; Prazeres, I.; Mestre, H.; Bento-Silva, A.; Rodrigues, M.J.; Duarte, N.; Serra, A.T.; Bronze, M.R.; Rijo, P.; Gaspar, M.M.; et al. A Newfangled Collagenase Inhibitor Topical Formulation Based on Ethosomes with *Sambucus Nigra* L. Extract. *Pharmaceuticals* **2021**, *14*, 467. [CrossRef]
22. Touitou, E.; Dayan, N.; Bergelson, L.; Godin, B.; Eliaz, M. Ethosomes—Novel Vesicular Carriers for Enhanced Delivery: Characterization and Skin Penetration Properties. *J. Control. Release* **2000**, *65*, 403–418. [CrossRef] [PubMed]
23. Arora, D.; Nanda, S. Quality by Design Driven Development of Resveratrol Loaded Ethosomal Hydrogel for Improved Dermatological Benefits via Enhanced Skin Permeation and Retention. *Int. J. Pharm.* **2019**, *567*, 118448. [CrossRef] [PubMed]
24. Bayer, I.S. Hyaluronic Acid and Controlled Release: A Review. *Molecules* **2020**, *25*, 2649. [CrossRef] [PubMed]
25. Bukhari, S.N.A.; Roswandi, N.L.; Waqas, M.; Habib, H.; Hussain, F.; Khan, S.; Sohail, M.; Ramli, N.A.; Thu, H.E.; Hussain, Z. Hyaluronic Acid, a Promising Skin Rejuvenating Biomedicine: A Review of Recent Updates and Pre-Clinical and Clinical Investigations on Cosmetic and Nutricosmetic Effects. *Int. J. Biol. Macromol.* **2018**, *120*, 1682–1695. [CrossRef] [PubMed]
26. Yang, S.; Liu, L.; Han, J.; Tang, Y. Encapsulating Plant Ingredients for Dermocosmetic Application: An Updated Review of Delivery Systems and Characterization Techniques. *Int. J. Cosmet. Sci.* **2020**, *42*, 16–28. [CrossRef]
27. Bryła, A. Encapsulation of Elderberry Extract into Phospholipid Nanoparticles. *J. Food Eng.* **2015**, *167*, 189–195. [CrossRef]
28. Shukla, R.; Tiwari, G.; Tiwari, R.; Rai, A.K. Formulation and Evaluation of the Topical Ethosomal Gel of Melatonin to Prevent UV Radiation. *J. Cosmet. Dermatol.* **2020**, *19*, 2093–2104. [CrossRef]
29. Khan, P.; Akhtar, N. Phytochemical Investigations and Development of Ethosomal Gel with *Brassica Oleraceae* L. (Brassicaceae) Extract: An Innovative Nano Approach towards Cosmetic and Pharmaceutical Industry. *Ind. Crop. Prod.* **2022**, *183*, 114905. [CrossRef]
30. Paiva-Santos, A.C.; Silva, A.L.; Guerra, C.; Peixoto, D.; Pereira-Silva, M.; Zeinali, M.; Mascarenhas-Melo, F.; Castro, R.; Veiga, F. Ethosomes as Nanocarriers for the Development of Skin Delivery Formulations. *Pharm. Res.* **2021**, *38*, 947–970. [CrossRef]
31. Malekar, S.A.; Sarode, A.L.; Bach, A.C.; Worthen, D.R. The Localization of Phenolic Compounds in Liposomal Bilayers and Their Effects on Surface Characteristics and Colloidal Stability. *AAPS PharmSciTech* **2016**, *17*, 1468–1476. [CrossRef]
32. Moulaoui, K.; Caddeo, C.; Manca, M.L.; Castangia, I.; Valenti, D.; Escribano, E.; Atmani, D.; Fadda, A.M.; Manconi, M. Identification and Nanoentrapment of Polyphenolic Phytocomplex from Fraxinus Angustifolia: In Vitro and in Vivo Wound Healing Potential. *Eur. J. Med. Chem.* **2015**, *89*, 179–188. [CrossRef] [PubMed]

33. Khan, B.A.; Mahmood, T.; Menaa, F.; Shahzad, Y.; Yousaf, A.M.; Hussain, T.; Ray, S.D. New Perspectives on the Efficacy of Gallic Acid in Cosmetics & Nanocosmeceuticals. *Curr. Pharm. Des.* **2018**, *24*, 5181–5187. [CrossRef]

34. Ramadon, D.; Wirarti, G.A.; Anwar, E. Novel Transdermal Ethosomal Gel Containing Green Tea *(Camellia Sinensis* L. Kuntze) Leaves Extract: Formulation and In Vitro Penetration Study. *JYP* **2017**, *9*, 336–340. [CrossRef]

35. Lukić, M.; Pantelić, I.; Savić, S.D. Towards Optimal PH of the Skin and Topical Formulations: From the Current State of the Art to Tailored Products. *Cosmetics* **2021**, *8*, 69. [CrossRef]

36. Fathalla, D.; Youssef, E.M.K.; Soliman, G.M. Liposomal and Ethosomal Gels for the Topical Delivery of Anthralin: Preparation, Comparative Evaluation and Clinical Assessment in Psoriatic Patients. *Pharmaceutics* **2020**, *12*, 446. [CrossRef] [PubMed]

37. Iizhar, S.A.; Syed, I.A.; Satar, R.; Ansari, S.A. In Vitro Assessment of Pharmaceutical Potential of Ethosomes Entrapped with Terbinafine Hydrochloride. *J. Adv. Res.* **2016**, *7*, 453–461. [CrossRef]

38. Magnano, G.C.; Sut, S.; Dall'Acqua, S.; Di Cagno, M.P.; Lee, L.; Lee, M.; Larese Filon, F.; Perissutti, B.; Hasa, D.; Voinovich, D. Validation and Testing of a New Artificial Biomimetic Barrier for Estimation of Transdermal Drug Absorption. *Int. J. Pharm.* **2022**, *628*, 122266. [CrossRef]

39. Zou, L.; Liu, W.; Liu, W.; Liang, R.; Li, T.; Liu, C.; Cao, Y.; Niu, J.; Liu, Z. Characterization and Bioavailability of Tea Polyphenol Nanoliposome Prepared by Combining an Ethanol Injection Method with Dynamic High-Pressure Microfluidization. *J. Agric. Food Chem.* **2014**, *62*, 934–941. [CrossRef]

40. Betz, G.; Aeppli, A.; Menshutina, N.; Leuenberger, H. *In Vivo* Comparison of Various Liposome Formulations for Cosmetic Application☆. *Int. J. Pharm.* **2005**, *296*, 44–54. [CrossRef]

41. Souto, E.B.; Müller, R.H. Cosmetic Features and Applications of Lipid Nanoparticles (SLN®, NLC®). *Int. J. Cosmet. Sci.* **2008**, *30*, 157–165. [CrossRef]

42. Salvioni, L.; Morelli, L.; Ochoa, E.; Labra, M.; Fiandra, L.; Palugan, L.; Prosperi, D.; Colombo, M. The Emerging Role of Nanotechnology in Skincare. *Adv. Colloid Interface Sci.* **2021**, *293*, 102437. [CrossRef] [PubMed]

43. Juncan, A.M.; Moisă, D.G.; Santini, A.; Morgovan, C.; Rus, L.-L.; Vonica-Țincu, A.L.; Loghin, F. Advantages of Hyaluronic Acid and Its Combination with Other Bioactive Ingredients in Cosmeceuticals. *Molecules* **2021**, *26*, 4429. [CrossRef]

44. Elena, O.B.; Maria, N.A.; Michael, S.Z.; Natalia, B.D.; Alexander, I.B.; Ivan, I.K. Dermatologic Gels Spreadability Measuring Methods Comparative Study. *Int. J. App. Pharm.* **2022**, *14*, 164–168. [CrossRef]

45. Xie, J.; Ji, Y.; Xue, W.; Ma, D.; Hu, Y. Hyaluronic Acid-Containing Ethosomes as a Potential Carrier for Transdermal Drug Delivery. *Colloids Surf. B Biointerfaces* **2018**, *172*, 323–329. [CrossRef]

46. Graça, A.; Gonçalves, L.; Raposo, S.; Ribeiro, H.; Marto, J. Useful In Vitro Techniques to Evaluate the Mucoadhesive Properties of Hyaluronic Acid-Based Ocular Delivery Systems. *Pharmaceutics* **2018**, *10*, 110. [CrossRef] [PubMed]

47. Leite, F.D.G.; Oshiro Júnior, J.A.; Chiavacci, L.A.; Chiari-Andréo, B.G. Assessment of an Anti-Ageing Structured Cosmetic Formulation Containing Goji Berry. *Braz. J. Pharm. Sci.* **2019**, *55*, e17412. [CrossRef]

48. Sallustio, V.; Chiocchio, I.; Mandrone, M.; Cirrincione, M.; Protti, M.; Farruggia, G.; Abruzzo, A.; Luppi, B.; Bigucci, F.; Mercolini, L.; et al. Extraction, Encapsulation into Lipid Vesicular Systems, and Biological Activity of *Rosa Canina* L. Bioactive Compounds for Dermocosmetic Use. *Molecules* **2022**, *27*, 3025. [CrossRef]

49. Brand-Williams, W.; Cuvelier, M.E.; Berset, C. Use of a Free Radical Method to Evaluate Antioxidant Activity. *LWT-Food Sci. Technol.* **1995**, *28*, 25–30. [CrossRef]

50. Marto, J.; Ascenso, A.; Gonçalves, L.M.; Gouveia, L.F.; Manteigas, P.; Pinto, P.; Oliveira, E.; Almeida, A.J.; Ribeiro, H.M. Melatonin-Based Pickering Emulsion for Skin's Photoprotection. *Drug Deliv.* **2016**, *23*, 1594–1607. [CrossRef] [PubMed]

51. Ma, H.; Guo, D.; Fan, Y.; Wang, J.; Cheng, J.; Zhang, X. Paeonol-Loaded Ethosomes as Transdermal Delivery Carriers: Design, Preparation and Evaluation. *Molecules* **2018**, *23*, 1756. [CrossRef]

52. Tzanova, M.M.; Randelov, E.; Stein, P.C.; Hiorth, M.; di Cagno, M.P. Towards a Better Mechanistic Comprehension of Drug Permeation and Absorption: Introducing the Diffusion-Partitioning Interplay. *Int. J. Pharm.* **2021**, *608*, 121116. [CrossRef] [PubMed]

53. Di Cagno, M.; Bibi, H.A.; Bauer-Brandl, A. New Biomimetic Barrier Permeapad™ for Efficient Investigation of Passive Permeability of Drugs. *Eur. J. Pharm. Sci.* **2015**, *73*, 29–34. [CrossRef] [PubMed]

54. Makuch, E.; Nowak, A.; Günther, A.; Pełech, R.; Kucharski, Ł.; Duchnik, W.; Klimowicz, A. The Effect of Cream and Gel Vehicles on the Percutaneous Absorption and Skin Retention of a New Eugenol Derivative With Antioxidant Activity. *Front. Pharmacol.* **2021**, *12*, 658381. [CrossRef] [PubMed]

55. Calonghi, N.; Farruggia, G.; Boga, C.; Micheletti, G.; Fini, E.; Romani, L.; Telese, D.; Faraci, E.; Bergamini, C.; Cerini, S.; et al. Root Extracts of Two Cultivars of Paeonia Species: Lipid Composition and Biological Effects on Different Cell Lines: Preliminary Results. *Molecules* **2021**, *26*, 655. [CrossRef]

Article

Niosomes for Topical Application of Antioxidant Molecules: Design and In Vitro Behavior

Maddalena Sguizzato [1], Alessia Pepe [2], Anna Baldisserotto [3], Riccardo Barbari [3], Leda Montesi [3], Markus Drechsler [4], Paolo Mariani [2] and Rita Cortesi [1,5,*]

[1] Department of Chemical, Pharmaceutical and Agricultural Sciences (DoCPAS), University of Ferrara, 44121 Ferrara, Italy
[2] Department of Life and Environmental Sciences, Polytechnic University of Marche, 60131 Ancona, Italy
[3] Department of Life Sciences and Biotechnology, University of Ferrara, 44121 Ferrara, Italy
[4] Bavarian Polymer Institute (BPI), Keylab "Electron and Optical Microscopy", University of Bayreuth, 95440 Bayreuth, Germany
[5] Biotechnology Interuniversity Consortium (C.I.B.), Ferrara Section, University of Ferrara, 44121 Ferrara, Italy
* Correspondence: crt@unife.it

Abstract: In the present study, gels based on xanthan gum and poloxamer 407 have been developed and characterized in order to convey natural antioxidant molecules included in niosomes. Specifically, the studies were conducted to evaluate how the vesicular systems affect the release of the active ingredient and which formulation is most suitable for cutaneous application. Niosomes, composed of Span 20 or Tween 20, were produced through the direct hydration method, and therefore, borate buffer or a micellar solution of poloxamer 188 was used as the aqueous phase. The niosomes were firstly characterized in terms of morphology, dimensional and encapsulation stability. Afterwards, gels based on poloxamer 407 or xanthan gum were compared in terms of spreadability and adhesiveness. It was found to have greater spreadability for gels based on poloxamer 407 and 100% adhesiveness for those based on xanthan gum. The in vitro diffusion of drugs studied using Franz cells associated with membranes of mixed cellulose esters showed that the use of a poloxamer micellar hydration phase determined a lower release as well as the use of Span 20. The thickened niosomes ensured controlled diffusion of the antioxidant molecules. Lastly, the in vivo irritation test confirmed the safeness of niosomal gels after cutaneous application.

Keywords: antioxidant molecules; niosomes; vesicles; hydrogels; topical application; nanoparticle solution X-ray scattering

Citation: Sguizzato, M.; Pepe, A.; Baldisserotto, A.; Barbari, R.; Montesi, L.; Drechsler, M.; Mariani, P.; Cortesi, R. Niosomes for Topical Application of Antioxidant Molecules: Design and In Vitro Behavior. *Gels* **2023**, *9*, 107. https://doi.org/10.3390/gels9020107

Academic Editor: Vitaliy V. Khutoryanskiy

Received: 30 December 2022
Revised: 20 January 2023
Accepted: 23 January 2023
Published: 26 January 2023

1. Introduction

Antioxidants molecules (AMs) represent our body's defense system against damage caused by free radicals: they are chemically defined as reducing agents capable of slowing down or preventing the oxidation chain reactions triggered by free radicals in the presence of oxidative stress.

Much of the inputs to the production of radicals occurs as a consequence of the body's metabolism and respiration, but also psycho-physical stress, environmental factors such as smog and smoke, and UV rays can increase the production of reactive elements inducing a cellular overload (oxidative stress) with altered functionality and consequent cutaneous disorders, e.g., dermatitis, tumors, psoriasis, wrinkles, alopecia, acne and skin aging.

For instance, some research has shown that a high concentration of the hydroxyl radical OH· is the cause of the onset of tumors, following an alteration of the genetic information due to the oxidation of the nitrogenous bases [1]. Instead, the attack of the radicals on the lipid substrates, the main components of the cell membranes, gives rise to allyl bond peroxidation reactions, resulting in cell wall degradation.

Therefore, antioxidants are the best allies in keeping the body healthy and for protecting the skin from the harmful effects of external factors. The mechanism of action of AMs in radical inhibition can be summarized in the following three points: (a) neutralization of reactive species, (b) sequestration of transition metal ions (chelating effect) and (c) enzymes inhibition involved in reactive oxygen species production [2]. From a physiological point of view, the biological system uses endogenous AMs such as catalase, glutathione peroxidase and superoxide dismutase to protect the body. When the endogenous defense systems may not be sufficient, the intake of exogenous AMs by diet (i.e., vitamin C and E), or even by topical or oral administration, becomes necessary.

However, it should be kept in mind that many of these substances are particularly insoluble and unstable at different pH and temperature levels and under environmental stress conditions; therefore, the most effective solution is the topical application through delivery systems such as phospholipid-based (liposomes) and niotenside-based (niosomes) vesicular systems [3] to ensure protection from degradation, thus opening up a vast field in the research and production of suitable cosmetic and pharmaceutical formulations.

In the present study, gallic (GA) and ferulic acid (FA) were considered as example AMs of natural origin due to their activity. Indeed, GA is a potent tyrosinase inhibitor with excellent antioxidant effects, together with antibacterial and anti-inflammatory activity [4–6]. Moreover, this natural triphenolic compound is an effective ingredient for the treatment of allergic contact dermatitis [7]. However, due to its poor water solubility, it has rarely been used for dermatologic applications. As for FA, its wide use in skin care products is corroborated by numerous studies demonstrating its protective role in skin structure. Furthermore, FA exhibits low toxicity and optimal absorption through the skin and tends to remain in the bloodstream longer than other phenolic compounds [8]. FA also performs powerful anti-inflammatory, antifibrotic—protective of the vascular endothelium—antiplatelet and antiapoptotic actions [9].

Finally, in order to obtain suitable systems for cutaneous application and delivery of AM embedded in niosomes, this research investigated the development and characterization of gels based on poloxamer 407 and xanthan gum.

2. Materials and Methods

2.1. Materials

Gallic acid (GA), ferulic acid (FA), cholesterol (CH), Tween 20 (T), Span 20 (S) the copolymers poly(ethylene glycol)-block-poly(propylene glycol)-block-poly(ethylene glycol) poloxamer 188 (P) and poloxamer 407 (pol) were from Sigma-Aldrich (St Louis, MO, USA). All other materials and solvents were from Merck Serono S.p.A. (Rome, Italy).

2.2. Niosomes Preparation

Niosomes were prepared by thin-layer hydration with slight modifications (Figure S1) [10]. In detail, non-ionic surfactant and cholesterol in a molar ratio 1:1 were solubilized in a methylene chloride/methanol mixture (1:1, v/v) and subjected to removal under vacuum of the organic solvent residue using a rotary evaporator (Rotavapor R-200, Büchi Italia, Cornaredo, Italy) obtaining a film on the glass wall that was hydrated using a 2 mg/mL AM aqueous solution, swirled and sonicated, at 60 °C, to give a final concentration of 25 mg/mL in terms of total vesicular components. Borate buffer (B) and Poloxamer 188 (P) solution (2.5% w/w) were alternatively used as a hydration medium. In Table 1 the detailed composition of each formulation is reported.

Table 1. Composition of the produced niosomes.

| Acronym | Composition | | | | | |
| | Molar Ratio | | | Aqueous Phase | Organic Phase (mg/mL) | AM (mg/mL) |
	Cholesterol	Span 20	Tween 20			
GA/B; FA/B	-	-	-	Borate Buffer (B)	-	2
GA/P; FA/P	-	-	-	Poloxamer 188 * (P)	-	2
NSB-GA; NSB-FA	1	1	-	Borate Buffer (B)	25	2
NSP-GA; NSP-FA	1	1	-	Poloxamer 188 * (P)	25	2
NTB-GA; NTB-FA	1	-	1	Borate Buffer (B)	25	2
NTP-GA; NTP-FA	1	-	1	Poloxamer 188 * (P)	25	2

* 2.5% *w/w*.

2.3. Niosomes Characterization

The morphology of niosomes was investigated using a Cryogenic Transmission Electron Microscope (Cryo-TEM). After vitrifying and transferring each sample to a Zeiss EM922Omega transmission electron microscope using a cryoholder (CT3500, Gatan Inc., Pleasanton, CA, USA) [11], specimens were examined with doses of about 1000–2000 e/nm^2 at 200 kV and images digitally recorded by CCD camera (UltraScan 1000, Gatan Inc., Pleasanton, CA, USA) using GMS 1.4 software (Gatan Inc., Pleasanton, CA, USA) to process the images. During the visualization, the temperature of the sample was maintained below $-175\ °C$.

Small-angle and wide-angle X-ray scattering (SAXS and WAXS) experiments were performed using the Offline Xeuss 3.0 Diamond-Leeds SAXS facility (DL-SAXS) at Diamond Light Source (Harwell, UK). A gallium Excillum metaljet (9.4 keV), coupled with a movable beamstop-less Eiger 2 R 1M placed at the two distances of 100 mm and 2000 mm (for WAXS and SAXS, respectively) was used. The final investigated Q-range, (Q being the modulus of the scattering vector, defined as $4\pi \sin \theta/\lambda$, where 2θ is the scattering angle) extended from 0.003 to 1.6 Å^{-1}. The experiment exploited the mail-in service. Niosome samples were prepared in 2 mm polycarbonate capillaries, loaded into a capillary ladder, and analyzed, at 36 °C. Each image had a 2 min collection time, with a 0.4 mm X-ray beam. Two-dimensional (2D) data were corrected for background, detector efficiency and sample transmission, then radially averaged to derive I(Q) vs. Q curves [12].

Niosomes dimension was measured on aqueous diluted fractions (1:10 by volume) using a Zetasizer Nano S90 (Malvern Instr., Malvern, UK) equipped with a 5 mW helium neon laser with a wavelength output of 633 nm. Plasticware, cleaned with detergent washed and rinsed with milliQ water, was used for measurements made at 25 °C at 90° angle and run time around 180 s. "CONTIN" method [13] was used for interpreting the obtained data.

2.4. AMs Content in Niosomes

The encapsulation efficiency (EE) in niosomes, expressed as AM content, was determined after 1 day from vesicles production on a ultracentrifuged 300 μl fraction of the formulation using a Microcon centrifugal filter unit YM-10 membrane (NMWCO 10 kDa, Sigma-Aldrich, St. Louis, MO, USA). Ultracentrifugation was conducted at 8000 rpm for 20 min on a Spectrafuge™ 24D Digital Microcentrifuge (Woodbridge, NJ, USA). After diluting 100 microliters of retentate with methanol (1:10, *v/v*) the samples were magnetically stirred for 30 min. Afterwards, the amount of AM was quantified by UV, and the encapsulation parameter was calculated using Equation (1).

$$EE = AM/T_{AM} \times 100 \tag{1}$$

where AM is the amount of GA or FA determined by UV and T_{AM} is the total AM amount weighed to prepare the formulation. AM quantification was performed by means of a double-ray UV/Vis spectrometer (Lambda19 UV/VIS/NIR Spectrometer, Perkin Elmer, Waltham, MA, USA) operating at 260 nm and 318 nm for GA and FA, respectively, in 1 mL quartz cuvettes. To estimate the AM content in the sample, a previously carried out calibration curve for each compound, was used as reference. The evaluation was performed on day 1, 7, 15 and 30 after production.

2.5. In Vitro Diffusion Experiments

AM diffusion was evaluated using Franz cells associated with the mixed cellulose esters membrane. Briefly, dried membranes were immersed in distilled water, at room temperature, for 1 h then mounted on Franz diffusion cells (LGA, Berkeley, CA, USA) exposing a surface area of 0.78 cm^2 (1 cm diameter orifice). The receptor compartment contained 5 mL of PBS 1X. This solution was stirred at 500 rpm, with the help of a magnetic bar, and thermostated, at 32 $\pm$ 1 °C, during all the experiments [14]. Then, 1 mL of each formulation was loaded in the donor compartment in contact with the membrane surface. At predetermined time points between 1 and 8 h, 300 uL of receptor phase was withdrawn and subjected to AM content determination by means of UV/Vis spectrometer. The volume of removed sample was replaced with an equivalent volume of fresh PBS. The AM content was measured in four separate experiments, and the mean values $\pm$ standard deviations were estimated. The mean values of AM diffused expressed as $\mu g/cm^2$, were then plotted as a function of time. Diffusion coefficients (J_s) were calculated from the linear portion of the accumulation curve and expressed as normalized fluxes (J_n) by dividing by AM concentration in the analyzed form, expressed in mg/mL.

2.6. Antioxidant Activity

DPPH radical-scavenging assay was utilized to rapidly evaluate the antioxidant capacity [15]. The DPPH assay estimates the hydrogen donation ability of an antioxidant to convert the stable DPPH free radical to 1,1-diphenyl-2-picrylhydrazyl, which is accompanied by a deep-purple-to-light-yellow colorimetric reaction, which can be evaluated by measuring the percentage reduction of the solution absorbance at 517 nm after the radical reaction with the products under examination. The percentage of radical scavenging capacity was obtained applying Equation (2), as follows:

$$\text{DPPH radical} - \text{scavenging capacity } (\%) = [1 - (A1 - A2)/A0] \times 100 \qquad (2)$$

where A0 is the absorbance of the control (without AM), A1 is the absorbance in the presence of the AM, and A2 is the absorbance without DPPH. AM (either solutions or niosomes) at different concentrations (0.750 mL) was added to a methanol solution of DPPH (1.5 mL). The absorbance at 517 nm was obtained by means of a UV–Vis spectrophotometer (Jenway 7305 Spectrophotometer, VWR International Srl, Milan, Italy), as previously described [16]. The IC$_{50}$ values were calculated from the results after applying a previously obtained calibration curve. The obtained values ($\mu g/mL$) were the mean $\pm$ standard deviation of at least three independent experiments.

The quantitative FRAP assay measures the plasma ferric ion reducing capacity as it is based on the reduction of ferric ions (Fe^{3+}) to ferrous ions (Fe^{2+}) under acidic conditions in the presence of 2,4,6-tripyridyl-s-triazine (TPTZ) [17]. Indeed, the presence of an AM facilitates the reduction of the Fe^{3+}–TPTZ complex to the ferrous form, leading to an intense blue coloration measured at 593 nm (absorption maximum wavelength). Trolox was employed to perform the calibration curves, and therefore, the FRAP antioxidant activity is given as μmol Trolox equivalent/g of compounds.

2.7. Niosomal Gel Preparation and Characterization

To obtain thickened formulations, xanthan gum and poloxamer 407 were added to niosomes (Figure S1). To be precise, xg (0.75 %, *w/w*) was added to niosomal formulation (xg-niosomes) and handily mixed for 10 min up to complete dispersion [18], while pol-niosomes were prepared by the "cold method" gradually adding an amount of poloxamer

407 (15 % w/w) to cold water (5–10 °C) magnetically stirred for 3 h and then kept at 5 °C for 12 h up to complete dispersion of poloxamer [19].

The spreading capacity of niosomal gels was evaluated 24 h from gel preparation, at 37 ± 1 °C, to mimic the body temperature conditions [20]. To be precise, 100 mg of formulation was placed on a Petri dish (3 cm diameter) and then subjected to pressure by a glass dish carrying a 10 g mass. The time employed by the gel to completely fill the dish was evaluated.

Spreadability assay was performed three times, and then the mean values ± standard deviations were calculated using Equation (3).

$$S = m \times l/t \tag{3}$$

where S is the spreadability of the niosomal gel, m is the weight (g) attached on the plate top, l is the diameter (cm) of the plate, and t is the time (s) employed by the gel to completely fill the plate diameter [20].

The in vitro leakage of niosomal gels was determined on rectangular agar slides prepared by adding agar (1.5% w/w) to a citrate buffer with pH 5.5 and stirred, at 95 °C, until solubilization. Then, the cooled gels agar slides were cut. An amount of 30 mg of gel was placed on the top of an agar slide placed in a Petri plate vertically positioned at 90° on a transparent box wall and maintained, at 37 ± 1 °C. After the gel placement, the time taken by the gel to run along the slide was measured. Gel leakage, expressed as the time traveled by the gel on the plate, was evaluated thrice, and the mean values ± standard deviations were determined.

2.8. Patch Test

An in vivo irritation test was performed in order to evaluate the effect of niosomal gels applied in a single dose on intact human skin. The occlusive patch test was conducted at the Cosmetology Center of the University of Ferrara following the basic criteria of the protocols for skin compatibility testing of potentially cutaneous irritant cosmetic ingredients on human volunteers (SCCNFP/0245/99) [21–23]. The protocol was approved by the Ethics Committee of the University of Ferrara, Italy (study number: 170583). The test was run on 20 healthy volunteers of both sexes, who gave written consent to the experimentation, excluding subjects affected by dermatitis, with a history of allergic skin reaction or under anti-inflammatory drug therapy (either steroidal or non-steroidal). An aluminum Finn chambers (Bracco, Milan, Italy) carrying 10 mg of xg-niosomes was applied onto the forearm or the back skin, safeguarded with self-sticking tape. The sample formulations were loaded onto the Finn chamber by an insulin syringe and left in contact with the skin surface for 48 h. At 15 min and 24 h after patch remotion and skin cleaning from formulation residues, skin irritative reactions (e.g., erythema and/or edema) were estimated. Specifically, erythematous reactions have been sorted out into three reaction degrees, namely, light, clearly visible, and moderate/serious erythema. Irritative reactions were expressed as percentage with respect to total reactions occurring in volunteers.

3. Results and Discussion

3.1. Production and Characterization of AMs-Loaded Niosomes

It is well known that the low bioavailability and stability of GA and FA hamper the administration on skin and limit the antioxidant potential of these molecules. In order to overcome these limitations, the encapsulation of GA and FA was investigated by the development of a preformulative study [24,25]. In particular, niosomes have been produced in order to obtain the optimal formulation for GA and FA physico-chemical stability, encapsulation efficiency, controlled release, antioxidant effect and administration on skin. Indeed, in niosomes, the accommodation of amphiphilic drugs can occur between the bilayer and the aqueous core of the vesicles. Niosomes were produced following the thin-layer hydration method with the composition reported in Table 1.

Then, the morphology and the inner structure of the produced niosomes were investigated and compared. In particular, the morphological characterization of the different formulations was addressed by Cryo-TEM visualization (Figure 1). Since the presence of the AMs did not alter the aspect of the vesicles, plain niosomes have been considered.

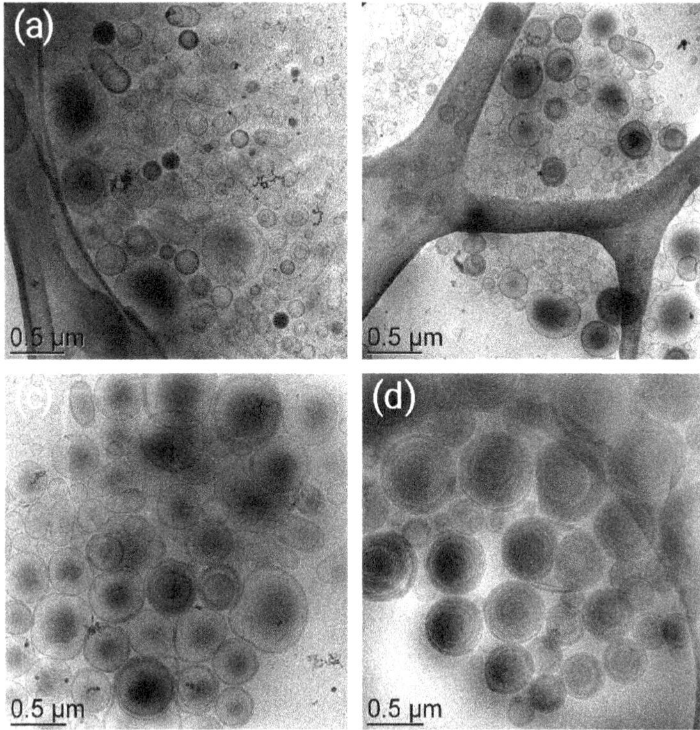

Figure 1. Cryo-TEM images of NSB (**a**), NSP (**b**), NTB (**c**) and NTP (**d**).

Vesicular systems with different peculiarities ascribable to the components used in the formulation have been obtained. Indeed, uni- or multilamellar non-ionic surfactant vesicles self-assembled in aqueous medium have been obtained [3].

Notably, both S (Figure 1a,b) and T (Figure 1c,d) gave rise to the formation of vesicles, although their lamellarity was affected by the combination between the type of surfactant and the hydration medium. For instance, when B is used, the coexistence of unilamellar, multilamellar and multivesicular systems is evidenced (Figure 1a,c). A similar morphological aspect was obtained for NSP (Figure 1b), while in the case of NTP, multilamellar vesicles were obtained (Figure 1d). This behavior could be ascribed to the interaction between the triblock copolymer p188 (P) and T, and it could be related to the ability of poloxamer micelles to stabilize the vesicular system by creating a "matryoshka" system, as reported in the literature [26].

In order to elucidate the internal structures of the different niosomal formulations, SAXS analysis was performed.

The results, reported in Figure 2, show similar scattering curves for all the samples, confirming that both S and T determine the formation of similar vesicles. In particular, at a very low angle, the scattering intensities appear to follow a Q^{-4}-law, suggesting the presence of very large particles (larger than 50–100 nm and probably polydisperse, as detected by Cryo-TEM and PCS), independent from niosomes composition. However, the vesicles show a different degree of lamellarity and a rather disordered nature. Indeed, the

Bragg peak observed at about 0.14 Å^{-1} in the scattering profiles confirms the lamellar inner structure of the niosomes, but its width and the absence of the characteristic higher-order peaks suggest that the positional correlation between adjacent bilayers is very small and/or that the degree of lamellarity is very low [27]. Interestingly, both the position and width of the peak appear related to the niosomes composition. As shown in Table 2, the unit cell parameter (which corresponds to the thickness of the vesicles layer plus the thickness of the aqueous layer separating two vesicles layers and which can be calculated from the position of the Bragg peak) depends on the used surfactant, while the degree of lamellarity, which can be indicatively derived by applying the Scherrer equation to the Bragg peak broadening under the assumption that other effects (such as stacking disorder or undulation effects) can be ignored [28], seems controlled by a combination of the type of surfactant and the hydration medium.

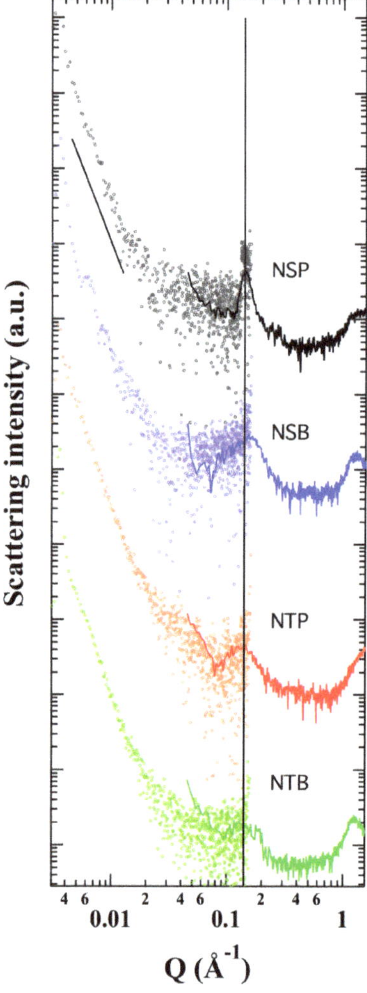

Figure 2. SAXS/WAXS profiles of NSP (black), NSB (blue), NTP (red) and NTB (green).

Table 2. Structural parameters for the uni-/multilamellar niosomes, as determined by SAXS/WAXS.

Niosome Acronym	Peak Position (2θ)	Peak Width (2θ)	Unit Cell *d* (nm)	Mean Crystallite Size *L* (nm)	Approximate Number of Interacting Bilayer *L/d*
NSP	1.299	0.129	4.41	41.7	9
NSB	1.290	0.554	4.44	9.7	2
NTP	1.180	0.388	4.86	13.9	3
NTB	1.191	0.681	4.81	7.9	2
errors			± 0.10	± 20%	

Table 2 shows that T probably induces a larger hydration of the vesicular layers (both in the presence of B and of P), so that the total thickness increases by about 10%. On the other side, the presence of P induces an ordering of the lamellar structure, which results in the enhancement of the mean crystallite size, particularly evident in the case of niosomes prepared with S as surfactant.

After production, the AMs-loaded formulations were characterized in terms of size and polydispersity. Afterwards, the dimensional stability was investigated for 30 days, as reported in Table 3.

Table 3. Dimensional behavior overtime of AMs-loaded niosomes, as determined by PCS.

Time (d)	NSB-GA Z-Ave (nm) *PdI*	NSP-GA Z-Ave (nm) *PdI*	NTB-GA Z-Ave (nm) *PdI*	NTP-GA Z-Ave (nm) *PdI*	NSB-FA Z-Ave (nm) *PdI*	NSP-FA Z-Ave (nm) *PdI*	NTB-FA Z-Ave (nm) *PdI*	NTP-FA Z-Ave (nm) *PdI*
1	549 ± 49 *0.13 ± 0.07*	456 ± 28 *0.19 ± 0.07*	436 ± 39 *0.15 ± 0.03*	610 ± 16 *0.25 ± 0.01*	419 ± 61 *0.27 ± 0.10*	594 ± 4 *0.36 ± 0.02*	495 ± 41 *0.27 ± 0.03*	862 ± 53 *0.39 ± 0.05*
7	614 ± 35 *0.11 ± 0.02*	573 ± 35 *0.30 ± 0.03*	451 ± 32 *0.26 ± 0.10*	686 ± 31 *0.40 ± 0.05*	373 ± 21 *0.31 ± 0.07*	499 ± 39 *0.27 ± 0.03*	473 ± 42 *0.28 ± 0.01*	895 ± 27 *0.35 ± 0.03*
15	613 ± 50 *0.17 ± 0.03*	604 ± 44 *0.24 ± 0.04*	501 ± 14 *0.41 ± 0.03*	825 ± 26 *0.28 ± 0.09*	377 ± 25 *0.22 ± 0.09*	456 ± 38 *0.23 ± 0.08*	498 ± 34 *0.30 ± 0.01*	932 ± 31 *0.33 ± 0.08*
30	601 ± 46 *0.21 ± 0.01*	779 ± 37 *0.26 ± 0.02*	546 ± 28 *0.31 ± 0.02*	1021 ± 48 *0.29 ± 0.07*	479 ± 29 *0.25 ± 0.03*	422 ± 26 *0.24 ± 0.01*	526 ± 23 *0.29 ± 0.02*	1082 ± 98 *0.32 ± 0.15*

The size distribution of niosomes after production ranged between 420 and 860 nm; thus, this parameter was strongly influenced by the composition of the vesicles. In particular, the higher dimensional increase was evidenced for niosomes comprising the combination of T and P as a surfactant and hydrating phase. This data agrees with the morphological and SAXS analyses, as well as with the literature [26], supporting the hypothesis that the interaction between the micelles and the bilayers could facilitate the formation of larger vesicles [26]. Indeed, as mentioned above, the resulting dimensions are influenced not only by the nature of T, responsible for the greater hydration of the layers, but also by the increase in lamellarity induced by P.

Moreover, no significant differences related to the alternative encapsulation of each AM have been evidenced.

Concerning dispersity indexes, the presence of heterogeneous populations has been denoted, in particular, one month after production, since PdI values higher than 0.2 indicate a wide size distribution [29].

3.2. Encapsulation Efficiency of Ams

The percentage of Ams content in the different vesicular systems was evaluated over time, and the chemical stability of the loaded drug was monitored up to 30 days after production. Figure 3 shows the drug content in niosomes, expressed as a percentage of the total amount used in the formulations.

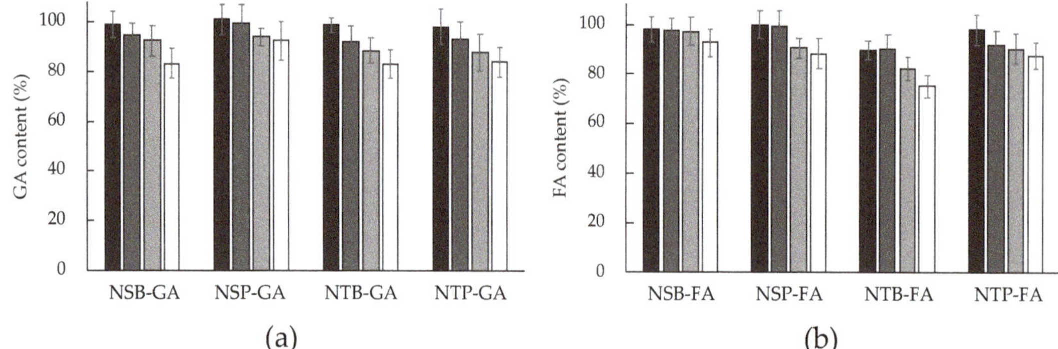

Figure 3. Percentage of loaded AMs in niosomes ((**a**), GA; (**b**), FA) at day 1 (black), 7 (dark grey), 15 (light grey) and 30 (white) after production. Data are the average of four independent experiments ± s.d.

The results reveal that niosomes allowed a quantitative encapsulation of AMs after production, reaching percentages between 90% and 100% independently of the type of formulation.

Specifically, in the case of GA, the composition of the formulations did not affect the encapsulation efficiency of the drug, which was maintained stable, reaching content values around 83% one month after production. The stability profile of GA content was similar for all the formulations. Concerning FA, NSB, NSP and NTP retained more than 85% of the drug after one month, while for NTB, the amount of drug decreased, reaching 75% of encapsulation. This behavior suggests that the presence of S, especially in combination with B as a hydration medium, ensured better stability to the system, as observed also in terms of size. Notably, the role of S as a destabilizing agent, when in aqueous solution, facilitates an increase in the entrapment efficiency of drugs, possibly explained by its HLB value being 8.6 [30].

3.3. In Vitro Diffusion Kinetics

The effect of the niosomes composition on AM diffusion has also been investigated by means of Franz-cell experiments. To be precise, to in vitro mimic the physiological conditions of AM diffusion through the skin, mixed cellulose esters membrane [31] and phosphate-buffered saline (pH 7.4) as receiving phase were employed. The amount of AM in the receiving phase was quantified by UV spectrometry, and the diffusion of the drug through the membrane was plotted against time, expressed as $\mu g/cm^2$ and hours, respectively. In Figure 4, the diffusion of GA (panel A) and FA (panel B) from niosomes and from two different reference solutions, namely, B and P, are displayed. The linear profile within 4 h was considered.

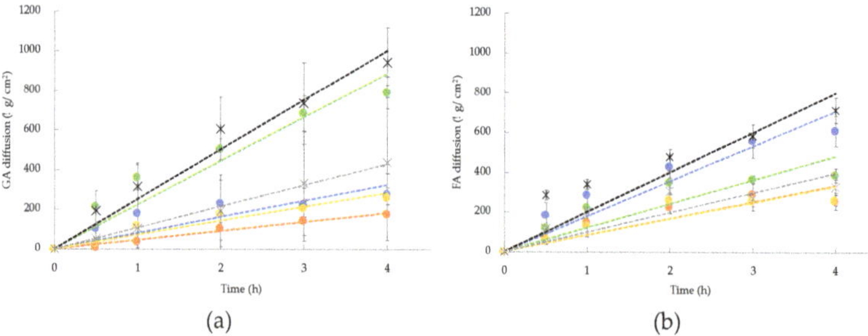

Figure 4. In vitro diffusion profiles of GA (**a**) and FA (**b**) from B (black cross), P (grey cross), NSB (blue), NSP (orange), NTB (green) and NTP (yellow). Data are the mean of four unrelated experiments ± s.d.

As expected, from the comparison between each reference solution (crosses) and the corresponding niosomal formulation (dots) in terms of the aqueous phase, a controlled diffusion of AM from the vesicles has been evidenced. This behavior was more pronounced for B as compared to P. However, also for P, the solution ensured a controlled diffusion, ascribable to the chemical nature of the co-polymer.

Furthermore, considering the hydration medium, it should be noticed that NSB and NTB displayed a faster drug diffusion profile with respect to those from NSP and NTP for both AMs. Indeed, the use of a micellar solution in the formulation conferred a "matryoshka" configuration to the system [26], responsible for the controlled release of the drug that can be loaded into the micelles and simultaneously into the internal core of vesicles. It is supposed that in this case, poloxamer 188, employed as a hydrating phase, retained the drug encapsulated in the micelles, allowing a slower diffusion of the drug through the membrane.

3.4. Antioxidant Activity

The use of GA and FA as antioxidants, in particular as radical-scavenging agents, is well demonstrated in the cosmetic and pharmaceutical fields [8,32–34]. In this study, DPPH assay was performed to compare the radical scavenging activity of the encapsulated AM by considering the IC_{50} values and FRAP assay to investigate their total antioxidant potential. Stability, intended as maintenance of the antioxidant effect, was monitored up to 30 days after production, and the results are showed in Figure 5.

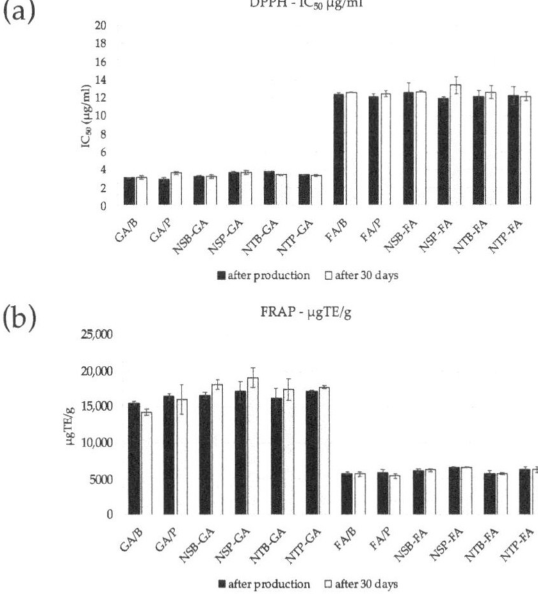

Figure 5. Antioxidant activity of GA and FA containing solutions and formulations as determined by DPPH (**a**) and FRAP (**b**) assays, 1 (black) and 30 (light gray) days after production.

Comparing the different niosomal formulations with the corresponding AM solutions, the data obtained showed that the encapsulation strategy was able to maintain the antioxidant potential of the molecules, confirming the suitability of niosomes as GA and FA delivery systems. Regarding the DPPH assay, as can be seen in Figure 5a, the activity profile is higher for GA (IC_{50} values in the range 2.87–3.77 µg/mL) than for FA (IC_{50} values in the range 11.83–12.51 µg/mL).

NTB-GA displayed higher antioxidant activity than GA/B solution (IC_{50} values were significantly different; $p < 0.0001$). Additionally, NSP-GA and NTP-GA showed higher and statistically significant IC_{50} values than GA/P ($p < 0.012$ and $p < 0.025$, respectively). In contrast, all FA-containing formulations possessed comparable antioxidant capacity to FA/B (non-statistically significant differences).

In general, the IC_{50} values monitored one month after production were maintained during the observation period without statistically significant changes, with the exception of NTB-GA, which showed a statistically significant decrease in activity ($p < 0.0001$).

Regarding the FRAP assay, the results corroborated what was shown by DPPH, indicating a greater antioxidant effect for GA (µmol TE/g ranging from 15,530.70–17,142.54) than FA (µmol TE/g ranging from 5714.88–6468.45) (Figure 5b). In particular, NSB-GA and NSP-FA showed higher antioxidant capacity than the reference GA/B and FA/B solutions (µgTE/g values were significantly different: $p < 0.023$ and $p < 0.046$, respectively). The 30-day stability of FRAP activity was maintained, except for GA/B, NSB-GA and NTP-GA, which showed a statistically significant decrease ($p < 0.019$, $p < 0.031$, and $p < 0.031$, respectively).

3.5. Niosomal Gel Production and Technological Behavior

Being niosomes dispersed in liquid form, the administration onto the skin is difficult and onsite permanence limited. Therefore, to obtain formulations with a certain grade of viscosity suitable for topical application, two different gelling agents, namely, xanthan gum and poloxamer 407 have been selected and added to niosome formulations. Indeed, these two biocompatible polymers are able to confer different adhesive properties to niosomes. Xanthan gum, an anionic polysaccharide obtained by bacterial fermentation [35], is a good thickening agent, while on the other hand, poloxamer 407 possesses peculiar thermo-reversible properties passing from a sol state at a low temperature to gel consistency at body temperature [31]. Spreadability and leakage were investigated in vitro to verify the suitability of the thickened formulations for topical administration. Indeed, these parameters are involved in many properties of the final formulation, such as gel extrusion from the package, skin coverage, patient compliance and drug therapeutic efficacy [18]. It has been demonstrated that presence of encapsulated drugs does not affect the technological performance of the thickened systems [18]; therefore, experiments were performed on empty formulations. The obtained results are reported in Table 4.

Table 4. Spreadability values and leakage behavior of niosomal gel obtained by the addition of xanthan gum (xg) and poloxamer 407 (pol) to the formulations.

Acronym	Spreadability (g·cm/s)	Leakage (s)
xg-NSB	3.10 ± 0.25	n.d.
xg-NSP	1.52 ± 0.09	n.d.
xg-NTB	3.02 ± 0.19	n.d.
xg-NTP	1.03 ± 0.07	n.d.
pol-NSB	8.93 ± 0.52	8.49 ± 0.38
pol-NSP	4.18 ± 0.29	7.16 ± 0.42
pol-NTB	9.77 ± 0.64	8.23 ± 0.48
pol-NTP	4.08 ± 0.22	7.36 ± 0.32

n.d.: not detectable.

As evidenced, the addition of xanthan gum led to the formation of a stiff gel formulation. In fact, concerning spreadability, the lowest values have been obtained on xg-niosomes, and no running across the agar plate has been observed in leakage tests. This aspect could ensure higher retention on the application site. Conversely, pol-niosomes showed higher values in terms of spreadability and leakage behavior.

In general, considering the hydration medium, niosomes composed of P resulted in thicker formulations, independently of the gelling agent used. This result could be related to the above-described role of the P micellar system in creating a rigid and complex structure.

Concerning the pH characteristics of niosomal gels, as summarized in Table 5, it emerges that the use of pol as a hydration medium gives rise to the formation of a pH around 5, but this is compatible anyway with skin characteristics. Taking into consideration the pKa values of both GA (4.41) and FA (4.58) we can hypothesize that the ratio at pH 5 between GA and gallate or FA and ferulate may be shifted towards the undissociated form. Indeed, from a micro and macroscopical point of view, no aggregation, precipitation, separation or changes in the aspect of the gelled formulation are evident. Therefore, we assume that the presence of these levels of dissociated form of AM is not able to induce changes in the solubility of the gelling polymers (i.e., poloxamer 407 and xanthan gum) and on the characteristics of the final topical formulations. Furthermore, many studies have shown no difference in antioxidant activity between ionized and non-ionized forms of the AMs [36–40].

Table 5. pH values of niosomal formulations before and after gelification obtained by the addition of xanthan gum (xg) or poloxamer 407 (pol).

Acronym	pH		
	Niosomes [1]	xg-Niosomes [2]	pol-Niosomes [3]
NSB-GA	4.7 ± 0.5	5.2 ± 0.3	5.4 ± 0.1
NSP-GA	4.9 ± 0.1	4.8 ± 0.1	4.9 ± 0.2
NTB-GA	5.0 ± 0.1	4.9 ± 0.1	5.9 ± 0.1
NTP-GA	4.7 ± 0.1	4.9 ± 0.2	5.4 ± 0.2
NSB-FA	6.0 ± 0.3	6.3 ± 0.3	6.5 ± 0.1
NSP-FA	5.2 ± 0.3	5.1 ± 0.2	5.6 ± 0.1
NTB-FA	6.0 ± 0.1	5.9 ± 0.2	6.0 ± 0.2
NTP-FA	5.5 ± 0.2	5.4 ± 0.2	5.3 ± 0.2

[1] niosomes formulation; [2] niosomes thickened with xg; [3] niosomes thickened with pol.

3.6. AMs Diffusion from Niosomal Gels

The AM diffusion from niosomal gels, obtained after addition of xg or pol, has been investigated and compared. In Figure 6, diffusion profiles of GA (panel A) and FA (panel B) from niosomal gels are shown considering the linear portion within 4 h.

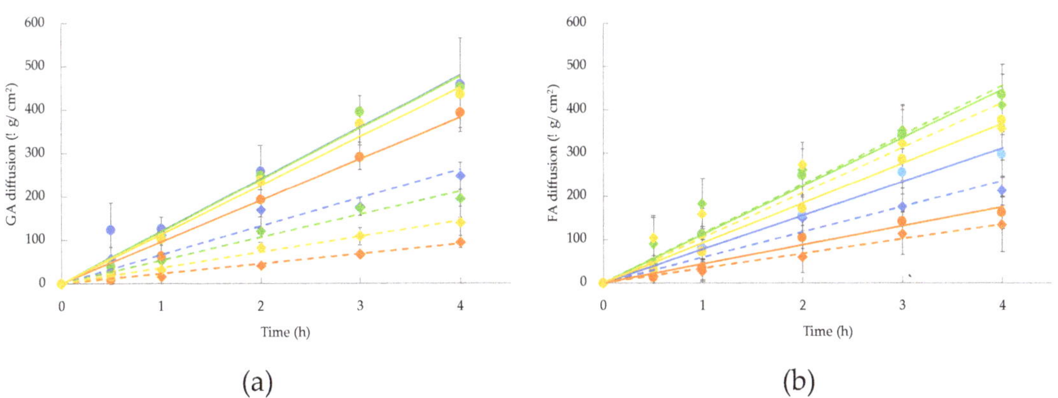

(a)　　　　　(b)

Figure 6. In vitro diffusion profiles of GA (**a**) and FA (**b**) from NSB (blue), NSP (orange), NTB (green) and NTP (yellow), when thickened with xg (dashed line, diamond) or pol (continue line, dot). Data represent the mean of four independent experiments ± s.d.

As depicted, the obtained diffusion profiles showed different behavior for GA and FA. In the case of GA, the passage of the drug through the membrane is mainly influenced by the thickening agent. In fact, the addition of xg provides a higher control on drug diffusion with respect to pol.

Concerning FA, the different behavior seems to be related to the nature of the surfactant employed for the formulation. In the case of NSB and NSP, xg induced a lowest FA passage, while for NTB and NTP no significant differences have been detected with the use of xg or pol. Furthermore, the diffusion kinetics of AMs were expressed as fluxes (Table 6), extrapolated from the slopes of diffusion profiles (J_s) and normalized in function of the experimental AM concentration in each formulation (J_n).

Table 6. Diffusion coefficients of GA and FA.

Acronym	GA		FA	
	J_s ($\mu g/cm^2 \cdot h$)	J_n ($cm^2 \cdot h$)	J_s ($\mu g/cm^2 \cdot h$)	J_n ($cm^2 \cdot h$)
NSB	81.0	40.5	176.7	88.3
xg-NSB	66.0	33.0	58.9	29.4
pol-NSB	120.0	60.0	78.2	39.1
NSP	45.3	22.7	83.3	41.6
xg-NSP	22.6	11.3	34.1	17.0
pol-NSP	95.7	47.8	43.8	21.9
NTB	220.7	110.4	119.5	59.8
xg-NTB	53.1	26.6	114.4	57.2
pol-NTB	119.2	59.6	112.1	56.1
NTP	70.3	35.1	82.4	41.2
xg-NTP	36.2	18.1	104.5	52.3
pol-NTP	112.9	56.5	92.3	46.2

Concerning GA, the pol-made gel led to a faster diffusion of the drug. In fact, the possible solubilization of GA in pol, thanks to the ability of the polymer in creating a micellar network, could be responsible for its increased passage through the membrane.

Considering the J_n values of FA diffusion, it was found that when T is used in the composition, the addition of a thickening agent did not affect a controlled passage of the drug with respect to formulations and data are superimposable. This behavior could be ascribed to the interaction between T and FA, and thus, it can be proposed that the diffusion is mainly governed by the vesicular system and less by the hydrophilic nature of the polymers, resulting in less structured gels.

In all cases, P as a hydration medium of niosomes plays a crucial role in GA diffusion, corroborating the influence of the "matryoshka" configuration on drug diffusion.

3.7. Patch Test

Being the topical administration the final target of these formulations, a patch test was performed in order to evaluate the safeness of the produced niosomal gels. Taking into account the results of spreadability and leakage, xg-niosomes have been selected for the patch test. To be precise, the potential irritation caused by applying xg-NSB, xg-NSP, xg-NTB and xg-NTP on skin have been evaluated on 20 healthy volunteers and the results expressed as a percentage of irritative reactions (Figure 7).

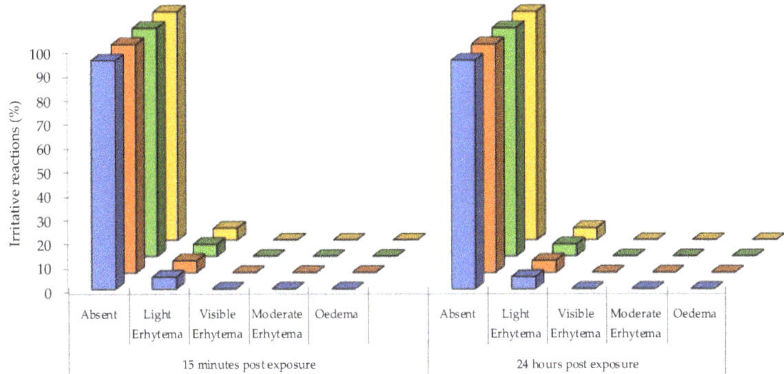

Figure 7. In vivo irritative reactions of xg-NSB (blue), xg-NSP (orange), xg-NTB (green) and xg-NTP (yellow) on 20 healthy volunteers, acquired by patch test.

It was found that the formulations exhibited the same results in terms of safeness; thus, all niosomal gels applied for 48 h under occlusive conditions can be classified as non-irritant. In fact, in 95% of cases, no irritative reactions have been detected, while 5% of cases led to negligible reactions, confirming the suitability of these systems for topical application.

4. Conclusions

The present investigation enabled the design of niosomal gels suitable for AMs loading and subsequent topical delivery. Specifically, this study highlighted the influence of a hydration medium (B or P) and a thickening agent (xg or pol) on the diffusion of AMs.

Strikingly, the results of the Franz-cell experiments and patch tests demonstrated that niosomal gel can deliver AMs in different ways depending on the interaction between the loaded drug, the non-ionic surfactant and the thickener agent. Indeed, in the case of FA, the presence of T facilitated an in vitro diffusion mainly governed by the vesicular system, while in the case of GA, an important role is due to the hydrophilic nature of the thickeners, corroborating the influence of the "matryoshka" configuration on drug diffusion. These data are in agreement with the cryo-TEM and SAXS analyses, showing the coexistence of unilamellar, multilamellar and multivesicular systems after hydration with B, or sole multilamellar vesicles, in the case of P hydration. However, an important contribution of the surfactant constituting the niosomes has been detected by means of SAXS analysis. Indeed, T seems to give rise to great hydration of the vesicle's layers both in the presence of B and of P, while in the case of niosomes prepared with S as the surfactant, the presence of P induces an ordering of the lamellar structure resulting in the enhancement of the mean crystallite size.

Moreover, no irritative reactions were detected during patch tests in 95% of cases, confirming the suitability of these systems for topical application.

Supplementary Materials: The following supporting information can be downloaded at: https://www.mdpi.com/article/10.3390/gels9020107/s1, Figure S1: Graphical diagram of the production steps of niosomes and niosomal gels.

Author Contributions: Conceptualization, R.C. and M.S.; methodology, R.C., M.S., A.B., L.M., P.M. and M.D.; investigation, M.S., A.B., L.M., R.B., A.P. and P.M.; data curation, M.S. and R.C.; writing—original draft preparation, M.S. and R.C.; writing—review and editing, R.C., M.S. and A.B.; visualization, M.D.; supervision, R.C.; funding acquisition, R.C. and M.S. All authors have read and agreed to the published version of the manuscript.

Funding: The research was funded by Piano Operativo Nazionale 2014–2020 (2021-PON-DM-1062-MS-RIC) and University of Ferrara (FAR 2020 and FAR 2021).

Data Availability Statement: Not applicable.

Acknowledgments: We acknowledge Stefano Manfredini of Unife for the availability of the antioxidant instrumentation, Irene Donelli of Unife for technical issue, Diamond Light Source and Leeds University for time on DL-SAXS facility under Proposal SM29797 and Piano Operativo Nazionale 2014–2020 Azione IV.6—Contratti di ricerca su tematiche Green (M.S.) and European Union–Next Generation EU, Project Code: ECS00000041—Project CUP: C43C22000380007 (P.M.).

Conflicts of Interest: The authors declare no conflict of interest.

References

1. Dizdaroglu, M.; Jaruga, P.; Birincioglu, M.; Rodriguez, H. Free Radical-Induced Damage to DNA: Mechanisms and Measurement. *Free. Radic. Biol. Med.* **2002**, *32*, 1102–1115. [CrossRef] [PubMed]
2. Van Tran, V.; Moon, J.-Y.; Lee, Y.-C. Liposomes for Delivery of Antioxidants in Cosmeceuticals: Challenges and Development Strategies. *J. Control. Release* **2019**, *300*, 114–140. [CrossRef] [PubMed]
3. Sguizzato, M.; Esposito, E.; Cortesi, R. Lipid-Based Nanosystems as a Tool to Overcome Skin Barrier. *Int. J. Mol. Sci.* **2021**, *22*, 8319. [CrossRef] [PubMed]
4. Oh, E.; Jeon, B. Synergistic Anti-Campylobacter Jejuni Activity of Fluoroquinolone and Macrolide Antibiotics with Phenolic Compounds. *J. Antibiot.* **2015**, *6*, 591–593. [CrossRef] [PubMed]
5. Shao, D.; Li, J.; Li, J.; Tang, R.; Liu, L.; Shi, J.; Huang, Q.; Yang, H. Inhibition of Gallic Acid on the Growth and Biofilm Formation of *Escherichia coli* and *Streptococcus mutans*. *J. Food Sci.* **2015**, *80*, M1299–M1305. [CrossRef]
6. Sorrentino, E.; Succi, M.; Tipaldi, L.; Pannella, G.; Maiuro, L.; Sturchio, M.; Coppola, R.; Tremonte, P. Antimicrobial Activity of Gallic Acid against Food-Related Pseudomonas Strains and Its Use as Biocontrol Tool to Improve the Shelf Life of Fresh Black Truffles. *Int. J. Food Microbiol.* **2018**, *266*, 183–189. [CrossRef]
7. Fu, R.; Zhang, Y.; Peng, T.; Guo, Y.; Chen, F. Phenolic Composition and Effects on Allergic Contact Dermatitis of Phenolic Extracts *Sapium sebiferum* (L.) Roxb. Leaves. *J. Ethnopharmacol.* **2015**, *162*, 176–180. [CrossRef]
8. Zduńska, K.; Dana, A.; Kolodziejczak, A.; Rotsztejn, H. Antioxidant Properties of Ferulic Acid and Its Possible Application. *Skin Pharmacol. Physiol.* **2018**, *31*, 332–336. [CrossRef]
9. Liu, Y.; Lin, Q.; Huang, X.; Jiang, G.; Li, C.; Zhang, X.; Liu, S.; He, L.; Liu, Y.; Dai, Q.; et al. Effects of Dietary Ferulic Acid on the Intestinal Microbiota and the Associated Changes on the Growth Performance, Serum Cytokine Profile, and Intestinal Morphology in Ducks. *Front. Microbiol.* **2021**, *12*, 698213. [CrossRef]
10. Puglia, C.; Bonina, F.; Rizza, L.; Cortesi, R.; Merlotti, E.; Drechsler, M.; Mariani, P.; Contado, C.; Ravani, L.; Esposito, E. Evaluation of Percutaneous Absorption of Naproxen from Different Liposomal Formulations. *J. Pharm. Sci.* **2010**, *99*, 2819–2829. [CrossRef]
11. Pattni, B.S.; Chupin, V.V.; Torchilin, V.P. New Developments in Liposomal Drug Delivery. *Chem. Rev.* **2015**, *115*, 10938–10966. [CrossRef] [PubMed]
12. Barbosa, L.R.S.; Ortore, M.G.; Spinozzi, F.; Mariani, P.; Bernstorff, S.; Itri, R. The Importance of Protein-Protein Interactions on the PH-Induced Conformational Changes of Bovine Serum Albumin: A Small-Angle X-Ray Scattering Study. *Biophys. J.* **2010**, *98*, 147–157. [CrossRef] [PubMed]
13. Pecora, R. Dynamic Light Scattering Measurement of Nanometer Particles in Liquids. *J. Nanopart. Res.* **2000**, *2*, 123–131. [CrossRef]
14. Andrade, L.M.; de Fátima Reis, C.; Maione-Silva, L.; Anjos, J.L.V.; Alonso, A.; Serpa, R.C.; Marreto, R.N.; Lima, E.M.; Taveira, S.F. Impact of Lipid Dynamic Behavior on Physical Stability, In Vitro Release and Skin Permeation of Genistein-Loaded Lipid Nanoparticles. *Eur. J. Pharm. Biopharm.* **2014**, *88*, 40–47. [CrossRef]
15. Blois, M.S. Antioxidant Determinations by the Use of a Stable Free Radical. *Nature* **1958**, *181*, 1199–1200. [CrossRef]
16. Wang, M.; Li, J.; Rangarajan, M.; Shao, Y.; LaVoie, E.J.; Huang, T.-C.; Ho, C.-T. Antioxidative Phenolic Compounds from Sage (*Salvia officinalis*). *Korean J. Pestic. Sci.* **1998**, *46*, 4869–4873. [CrossRef]
17. Benzie, I.F.F.; Strain, J.J. The Ferric Reducing Ability of Plasma (FRAP) as a Measure of "Antioxidant Power": The FRAP Assay. *Anal. Biochem.* **1996**, *239*, 70–76. [CrossRef]
18. Esposito, E.; Sguizzato, M.; Bories, C.; Nastruzzi, C.; Cortesi, R. Production and Characterization of a Clotrimazole Liposphere Gel for Candidiasis Treatment. *Polymers* **2018**, *10*, 160. [CrossRef]
19. Sguizzato, M.; Mariani, P.; Ferrara, F.; Drechsler, M.; Hallan, S.S.; Huang, N.; Simelière, F.; Khunti, N.; Cortesi, R.; Marchetti, N.; et al. Nanoparticulate Gels for Cutaneous Administration of Caffeic Acid. *Nanomaterials* **2020**, *10*, 961. [CrossRef]
20. Esposito, E.; Drechsler, M.; Mariani, P.; Panico, A.M.; Cardile, V.; Crascì, L.; Carducci, F.; Graziano, A.C.E.; Cortesi, R.; Puglia, C. Nanostructured Lipid Dispersions for Topical Administration of Crocin, a Potent Antioxidant from Saffron (*Crocus sativus* L.). *Mater. Sci. Eng. C* **2017**, *71*, 669–677. [CrossRef]
21. Robinson, M.K.; Cohen, C.; de Brugerolle de Fraissinette, A.; Ponec, M.; Whittle, E.; Fentem, J.H. Non-Animal Testing Strategies for Assessment of the Skin Corrosion and Skin Irritation Potential of Ingredients and Finished Products. *Food Chem. Toxicol.* **2002**, *40*, 573–592. [CrossRef] [PubMed]
22. Nohynek, G.J.; Antignac, E.; Re, T.; Toutain, H. Safety Assessment of Personal Care Products/Cosmetics and Their Ingredients. *Toxicol. Appl. Pharmacol.* **2010**, *243*, 239–259. [CrossRef] [PubMed]

23. The Scientific Committee on Cosmetic Products and Non-Food Products Intended for Consumers. *Opinion Concerning Basic Criteria of the Protocols for the Skin Compatibility Testing of Potentially Cutaneous Irritant Cosmetic Ingredients or Mixtures of Ingredients on Human Volunteers*; SCCNFP/0245/99; SCCNFP: Brussels, Belgium, 1999.
24. Badhani, B.; Sharma, N.; Kakkar, R. Gallic Acid: A Versatile Antioxidant with Promising Therapeutic and Industrial Applications. *RSC Adv.* **2015**, *5*, 27540–27557. [CrossRef]
25. Ou, S.; Kwok, K.-C. Ferulic Acid: Pharmaceutical Functions, Preparation and Applications in Foods. *J. Sci. Food Agric.* **2004**, *84*, 1261–1269. [CrossRef]
26. Franzè, S.; Musazzi, U.M.; Minghetti, P.; Cilurzo, F. Drug-in-Micelles-in-Liposomes (DiMiL) Systems as a Novel Approach to Prevent Drug Leakage from Deformable Liposomes. *Eur. J. Pharm. Sci.* **2019**, *130*, 27–35. [CrossRef] [PubMed]
27. Andreozzi, P.; Funari, S.S.; La Mesa, C.; Mariani, P.; Ortore, M.G.; Sinibaldi, R.; Spinozzi, F. Multi- to Unilamellar Transitions in Catanionic Vesicles. *J. Phys. Chem. B* **2010**, *114*, 8056–8060. [CrossRef]
28. Rappolt, M.; Gregorio, G.M.D.; Almgren, M.; Amenitsch, H.; Pabst, G.; Laggner, P.; Mariani, P. Non-Equilibrium Formation of the Cubic *Pn 3 m* Phase in a Monoolein/Water System. *Europhys. Lett.* **2006**, *75*, 267–273. [CrossRef]
29. Danaei, M.; Dehghankhold, M.; Ataei, S.; Hasanzadeh Davarani, F.; Javanmard, R.; Dokhani, A.; Khorasani, S.; Mozafari, M. Impact of Particle Size and Polydispersity Index on the Clinical Applications of Lipidic Nanocarrier Systems. *Pharmaceutics* **2018**, *10*, 57. [CrossRef]
30. Mahale, N.B.; Thakkar, P.D.; Mali, R.G.; Walunj, D.R.; Chaudhari, S.R. Niosomes: Novel Sustained Release Nonionic Stable Vesicular Systems—An Overview. *Adv. Colloid Interface Sci.* **2012**, *183–184*, 46–54. [CrossRef]
31. Sguizzato, M.; Valacchi, G.; Pecorelli, A.; Boldrini, P.; Simelière, F.; Huang, N.; Cortesi, R.; Esposito, E. Gallic Acid Loaded Poloxamer Gel as New Adjuvant Strategy for Melanoma: A Preliminary Study. *Colloids Surf. B Biointerfaces* **2020**, *185*, 110613. [CrossRef]
32. Marino, T.; Galano, A.; Russo, N. Radical Scavenging Ability of Gallic Acid toward OH and OOH Radicals. Reaction Mechanism and Rate Constants from the Density Functional Theory. *J. Phys. Chem. B* **2014**, *118*, 10380–10389. [CrossRef] [PubMed]
33. Yang, J.; Chen, J.; Hao, Y.; Liu, Y. Identification of the DPPH Radical Scavenging Reaction Adducts of Ferulic Acid and Sinapic Acid and Their Structure-Antioxidant Activity Relationship. *LWT* **2021**, *146*, 111411. [CrossRef]
34. Liu, C.; Chen, C.; Ma, H.; Yuan, E.; Li, Q. Characterization and DPPH Radical Scavenging Activity of Gallic Acid-Lecithin Complex. *Trop. J. Pharm. Res.* **2014**, *13*, 1333. [CrossRef]
35. Witczak, Z.J. Polysaccharides in Medicinal Applications. Edited by Severian Dumitriu, Marcel Dekker, Inc. New York, ISBN 0-8247-9540-7. 1996, 794 pp. $195.00. *J. Carbohydr. Chem.* **1997**, *16*, 245–247. [CrossRef]
36. Minnelli, C.; Moretti, P.; Fulgenzi, G.; Mariani, P.; Laudadio, E.; Armeni, T.; Galeazzi, R.; Mobbili, G. A Poloxamer-407 Modified Liposome Encapsulating Epigallocatechin-3-Gallate in the Presence of Magnesium: Characterization and Protective Effect against Oxidative Damage. *Int. J. Pharm.* **2018**, *552*, 225–234. [CrossRef]
37. Avadhani, K.S.; Manikkath, J.; Tiwari, M.; Chandrasekhar, M.; Godavarthi, A.; Vidya, S.M.; Hariharapura, R.C.; Kalthur, G.; Udupa, N.; Mutalik, S. Skin Delivery of Epigallocatechin-3-Gallate (EGCG) and Hyaluronic Acid Loaded Nano-Transfersomes for Antioxidant and Anti-Aging Effects in UV Radiation Induced Skin Damage. *Drug Deliv.* **2017**, *24*, 61–74. [CrossRef]
38. Xu, X.; Xiao, H.; Zhao, J.; Zhao, T. Cardioprotective Effect of Sodium Ferulate in Diabetic Rats. *Int. J. Med. Sci.* **2012**, *9*, 291–300. [CrossRef]
39. Karamać, M.; Koleva, L.; Kancheva, V.; Amarowicz, R. The Structure–Antioxidant Activity Relationship of Ferulates. *Molecules* **2017**, *22*, 527. [CrossRef]
40. Zhang, J.; Huang, X.; Huang, S.; Deng, M.; Xie, X.; Liu, M.; Liu, H.; Zhou, X.; Li, J.; Ten Cate, J.M. Changes in Composition and Enamel Demineralization Inhibition Activities of Gallic Acid at Different PH Values. *Acta Odontol. Scand.* **2015**, *73*, 595–601. [CrossRef]

Article

Biopolymer Lipid Hybrid Microcarrier for Transmembrane Inner Ear Delivery of Dexamethasone

Maximilian George Dindelegan [1,2], Violeta Pașcalău [3,*], Maria Suciu [4,5], Bogdan Neamțu [3], Maria Perde-Schrepler [6], Cristina Maria Blebea [2], Alma Aurelia Maniu [2], Violeta Necula [2], Anca Dana Buzoianu [1], Miuța Filip [7], Alexandra Csapai [3] and Cătălin Popa [3]

[1] Department of Clinical Pharmacology, "Iuliu Hatieganu" University of Medicine and Pharmacy, 23 Gh. Marinescu Street, 400337 Cluj-Napoca, Romania; maximilian.dindelegan@gmail.com (M.G.D.); abuzoianu@umfcluj.ro (A.D.B.)

[2] Department of Otorhinolaringology, "Iuliu Hatieganu" University of Medicine and Pharmacy, 4-6 Clinicilor Street, 400006 Cluj-Napoca, Romania; cristina_blebea@yahoo.com (C.M.B.); almacjro@yahoo.com (A.A.M.); neculav@yahoo.com (V.N.)

[3] Department of Materials Science and Engineering, Technical University of Cluj-Napoca, 28 Memorandumului Street, 400114 Cluj-Napoca, Romania; bogdan.neamtu@stm.utcluj.ro (B.N.); alexandra.csapai@stm.utcluj.ro (A.C.); catalin.popa@stm.utcluj.ro (C.P.)

[4] Electron Microscopy Center "C. Craciun", Biology and Geology Faculty, Babes-Bolyai University, 5-7 Clinicilor Street, 400006 Cluj-Napoca, Romania; suciu.maria@ubbcluj.ro

[5] National Institute for Research and Development of Isotopic and Molecular Technologies, 67-103 Donath Street, 400293 Cluj-Napoca, Romania

[6] Institute of Oncology "Prof Dr. Ion Chiricuta", 34-36 Republicii Street, 400015 Cluj-Napoca, Romania; pmariaida@yahoo.com

[7] "Raluca Ripan" Institute for Research in Chemistry, Babes-Bolyai University, 30 Fantanele Street, 400294 Cluj-Napoca, Romania; miuta.filip@ubbcluj.ro

* Correspondence: violeta.pascalau@stm.utcluj.ro

Citation: Dindelegan, M.G.; Pașcalău, V.; Suciu, M.; Neamțu, B.; Perde-Schrepler, M.; Blebea, C.M.; Maniu, A.A.; Necula, V.; Buzoianu, A.D.; Filip, M.; et al. Biopolymer Lipid Hybrid Microcarrier for Transmembrane Inner Ear Delivery of Dexamethasone. *Gels* **2022**, *8*, 483. https://doi.org/10.3390/gels8080483

Academic Editors: Maddalena Sguizzato, Rita Cortesi and Rachel Yoon Chang

Received: 12 July 2022
Accepted: 27 July 2022
Published: 1 August 2022

Publisher's Note: MDPI stays neutral with regard to jurisdictional claims in published maps and institutional affiliations.

Abstract: Dexamethasone is one of the most often used corticosteroid drugs for sensorineural hearing loss treatment, and is used either by intratympanic injection or through systemic delivery. In this study, a biopolymer lipid hybrid microcarrier was investigated for enhanced local drug delivery and sustained release at the round window membrane level of the middle ear for the treatment of sensorineural hearing loss (SNHL). Dexamethasone-loaded and dexamethasone-free microparticles were prepared using biopolymers (polysaccharide and protein, pectin and bovine serum albumin, respectively) combined with lipid components (phosphatidylcholine and Dimethyldioctadecylammonium bromide) in order to obtain a biopolymer–liposome hybrid system, with a complex structure combining to enhance performance in terms of physical and chemical stability. The structure of the microparticles was evaluated by FTIR, XRD, thermal analysis, optical microscopy, and scanning electron microscopy (SEM). The encapsulation efficiency determination and the in vitro Dexamethasone release study were performed using UV-Vis spectroscopy. The high value of encapsulation efficiency and the results of the release study indicated six days of sustained release, encouraging us to evaluate the in vitro cytotoxicity of Dexamethasone-loaded microparticles and their influence on the cytotoxicity induced by Cisplatin on auditory HEI-OC1 cells. The results show that the new particles are able to protect the inner ear sensory cells.

Keywords: microparticles; Lipoid S 100; pectin; BSA; hydrogel

1. Introduction

Hearing loss represents a worldwide health issue. More than 1.5 billion people are affected by hearing loss, according to the "World Report on Hearing" published by the World Health Organization in 2021. A large number, approximately 430 million, suffer from moderate or worse hearing loss levels, which impact their quality of life and daily activities [1]. Sensorineural hearing loss can be genetically inherited or acquired. Acquired SNHL follows sensory inner ear cell destruction. The causes of inner ear cell destruction are represented by infectious pathogens (viral or bacterial), ototoxic drugs (aminoglycoside antibiotics, loop diuretics, platinum-based chemotherapeutics), traumatic causes, noise-induced hearing loss, and idiopathic hearing loss.

Sudden sensorineural hearing loss (SSNHL) is defined as SNHL of at least 30 dB over at least three contiguous frequencies occurring in a maximum of 72 h [2]. SSNHL has an incidence of 5–20 per 100,000 persons. In the majority of patients the etiology remains unknown, and thus their hearing loss is considered idiopathic [3]. About 98% of otolaryngologists in the U.S. state that they treat idiopathic SSNHL with oral steroids, and 8% reported using intratympanic administration of steroid drugs [4].

Platinum-based chemotherapeutics, such as Cisplatin, cause bilateral high-frequency sensorineural hearing loss [5]. Many patients who are undergoing treatment with cisplatin chemotherapy experience hearing loss, with studies suggesting that 40–80% are affected and experience permanent hearing loss [6]. Cisplatin induces hearing loss through an imbalance of the antioxidant defense system; increasing the reactive oxygen species increase the early immediate release and synthesis of pro-inflammatory cytokines, which induces apoptosis in inner ear cells [7]. Experimental studies on mice have shown that cisplatin-induced ototoxicity can be avoided by daily intratympanic dexamethasone injection for all sound frequencies except 32 kHz, where little protection was observed [8].

The most widely accepted and used treatment for acquired SNHL is represented by corticotherapy [9]. Corticosteroids can be administered through systemic delivery or local delivery. Systemic delivery of steroid drugs into the inner ear is restricted because of the presence of the blood–labyrinth barrier, which isolates the inner ear from the bloodstream [10], and can lead to important systemic side effects because of the high doses needed to reach therapeutic concentration [11]. The other option used in clinical practice for reaching therapeutic drug concentrations in the inner ear is represented by intratympanic delivery of substances. Injection of a drug through the tympanic membrane into the middle ear allows it to diffuse through the round window membrane and oval window membrane into the inner ear, effectively bypassing the blood–labyrinth barrier and avoiding the systemic side effects of the molecules [12]. The main disadvantages of transtympanic delivery are represented by clearing of the substance through the eustachian tube, resorption through the middle ear mucosa, the risk of otitis media, and persistent perforation of the tympanic membrane [13]. The clearance of the substances from the middle ear means that the intratympanic injection must be repeated multiple times throughout the treatment schedule. To counter clearing of the drugs from the level of the middle ear, various prolonged drug delivery systems have been proposed [14,15]. Developing a stable and safe prolonged drug delivery platform at the level of the inner ear would lead to a paradigm shift for corticosteroid treatment in clinical practice, effectively removing the need for repeated injections through the tympanic membrane into the middle ear.

Dexamethasone (Dexa) is a corticosteroid drug that has been used in the treatment of sudden sensorineural hearing loss both by systemic administration and by intratympanic injection in clinical settings [16,17]. Compared to prednisone and prednisolone, it has minimal mineralocorticoid effect and a longer duration of action [18,19].

Developing microcarriers capable of prolonged local delivery of corticosteroid drugs into the middle ear and from there into the inner ear could represent a viable solution for the treatment of patients suffering from treatable acquired SNHL (ototoxic hearing loss, SSNHL).

Our approach is based on a liposomal-type microcarrier with enhanced structure for Dexa loading, delivery, and sustained release, and with major components that are natural. Liposomes are self-assembled phospholipid vesicles composed of an aqueous compartment surrounded by one or more phospholipid bilayer membranes. Although liposomes offer several advantages over other carrier systems, such as the ability to carry both hydrophobic and hydrophilic compounds, high encapsulation efficiency, biocompatibility, and transport ability through cell membranes, they are thermodynamically unstable systems prone to aggregation, fusion, degradation, or hydrolyzation, resulting in leakage of the entrapped compounds. The main limitations of liposome application in drug delivery systems are their physical and chemical instability [20,21] and their retention time after both systemic and local application [22]. Hydrophobic interactions are responsible for their physical instability, while degradative lipid oxidation determines their chemical instability. Our intent was to improve the physical stability of liposomes by introducing both a polysaccharide and a protein in the liposome structure to increase repulsion forces through both steric and electrostatic interactions.

The use of biopolymers such as polysaccharides and proteins has led to a second generation of liposomes, namely, biopolymer lipid hybrid (PLH) systems, which have improved in vitro and in vivo performance based on the cumulative benefits of the polymer and lipid together [23]. From the many types of PLH, we have chosen ones with a modified surface in order to improve the system's structural stability in the case of changes in shape, size, surface charge, or lipid chain ordering [24]. We synthesized a polysaccharide protein–lipid hybrid system which can offer advantages of biodegradability, biocompatibility, and lower toxicity. Lipids, one of the major human nutrients, are biocompatible and biodegradable. The safety and diversity of lipids have attracted interest to their applications in drug delivery.

The lipid phosphatidylcholine contained soy lecithin (Lipoid S100) (L), and Dimethyl-dioctadecylammonium bromide (DDAB) was used for the synthesis of the surface coated liposomes. Phosphatidylcholine is one of the most widely used lipids in liposome preparation, and DDAB is another lipid component which contributes to increasing the electrostatic interactions with the other components of the system.

As a polysaccharide, we chose pectin (P), an anionic polysaccharide composed of galacturonic acid units linked by α-1,4 bonds. It has been demonstrated that interaction of P with egg lecithin liposomes can regulate the freedom of lipid molecules and enhance their rigidity and mechanical strength [25]. Natural P is characterized by different degrees of carboxylic group methoxylation, being conventionally designated as low (LMP) and high methoxylated (HMP) P, respectively. Studies have compared the performance of the two kinds of P on the stability of liposomes. The reported results highlight higher stability for LMP-coated liposomes, which is provided by the higher amount of carboxylic groups; these make the surface more negative, increasing electrostatic repulsion between particles and allowing LMP to better bind to the liposome surface through hydrogen bonding. In addition, LMP has increased availability for participation in crosslinking reactions with metal cations, forming stable hydrogel networks on the coated liposome surfaces [26,27]. P is a water-soluble and edible polysaccharide, and in addition to its emulsifying, suspending, and hydrogel-forming properties it acts as a mucoadhesive agent as well, sustaining drug release for rectal and nasal drug delivery [28,29].

The choice of Bovine serum albumin (BSA) as a protein candidate for building the microcarrier was based on its intrinsic biodegradability, biocompatibility, and lack of toxicity and immunogenicity [30,31], being intensively used as a protein model for transporting different hydrophobic macromolecules or small drug molecules to target sites [32,33]. BSA is an ampholyte with a pH-dependent charge, providing reversible sites to bind and release drugs [34,35].

Our approach is based on a modified dispersion technique for liposome preparation, which has the advantage of being the most well suited for passive loading of hydrophobic drugs with high encapsulation efficiency [36,37]. We studied and selected the optimal parameters for processing free (Lmp) and Dexa-loaded (Lmp/Dexa) Lipid/pectin/BSA microparticulate systems, considering the concentrations and sequence of adding the components, stirring speed, temperature, incubation time, separation, and storage conditions, as well as samples preparation for physical chemical characterization. On the other hand, the choice of the Dexa encapsulation pathway in the microparticle synthesis stage has the advantage of allowing a high degree of encapsulation and represents a way of ensuring the protection of the drug while not subjecting it to several stages of processing, which would affect its structure. The use of both polysaccharides and proteins to improve the system's stability and sustain drug delivery by increasing repulsion forces through steric and electrostatic interactions and by building a protective shell against degradative lipid oxidation is the novelty of our study. The results achieved in physical chemical characterization of the obtained microparticles confirm the complex structure as well as the stability of the developed system. The in vitro release process of Dexa and in vitro cytotoxicity on the House Ear Institute-Organ of Corti 1 (HEI-OC1) cell line and the spontaneously immortalized human keratinocyte cell line HaCaT, along with the influence on the cytotoxicity induced by Cisplatin in HEI-OC1 cells of the new system, were all assessed.

The complex architecture of Lmp/Dexa as a hybrid biopolymer lipid system offers both stability and multifunctionality, making the microcarrier intelligent and suitable for specific local applications of Dexa. To the best of our knowledge, this is the first report to propose this approach.

2. Results and Discussion

2.1. Encapsulation Efficiency of Dexa (%) and Dexa Loading Efficiency (%)

The hydrophobicity of Dexa was the main reason for choosing to build the above biopolymer lipid hybrid architecture. Dexa loading in Lmp was achieved in the same stage as microparticle processing by dispersing the ethanolic solution in which Dexa and the two lipids were dissolved into the aqueous solution containing the hydrophilic components (P–Calcium crosslinked hydrogel and BSA). The Dexa-loaded Lmp is denoted Lmp/Dexa. Table 1 shows the mean values of E_{ef} (%) and Dexa Load$_{ef}$ (%) (g/100g), calculated based on Equation (1) for all three variants of microparticles (Lmp/Dexa) achieved, taking into account three sample preparations. It should be noted that the efficiency of the encapsulation is influenced by the Dexa concentration used in the preparation, which allows variants of microparticles with Dexa content suitable for specific applications. As the experimental data show, one way to increase the loading efficiency is to use a higher concentration of Dexa in the preparation phase of the microparticles; the process parameters can be optimized as well. A possible option to increase the loading efficiency could be the use of a variety of water-soluble Dexa, such as Disodium Dexa Phosphate, which would allow much higher Dexa concentrations. This will be one of our concerns in the future.

Table 1. Values of Dexa Encapsulation efficiency (%) and Dexa Loading efficiency for different ratios of Dexa:Lipoid S100:DDAB used in Lmp/Dexa preparation.

Dexa:Lipoid S100:DDAB Ratio	Dexa Encapsulation Efficiency (%) $\pm$ SD	Dexa Loading Efficiency (%) $\pm$ SD
2:10:1	83.07 $\pm$ 1.35	0.22 $\pm$ 0.007
3:10:1	89.86 $\pm$ 0.42	0.36 $\pm$ 0.009
4:10:1	92.5 $\pm$ 0.61	0.50 $\pm$ 0.01

The high value of E_{ef} (%) should be due to the hydroxy groups of Dexa involved in hydrogen bonds with the hydroxy groups of P and BSA, as well as to increased electrostatic interactions with Phosphatidylcholine (Lipoid S100) and DDAB. In this respect, Dexa loading in Lmp was favored in both the biopolymer coating and in the core of the hybrid system. The first variant was chosen, with the ratio 2:10:1, and throughout the paper we refer to it under the name Lmp/Dexa.

2.2. Lmp and Lmp/Dexa Structure Characterization

2.2.1. Fourier-Transform Infrared Spectroscopy

Fourier-transform infrared (FTIR) spectra of the precursors (Lipoid S 100, P, BSA, DDAB) as well as the free (Lmp) and Dexa-loaded (Lmp/Dexa) formulations achieved in the experimental section are shown in Figure 1. Interpretation of the spectra can provide information about the chemical structure of the synthesized microstructures and the Dexa encapsulation process. The FTIR spectra may highlight, where applicable, the new bonds formed between Dexa and Lmp and the disappearance of bonds from the precursors. As can be seen in Figure 1a, the Lmp spectrum shows characteristic bands of P, Phosphatidyl choline, BSA, and DDAB. The vibration of the cyclic etheric band at 1011 cm^{-1} from P overlaps with those of alkyl ether from Phosphatidyl choline at 1073 cm^{-1}, the vibration bands of OH and C–H bonds, which appear at 3313, 2860, and 2921 cm^{-1}, respectively, and specific bands of P and Phosphatidyl choline such as the asymmetric and symmetric stretch of COO– at 1611 cm^{-1} and 1464 cm^{-1} and the stretch vibration due to the C=O carboxylic acid from P and C=O of ketone from Lipoid S100 at 1734 cm^{-1}.

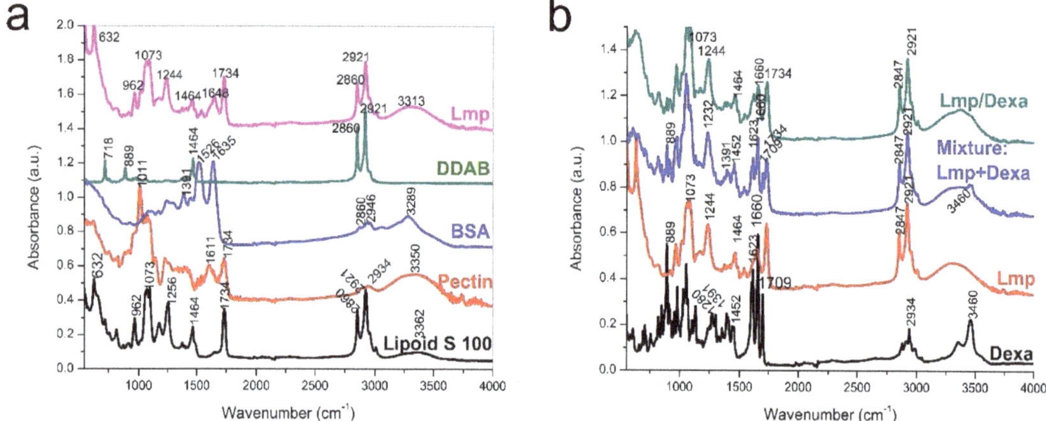

Figure 1. Overlapped FTIR spectra of: Lmp and the precursors (Lipoid S 100, P, BSA and DDAB) used for Lmp synthesis (**a**); pure Dexa, Lmp, Physical mixture of Lmp + Dexa and Lmp/Dexa (**b**).

The two amide-specific bands of BSA (particularly the in-plane bend vibration band at 1528 cm^{-1} and the stretching vibration band of the C=O bond at 1635 cm^{-1}) can both be observed in the BSA, and are superimposed with those of the carboxylate bands into a broad band in the Lmp spectrum, the a result of low BSA content of in the structure. The FTIR spectrum of pure Dexa (Figure 1b) shows multiple bands due to the vibrations of the chemical structure with many functional groups in the molecule: –C–F stretch—1432 cm^{-1}; ketone C=O stretch—1709 cm^{-1}; C–H stretch—2934 cm^{-1}; HO– stretch—3460 cm^{-1}. The FTIR spectrum of Lmp/Dexa, shows bands provided by both Lmp and Dexa, most of them being superimposed. No major differences in the FTIR spectra of physical mixture Lmp + Dexa and Lmp/Dexa were observed, thus, the FTIR tests confirmed that there are no new bonds created between pure Dexa and the functional groups of Lmp.

2.2.2. X-ray Diffraction

The XRD patterns of Lmp and Lmp/Dexa samples are shown in Figure 2 alongside those of the precursors used for their synthesis, the samples of Dexa, Lmp, and Lmp/Dexa, respectively, treated with PBS, and the Lmp + Dexa physical mixture. As can be seen in Figure 2a, Lmp is in amorphous state, with the diffraction pattern showing only a characteristic halo, although some of the precursors are in a crystalline state. In the same figure, the X-ray pattern of the pure Dexa reveals a crystalline state. The XRD pattern of Lmp/Dexa samples indicates the presence of two phases, one amorphous and the other crystalline. The crystalline phase can be assigned to Dexa because the diffraction peaks of Lmp/Dexa appeared at the same positions as in the case of the pure Dexa. However, certain intensities are not identical, indicating a preferential orientation of the recrystallized Dexa crystals due to spatial constraints, as reported by other authors [38,39]. The presence of Dexa in the Lmp/Dexa samples confirms the success of the loading process. In order to study the release of Dexa from Lmp/Dexa in PBS medium, the diffraction patterns of pure Dexa and Lmp/Dexa samples recovered from their suspensions in PBS were examined and compared with those of dried PBS, pure Dexa, and Dexamethasone sodium phosphate. The results are shown in Figure 2b. In the XRD pattern of the Dexa + PBS sample a new maximum can be seen, indicating that Dexa has undergone a transformation in its crystalline structure. As a result, the newly formed crystalline species, which could be a phosphate ester of Dexa, absorbs UV-Vis radiation corresponding to a wavelength other than that of pure Dexa, and measurement of the concentration of the new species using the calibration curve recorded for pure Dexa is no longer possible. These results confirm the experimental observations, and constitute an argument in favor of the decision to avoid conducting the study of release in PBS and instead conduct it in water containing a minimum amount of ethanol. Figure 2c compares the diffraction patterns of the Lmp + Dexa physical mixture with the Lmp/Dexa synthesized hybrid system. The two diffractograms are similar, leading to the conclusion that during the encapsulation process Dexa does not undergo polymorphic transformation.

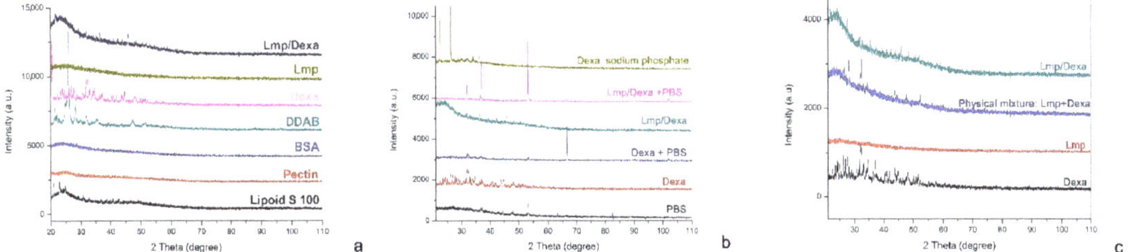

Figure 2. X-ray diffraction patterns of: Lmp and Lmp/Dexa and the precursors (Lipoid S 100, P, BSA and DDAB, Dexa) used for Lmp and Lmp/Dexa synthesis (**a**); dried PBS, Dexa, Dexa recovered from suspension in PBS, Lmp/Dexa, and Lmp/Dexa recovered from suspension in PBS, and Dexa sodium phosphate (**b**); pure Dexa, Lmp and the physical mixture of Lmp+Dexa, and Lmp/Dexa (**c**).

2.2.3. Thermal Analysis

The thermograms of pure Dexa, the free Lmp and Dexa encapsulated system, and the physical mixture between Lmp and Dexa are shown in Figure 3. The interaction between Dexa and the lipidic components of Lmp/Dexa was demonstrated by the decrease of the melting temperature (the endothermic peak), as can be seen in Figure 3, from 158 °C for Lmp to 155 °C for Lmp/Dexa and to 157 °C for the physical mixture between Lmp and Dexa. The Dexa molecules seem to be dissolved in the melted Lmp matrix, justifying the absence of the endothermic peak from 260 °C in the DSC thermograms of Lmp/Dexa and the physical mixture of Lmp + Dexa.

2.2.4. Optical Microscopy

Optical microscopy was used to observe the shape and size of both Lmp and Lmp/Dexa. The optical microscopy images (Figure 4) of the samples prepared without Fluorescein for (a) Lmp and (e) Lmp/Dexa and with Fluorescein for (b–d) Lmp and (f–h) Lmp/Dexa show particles of spherical shape, around 25 μm size in average, and with or without Dexa hexagonal crystals on the surface. Moreover, the free Dexa crystals (35 μm length) recovered from the supernatant resulting from Lmp/Dexa without Fluorescein (i) and Lmp/Dexa with Fluorescein preparation (j–l) are shown. Considering the fact that P makes up the majority of the shell that surrounds the lipid part of the microsystem, the charge is expected to be negative. Images of fluorescent samples suggest the presence of pectin hydrogel, capable of fixing fluorescein in the microparticle shell. This information is relevant for the stability profile of the microparticles as well as for their swelling behavior in aqueous environments. As can be seen, free Dexa crystals do not change their appearance when preparing microparticles with fluorescein.

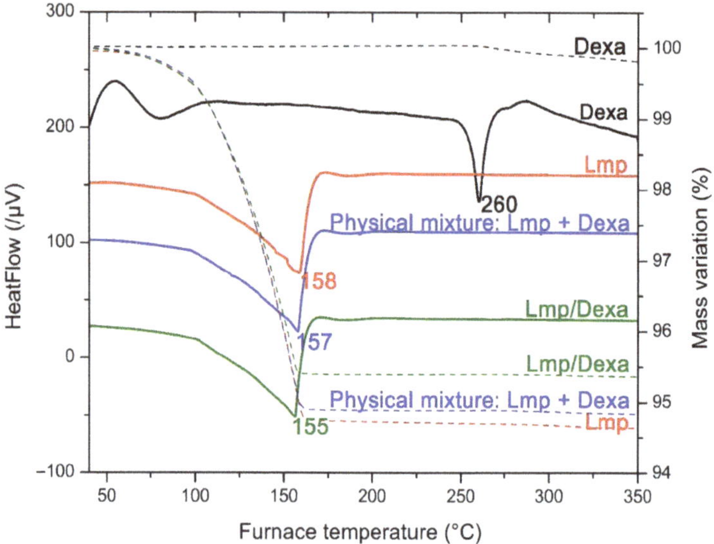

Figure 3. Overlapped DSC and TG thermograms of pure Dexa, Lmp, the physical mixture of Lmp+Dexa, and Lmp/Dexa.

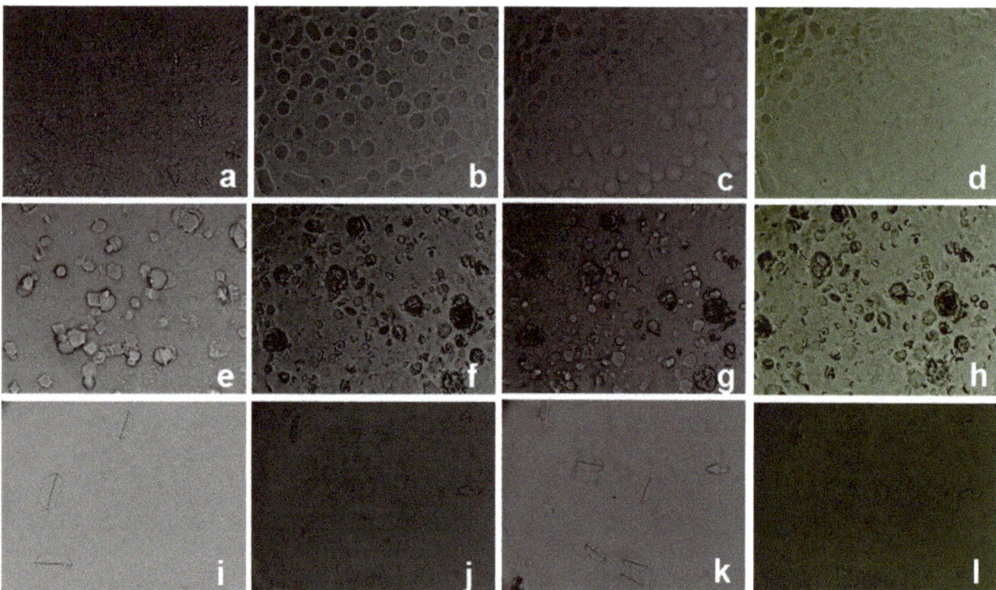

Figure 4. Optical images of: Lmp without Fluorescein (**a**) and Lmp with Fluorescein Blue filter (**b**), Bright field (**c**), and Green filter (**d**) (40×); Lmp/Dexa without Fluorescein (**e**) and Lmp/Dexa with Fluorescein Blue filter (**f**), Bright field (**g**), and Green filter (**h**) (40×); Free Dexa crystals recovered from the supernatant resulting from Lmp/Dexa without Fluorescein (**i**) and Lmp/Dexa with Fluorescein preparation Blue filter (**j**), Bright field (**k**), and Green filter (**l**) (40×).

2.2.5. SEM

The SEM micrographs (Figure 5) provide information about the shape and size of the examined particles as well as the presence of drug crystals on the surface. The SEM images highlight core-shell spherical microparticles around 20 μm in diameter. The shell seems to contain drug crystals.

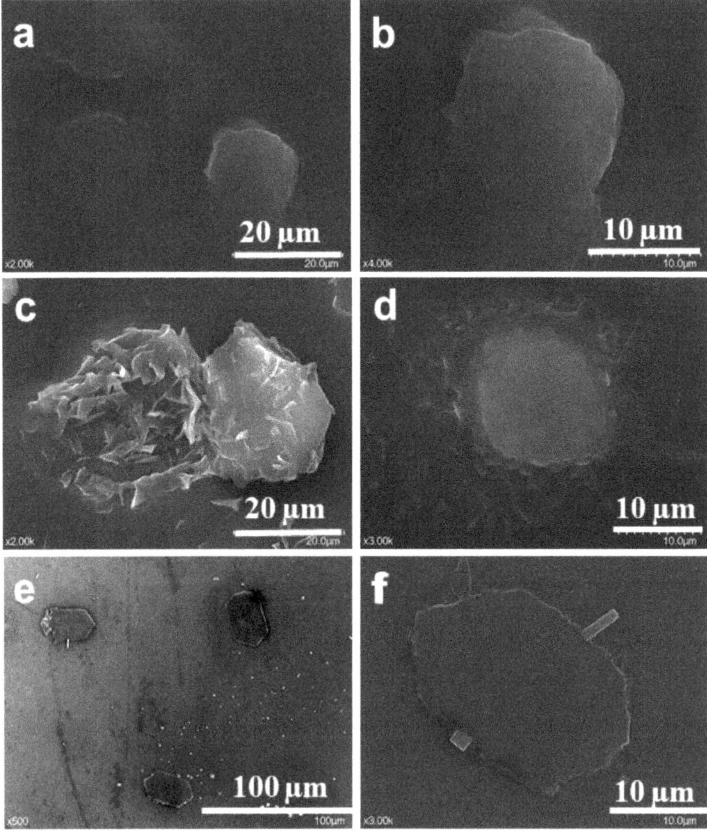

Figure 5. SEM microimages of Lmp (**a**,**b**), Lmp/Dexa (**c**,**d**), and Dexa crystals (**e**,**f**) from the supernatant separated after Lmp/Dexa sedimentation.2.3. Lmp/Dexa Charateristics.

2.3. Lmp/Dexa Charateristics

2.3.1. In Vitro Release Study

The release study of Dexa from Lmp/Dexa was performed in both water and PBS pH 7.4, each containing a small amount of ethanol (1%), which has the role of dissolving the released Dexa molecules. The supply of ethanol required for the continuous dissolution of Dexa molecules was ensured by replacing the volumes of samples removed from the release medium with identical volumes of fresh medium containing ethanol. Based on the Dexa concentrations determined from the collected samples from the release medium at different time intervals, the cumulative amounts of Dexa released were calculated, and taking into account the Dexa content of the Lmp/Dexa sample subjected to release, the cumulative release (%) values were calculated and plotted against time (Figure 6). The release profile of Dexa from pure Dexa solution of the same Dexa concentration as in the microparticles was performed, and is included in Figure 6. As expected, Dexa release from free Dexa solution has a different release profile than Lmp/Dexa release, being released almost completely in 2 h. As indicated in the release profile, Dexa is released from Lmp/Dexa continuously for six days, with a burst effect in the first 20 h in water. The release process seems to be favoured in PBS, with a decreased burst effect. After five days we observed the appearance of particles in suspension in the release medium, both in water and in PBS, which indicates a process of degradation of the studied system. This process could be caused by the multiple openings of the system in order to remove the samples taken from the release medium, and would of course be favored by the aqueous environment and the temperature of 37 °C. The release of Dexa into water is favored by the presence of the pectin hydrogel crosslinked with Ca^{2+} ions in the particle shell, as their network hydrates, swells, and dilates the pores, thus allowing small Dexa molecules to diffuse into the release medium. Although the ethanol molecules have the opposite effect of dehydrating the hydrogel network by competing for the water molecules with which they form stable hydrogen bonds, the very low ethanol content in the release medium can only slightly counterbalance the hydration process, swelling, and enlarging of the network pores. The presence of ethanol can lead to a slight prolongation of the release process. Comparing the process of release in diluted ethanol solution with that of release in pure water is not possible, because the released Dexa molecules do not dissolve in water, and thus cannot be quantified by the above-mentioned method. The results of the study certainly indicate a prolonged release of Dexa from Lmp/Dexa for six days, which recommends the use of the system studied in the in vitro and in vivo experiments for the development of a suitable and precisely controlled delivery system for inner ear therapy.

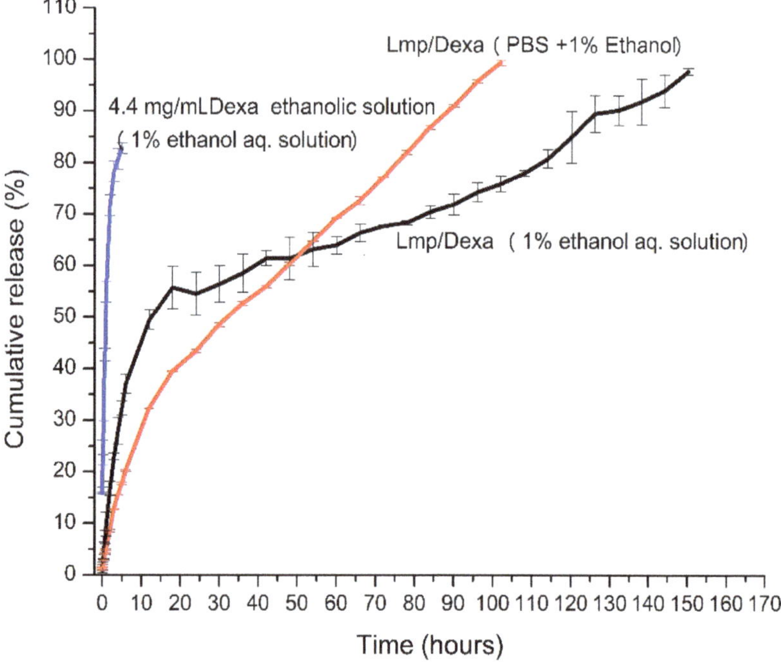

Figure 6. Cumulative release (%) versus time of Dexa released from Lmp/Dexa in 1% ethanol aqueous solution and in PBS containing 1% ethanol, and from Dexa ethanolic solution in 1% ethanol aqueous solution.

The time length of sustained drug release can vary depending on the nature of the carrier particles, their size, the degree of drug loading, and the nature and intensity of the physical or chemical interactions in which the active compound is involved. In general, this range is shorter for nanoparticles and longer for microparticles used as carriers [40–42].

There are exceptions if the drug is chemically bonded to the functional groups belonging to the carrier matrix or to a high density of cationic charges that form strong electrostatic interactions with the anionic charges of the drug.

The Dexa release profile from Lmp/Dexa indicates an extended release of up to six days with a burst effect at the beginning of the time interval, which is in agreement with other reported results [43]. This may be due to the high value of loading, as well as to the interactions between Dexa and BSA, which is an ampholyte with a pH-dependent charge, presenting reversible sites to bind and release drugs as well as P and the lipid components.

Further studies on improving of the protective coating of the microparticles through the formation of polyelectrolytic complexes between anionic and cationic polysaccharides used in the shell building process are able to provide extra stability and sustain a longer release time.

2.3.2. Cytotoxicity Assay of Lmp and Lmp/Dexa
Cell Viability Assays

The results regarding the cell viability assays after treatment of HEI-OC1 cells and HaCaT cells with Lmp and Lmp/Dexa, respectively, are illustrated in the Figure 7. Lmp and Lmp/Dexa both reduced the percentage of viable HEI-OC1 cells in a concentration-dependent manner. Concentrations below 10 µg/mL showed no significant reduction in viability compared to the control cells.

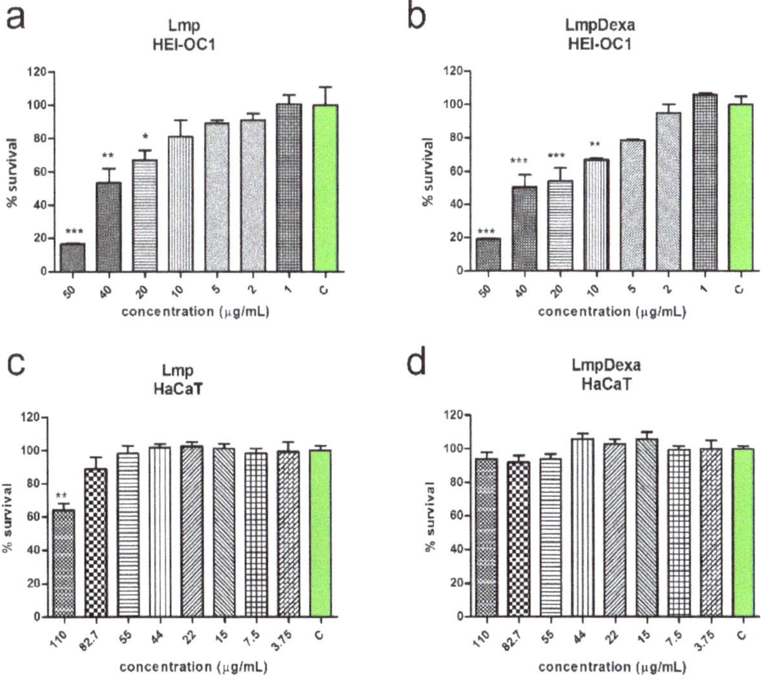

Figure 7. Concentration-dependent reduction of viability in HEI-OC1 and HaCaT cells after treatment with Lmp (**a,c**) and Lmp/Dexa (**b,d**) (*** $p < 0.001$; ** $p < 0.01$; * $p < 0.5$; one-way ANOVA, Dunnett's multiple comparisons test).

When tested on HaCaT keratinocytes, none of the Lmp samples (with or without Dexa) showed significant toxicity except when the highest concentrations were applied (110 and 82.7 µg/mL).

Our tests showed significant differences between the effects on the viability of the two cell lines used.

These differences in behavior may be explained by different physiological functions and differences in cellular uptake [44]. Studying 23 types of NPs in ten different cell lines, Kroll et al. reported differences in the sensitivity of the cell lines [45].

Another difficulty in determining whether a drug is cytotoxic lies in the sensitivity of the in vitro system. In vivo, in addition to the metabolism of a drug there are many other factors which determine its circulatory levels, such as binding to serum proteins, renal excretion, etc., and one cannot predict the final concentration that a cell will meet. In vitro, if the concentrations of most drugs under investigation is raised there is a level at which most, if not all, become cytotoxic [46].

Our results demonstrate that the sensitivity towards nanomaterial exposure is not only cell-type-specific, but also depends on the particle type used [41].

Protective Effect of Dexamethasone and Dexamethasone-Loaded Lmp against Cisplatin

Furthermore, we analyzed whether the new particles could protect inner ear cells against the cytotoxicity of Cisplatin. As reported in several previous studies, the treatment of HEI-OC1 cells with cisplatin induced a significant dose-dependent decrease in viability [47]. The IC50 of the concentration, reducing cell viability by 50%, was 65.79 µM.

In our experiments, we used a concentration of 100 µM Cis, which is above the IC50 concentration, in order to induce cell toxicity. We then treated the HEI-OC1 cells with Dexamethasone in different concentrations, followed by 100 µM Cis. We obtained a reduction in toxicity for Dexamethasone at the concentrations of 50 and 25 µg/mL, respectively (Figure 8).

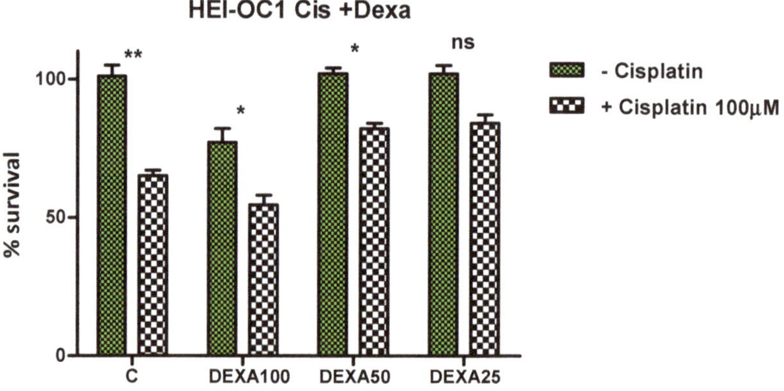

Figure 8. Protection of HEI-OC1 cells by pretreatment with Dexamethasone. Two-way ANOVA, Bonferroni post-tests. (** $p < 0.01$; * $p < 0.5$).

Dexamethasone is a synthetic fluorinated corticosteroid with multiple effects on practically every auditory cell, and it is frequently used in clinics to protect the auditory organ against inflammatory responses induced by noise, drugs, and other ototoxic stimuli. As with other drugs investigated in a study by Kalinec et al. [48], dexamethasone induced a decrease in cell viability without any significant change in cell death or caspase activation.

When we treated HEI-OC1 cells with Lmp in concentrations that were non-toxic (below 20 µg/mL) followed by 100 µM Cisplatin at concentrations of 5, 2, 1, and 0.1 µg/mL, no significant decreases in cell viability were observed compared to the non-treated cells for both types of microcapsules. (one-way ANOVA, Dunnett post-test) (Figure 9).

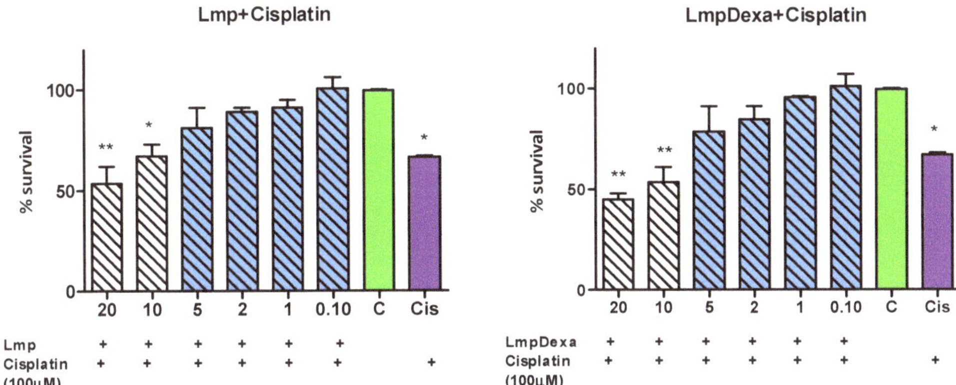

Figure 9. Lmp and Lmp/Dexa effect on cisplatin treated HEI-OC1 cells. (One-way ANOVA, Dunnett post-test). (** $p < 0.01$; * $p < 0.5$).

Our results show that the new particles, Lmp and Lmp/Dexa, are able to protect the inner ear neurosensorial cells from cisplatin toxicity in an in vitro model represented by the HEI-OC1 cell line. Dexa, as shown in our study, provides protection from cisplatin-induced toxicity in vitro. Particles that can continuously release Dexa in a controlled and prolonged manner could deliver a more efficient treatment option for different types of sensorineural hearing loss. These results point to a promising role for the newly described particles in hearing loss treatment. Further functional and anatomical in vivo studies are needed in order to test the otoprotection provided by Lmp/Dexa before the drug-delivery system can be considered a viable solution in clinical practice.

3. Conclusions

In our study, the main objective of developing a biopolymer lipid hybrid microcarrier for enhanced local Dexamethasone delivery and sustained release for the treatment of SNHL was accomplished. Dexamethasone-loaded and Dexamethasone-free microparticles were prepared using biopolymers (polysaccharide–protein and pectin–bovine serum albumin, respectively) combined with the lipid components (Lipoid S 100 and Dimethyldioctadecy-lammonium bromide) into a biopolymer–liposome hybrid system with a complex structure that combines for enhanced performance in terms of both physical and chemical stability. Structural characterization was achieved and in vitro study of Dexa release and cytotoxicity on the HEI-OC1 and HaCaT cell lines was carried out. The FTIR tests confirmed no new bonds are created between pure Dexa and the functional groups of Lmp. In addition, XRD patterns highlight that during the encapsulation process Dexa does not undergo polymorphic transformation. Images of fluorescent samples suggest the presence of pectin hydrogel in the shell of the microparticles, responsible for the stability profile of the microparticles as well as for their swelling behavior in aqueous environments. The results obtained in synthesis and characterization of the new microparticles, Lmp and Lmp/Dexa, including sustained release for up to six days and the lack of cytotoxicity to the HEI-OC1 cell line in the in vitro studies, are encouraging for the initiation of further work regarding the use of Lmp/Dexa in an in vivo functional hearing study. The sustained release of Dexa from Lmp/Dexa makes Lmp/Dexa more eligible for local delivery in the middle ear than a pure Dexa solution. Furthermore, the in vitro studies performed here will be followed by prospective in vivo studies on experimental animals in order to demonstrate the suitability of this microcarrier alone or when included in a hydrogel network as a potential vehicle for Dexa in the treatment of sensorineural hearing loss.

4. Materials and Methods

4.1. Materials

Phosphatidylcholine from Soybean Lipoid S 100 was kindly provided by Lipoid Gmbh (Ludwigshafen, Germany) Lipoid GmbH, Germany. Dexamethasone, Dexamethasone sodium phosphate, Dimethyldioctadecildiammonium bromide $\geq$ 98%, bovine serum albumin (BSA), albumin fraction V for biochemistry, Calcium chloride, and pectin from citrus peel $\geq$ 74% galacturonic acid were all purchased from Merck (Darmstadt, Germany). Sterile filtered and endotoxin tested Dulbecco's phosphate buffered saline (PBS) and all the reagents used in biological studies were purchased from Sigma Life Science/Merck (Darmstadt, Germany)SIGMA Life Science. Ethanol 96% was purchased from S.C. Nordic Chemicals SRL, Cluj-Napoca, Romania.

4.2. Methods

4.2.1. Preparation of Lipid/Pectin/BSA Microparticles (Lmp)

For Lmp preparation, we used the modified solvent dispersion technique [37,42,49,50]. A 0.2% P solution obtained by P dissolution in distilled water under vigorous magnetic stirring (1000 rpm) with heating up to 90 °C, filtrated on yellow filter paper and cooled at room temperature, was mixed with 0.1% BSA water solution at a ratio of 2:1 (*w/w*) until it homogenized. The mixture was incubated at 30 °C under slow magnetic stirring (200 rpm). A 1% CaCl$_2$ (10% CaCl$_2$ relative to the solid P) aqueous solution was added under continuous stirring. An ethanolic solution previously made by successively adding Lipoid S 100 and DDAB at a ratio of 1:10 (*w/w*) was injected with a 2 mL syringe. Incubation was maintained at 30 °C without stirring for 15 min. P and Lipoid S100 were in a 1:5 ratio (*w/w*). The obtained suspension containing Lmp was stored at 4 °C for minimum of 24 h before use. After removal of the supernatant, the sediment containing Lmp was stored at 4 °C before use.

4.2.2. Preparation of Dexa-Loaded Lipid/Pectin/BSA Microparticles (Lmp/Dexa)

For Dexa-loaded microparticles (Lmp/Dexa), the same procedure described above was used, with Dexa being introduced together with the Lipoid S 100 and DDAB ethanolic solution. Three different concentrations of Dexa were used in order to determine the influence of its concentration on the microparticles' encapsulation efficiency. In this regard, a multicomponent ethanolic solution was prepared by successively adding the components, after each complete dissolution, in the following order: Dexa, Lipoid S100, and DDAB in Ethanol 96%, under agitation and at room temperature, in a ratio of 2:10:1; 3:10:1 and 4:10:1 Dexa:Lipoid and S100:DDAB (*w/w*). Dexa and P were used 1:1 (*w/w*). Similarly, the addition of the ethanolic solution in the mixture of P and BSA solutions was carried out by incubation at 30 °C and 200 rpm, drop-wise, using a 2 mL syringe while maintaining incubation and without stirring after complete addition. The suspension containing Lmp/Dexa was kept at 4 °C for a minimum of 24 h for deposition. After removal of the supernatant, the sediment containing Lmp/Dexa was stored at 4 °C before use. The supernatants collected from each of the three prepared Lmp/Dexa variants were used to determine the free Dexa concentration, which was used for the further calculation of its encapsulation efficiency.

4.2.3. Fluorescent Lmp and Lmp/Dexa Samples Preparation

In order to obtain information about the structure of the new developed hybrid biopolymer lipid microparticles, fluorescent microparticles were prepared using water-soluble Disodium Fluoresceinum. In this way, it is be possible to observe the position of the pectin hydrogel layer via fluorescence optical microscopy images, as it becomes fluorescent relative to the lipid layer whether inside or in the coating of the microparticles, thus providing information about the structure of the microparticles. The 1 mg/mL Fluorescein aqueous solution was introduced during the processing of the microparticles, more precisely, in the aqueous phase before the addition of the ethanolic solution of lipids or the Dexa or lipids (5% Disodium Fluoresceinum relative to the solid P). The resulting suspensions were kept at 4 °C in the dark.

4.2.4. Suspensions of Lmp and Lmp/Dexa in PBS and Physical Mixture of Pure Dexa with Lmp

Preliminary tests on the behavior of pure Dexa, Lmp, and Lmp/Dexa in PBS were performed in order to study of the release of Dexa from Lmp/Dexa in PBS. In this regard, samples of pure Dexa, Lmp, and Lmp/Dexa were dispersed in PBS. The samples of Dexa–PBS, Lmp–PBS, and Lmp/Dexa–PBS recovered from the PBS were analyzed by FTIR and XRD and compared to pure Dexa, Lmp and Lmp/Dexa. In addition, a physical mixture of pure Dexa with Lmp in the ratio corresponding to the content of Dexa in Lmp/Dexa was prepared and analyzed in comparison to Lmp/Dexa.

4.2.5. Encapsulation Efficiency of Dexa and Dexa Loading Efficiency in Lmp/Dexa

In order to calculate the encapsulation efficiency (E_{ef}), the UV-Vis spectrophotometry quantitative method (SPECORD 250 PLUS spectrophotometer equipped with WinAspect software) was used. Dexa concentration was determined at λ = 243 nm based on the calibration curve, registered using five diluted solutions of pure Dexa in 50% (*v/v*) ethanol solution with concentrations between 0.001 mg/mL and 0.02 mg/mL.

After 24 h at 4 °C, the suspension containing Lmp/Dexa separated into two layers, a clear supernatant solution and a layer of white sediment. After removing the supernatant, both layers were measured considering the net volume of each fraction. The measured supernatant was used to evaluate the free Dexa concentration and to calculate E_{ef} (%). The supernatant was formed from the excess aqueous solution of P, to which was added the excess ethanolic solution containing dissolved free Dexa. Considering that Dexa is not water soluble, if the clear supernatant solution contains Dexa molecules they were certainly from the ethanolic solution in the synthesis stage. The measured volume of sediment was employed in the following in vitro studies. The sedimented Lmp/Dexa was washed with a small amount of distilled water and recovered. Using the Dexa calibration curve, the Dexa concentration in the removed supernatant solution containing Dexa remaining from the Dexa loading process (free Dexa) was determined. The E_{ef} for Dexa was estimated according to Equation (1) [51]:

$$E_{ef} (\%) = (\text{Total feeding Dexa} - \text{Free Dexa}) \times 100/\text{Total feeding Dexa} \qquad (1)$$

The Dexa loading Efficiency in Lmp/Dexa ($Dexa_{loadef}$ (%)) was calculated according to Equation (2):

$$Dexa_{loadef} (\%) = \text{Total loaded Dexa} \times 100/\text{Total Lmp} \qquad (2)$$

where

$$\text{Total loaded Dexa} = \text{Total feeding Dexa} - \text{Free Dexa} \qquad (3)$$

4.2.6. Methods for Structural Characterization of Lmp and Lmp/Dexa
FTIR Spectroscopy

FTIR spectroscopy was used to identify the functional groups of the Lmp and Lmp/Dexa by FT-IR spectra recorded using a JASCO-FTIR 610 spectrometer (Jasco Europe s.r.l. Cremella, Italy) equipped with an ATR (attenuated total reflectance) accessory with a horizontal ZnSe crystal (Jasco PRO400S). The spectra resolution was 4 cm^{-1}, and scans were repeated 100 times. An appropriate amount of the samples was placed on the ZnSe crystal, then the FTIR spectrum was recorded in the range 4000–500 cm^{-1}.

X-ray Diffraction

The influence of the entrapment process on the crystalline structure of Dexa was studied using X-ray diffraction (XRD). The diffraction patterns of the precursors, Lmp and Lmp/Dexa, were registered by an INEL EQUINOX 3000 diffractometer with CoKα radiation (λ = 1.7903Å) and a 2-theta angular range of 20–110°.

Thermal Analysis

The thermal stability of Lmp/Dexa compared with pure Dexa, Lmp, and Lmp + Dexa (the physical mixture) samples was investigated by Differential Scanning Calorimetry (DSC) and Thermogravimetric (TG) analyses using a DSC-TG Labsys Setaram Instrument. The samples were analyzed at a heat flow rate of 10 °C/min under Argon purging within the temperature range of 25 °C–350 °C using an aluminum crucible. A high purity alumina powder was used as a reference in the DSC measurements. DSC and TG measurements were carried out to establish possible interaction between the biopolymers and lipid components and to evaluate the stability of the Dexa encapsulation into the carriers. All experiments were run in triplicate.

Optical Microscopy

Optical microscopy can be a rapid, comfortable, brief, and non-resource intensive technique, and accordingly was used to follow up formation of the microparticles in terms of their shape and size and to control their stability over time. Samples of Lmp and Lmp/Dexa suspensions were deposited drop-wise onto cover glass and dried at room temperature. Images were captured using an Optika B-383FL microscope.

In order to highlight the parts of the Lmp system, we used an optical microscope equipped with a fluorescence illuminator. For this purpose, both samples, the one with intrinsic fluorescence due to the presence of BSA and the samples processed with disodium fluoresceinum, were used. The B filter (excitation at 475 nm and emission at 515 nm) and G filter (excitation at 530 nm and emission at 590 nm) f were used or fluorescence reflected light observation and compared with Bright field (BF) empty for transmitted light observation.

Scanning Electron Microscopy (SEM)

Although SEM is not usually used to characterize the morphology of liposomes because they do not withstand working conditions, degrading even more, obtaining relevant images of the proposed hybrid system confirms the stability of its structure. SEM micrographs were used to highlight the changes in the surface morphology of Lmp after Dexa loading. In this respect, samples of both Lmp and Lmp/Dexa suspensions were deposited drop-wise, without spreading the layer evenly, onto a cover glass and dried at room temperature. The dried samples were mounted on carbon sticky tabs and then covered with a 10 nm Au layer. Samples were analyzed using a Hitachi SU8230 scanning electron microscope at 30 kV, 10 µA and 8 mm working distance.

4.2.7. Methods for the Study of Lmp/Dexa Characteristics
In Vitro Release Study

The in vitro release of Dexa from Lmp/Dexa was conducted into both water and PBS pH 7.4, each containing a small amount of ethanol (1%) to allow for Dexa solubility. The study of release from pure Dexa solution of the same Dexa concentration as in the microparticles was performed in 1% ethanol aqueous solution. The assessment of Dexa concentration in the release medium using UV-Vis Quant mode based on the calibration curve was performed. The Dexa release profile from Lmp/Dexa was developed using a previously reported dialysis method [52]. The study was conducted at 37 °C, under slow magnetic stirring at 100 rpm. A 1mL Lmp/Dexa sample and 1 mL 1% Ethanol solution/PBS pH 7.4 containing 1% ethanol were introduced in a dialysis bag (MWCO:3 kDa) and the sealed bag was immersed in 19 mL 1% Ethanol solution/PBS ph 7.4 containing 1% ethanol, the release medium, being placed into a 100 mL Erlenmeyer flask with a flat bottom and closed with ground glass stopper and left for 144 h. At certain time intervals, three samples of 1 mL release medium each were removed and replaced with the same volume of fresh medium. The sample was further incubated. The collected samples were diluted with 50% ethanol solution (v/v), homogenized, and then introduced into the spectrometer measuring cuvette for Dexa quantification at $\lambda = 243$ nm against the adequate calibration curve previously registered. All studies were performed in triplicate. The Dexa concentrations at any specified time $t_{n\ (1-32)}$ ($t_1 = 5$ min, $t_2 = 10$ min, $t_3 = 15$ min, $t_4 = 30$ min, $t_5 = 45$ min and $t_6 = 1$ h, $t_7 = 2$ h, $t_8 = 3$ h, $t_9 = 4$ h, $t_{10} = 5$ h, $t_{11} = 6$ h, $t_{12} = 12$ h, $t_{13} = 24$ h, $t_{14} = 30$ h, $t_{15} = 36$ h, $t_{16} = 42$ h, $t_{17} = 48$ h, $t_{18} = 54$ h, $t_{19} = 60$ h, $t_{20} = 66$ h, $t_{21} = 72$ h, $t_{22} = 78$ h, $t_{23} = 84$ h, $t_{24} = 90$ h, $t_{25} = 102$ h, $t_{26} = 108$ h, $t_{27} = 114$ h, $t_{28} = 120$ h, $t_{29} = 126$ h, $t_{30} = 132$ h, $t_{31} = 138$ h and $t_{32} = 144$ h) (C_n) were determined.

The cumulative Dexa percent was calculated as the percent of released Dexa at a specific time n against the amount of Dexa estimated in the sample subjected to release, based on Dexa Loading$_{ef}$ (%).

The cumulative release (%) of Dexa from the total Dexa before releasing was further plotted against the release time.

Cytotoxicity Assay of Lmp and Lmp/Dexa

1.   Cell Lines, Cultures, and Treatments

For the in vitro study, we used the House Ear Institute-Organ of Corti 1 (HEI-OC1) cell line, derived from the auditory organ of a transgenic mouse "Immortomouse" [53]. The cells were cultured in Dulbecco's Modified Eagles' Medium (DMEM) supplemented with 10% fetal bovine serum (FBS) in permissive conditions of 33 °C and 10% CO_2. These conditions allowed for the expression of an immortalizing gene that triggers de-differentiation and accelerated proliferation [54]. The HEI-OC1 cell line was a gift from Professor Federico Kalinek of the House Ear Institute (Los Angeles, CA, USA). This cell line is considered adequate for investigating the cellular and molecular mechanisms involved in ototoxicity and for screening of the potential ototoxicity or otoprotective properties of new pharmacological drugs [55].

Another cell line used in our experiments was the spontaneously immortalized human keratinocyte cell line HaCaT, purchased from the Cell Line Service of the German Cancer Research Centre in Heidelberg. The keratinocytes were cultured in high-glucose DMEM supplemented with 10% d FBS, 2 mM glutamine, 50 UI/mL penicillin, and 50 mg/mL streptomycin. The reason we used the HaCaT cell line was to compare the sensitivity towards the tested particles, having in mind that the HEI-OC1 cell line is derived from mouse cells and the HaCat cell line represents a human cellular line. We wanted to test the toxicity of the Lmp and Lmp/Dexa on a human cellular line, because the microcarriers have the end goal of treating acquired SNHL in human patients.as.

All media and substances were purchased from Sigma-Aldrich Chemie GmbH, Taufkirchen, Germany.

For the cytotoxicity experiments, the cells were plated in 96-well plates, (Nunc, Thermo Scientific, Waltham, MA, USA) 2×10^4 cells/well. Treatments were carried out with serial dilutions of Lmp and Lmp/Dexa. For HEI-OC1 cells, the following concentrations were used: 50, 40, 20, 10, 5, 2, 1, and 0.1 µg/mL. The concentrations are expressed as the concentration of Dexamethasone contained in the microcapsules. For HaCaT cells, the concentrations were 100, 80, 50, 40, 20, 15, 7.5, and 3.75 µg/mL. For each experiment, cells in a minimum of three wells were left untreated as a control.

For the protection experiments on HEI-OC1 cells, 100 µM Cisplatin was added to each well 30 min after treatment with Lmp and Lmp/Dexa, a concentration close to the IC50 of Cisplatin. In addition, for comparison of the protection offered by Lmp/Dexa with a well-known protective agent, two concentrations of Dexamethasone (25 and 50 µg/mL) were used.

2.   In Vitro Release Study

The viability of the cells after exposure to Lmp and Lmp/Dexa was assessed by the alamar Blue fluorimetric test from Molecular Probes (Invitrogen, Carlsbad, CA, USA). At

24 h after the treatments, 20 μL of alamarBlue reagent was added to each well and incubated in standard cell culture conditions for 1 h. The emitted fluorescence was read on a BioTek Synergy 2 microplate reader at 570 nm excitation and 585 nm emission. At each concentration, the surviving fraction was calculated as the fluorescence of the sample/fluorescence of control non treated cells × 100. Each experiment was repeated three times.

Another viability test used was the MTT (3-(4,5-dimethylthiazol-2-yl)-2,5-diphenyl-2H-tetrazolium bromide) test. The absorbance was read with a Tecan Sunrise microplate reader at 570 nm.

4.2.8. Statistical Analysis

The results were analyzed using GraphPad Prism 5 Software. (San Diego, CA, USA) and Microsoft Excel. One-way ANOVA analysis of variance and Dunnett's test were used to compare the means of different groups to the control (non-treated) group. A significant effect was considered for p values below 0.5.

Author Contributions: Conceptualization, M.G.D., V.P. and C.P.; Synthesis V.P. and A.C.; Physical chemical analyses B.N., M.F., C.M.B. and M.S.; writing—original draft preparation M.G.D., V.P. and M.P.-S.; writing—review and editing, V.N. and A.A.M.; supervision A.D.B. All authors have read and agreed to the published version of the manuscript.

Funding: This research was partial financially supported by the Project "Entrepreneurial competences and excellence research in doctoral and postdoctoral programs-ANTREDOC", a project co-funded by the European Social Fund financing agreement no. 56437/24.07.2019, and by The Executive Agency for Higher Education, Research, Development, and Innovation Funding, Romania (project number PN-III-P2-2.1-PED-2019-3813).

Institutional Review Board Statement: Not applicable.

Informed Consent Statement: Not applicable.

Data Availability Statement: Not applicable.

Conflicts of Interest: The authors declare no conflict of interest.

References

1. World Health Organization. World Report on Hearing. 2021. Available online: https://www.who.int/publications/i/item/world-report-on-hearing (accessed on 25 July 2022).
2. Wilson, W.R.; Byl, F.M.; Laird, N. The Efficacy of Steroids in the Treatment of Idiopathic Sudden Hearing Loss: A Double-blind Clinical Study. *Arch. Otolaryngol.* **1980**, *106*, 772–776. [CrossRef] [PubMed]
3. Kuhn, M.; Heman-Ackah, S.E.; Shaikh, J.A.; Roehm, P.C. Sudden Sensorineural Hearing Loss: A Review of Diagnosis, Treatment, and Prognosis. *Trends. Amplif.* **2011**, *15*, 91–105. [CrossRef]
4. Shemirani, N.L.; Schmidt, M.; Friedland, D.R. Sudden sensorineural hearing loss: An evaluation of treatment and management approaches by referring physicians. *Otolaryngol. Neck. Surg.* **2009**, *140*, 86–91. [CrossRef]
5. Waissbluth, S.; Peleva, E.; Daniel, S.J. Platinum-induced ototoxicity: A review of prevailing ototoxicity criteria. *Eur. Arch. Oto-Rhino-Laryngol.* **2017**, *274*, 1187–1196. [CrossRef]
6. Breglio, A.M.; Rusheen, A.E.; Shide, E.D.; Fernandez, K.A.; Spielbauer, K.K.; McLachlin, K.M.; Hall, M.D.; Amable, L.; Cunningham, L.L. Cisplatin is retained in the cochlea indefinitely following chemotherapy. *Nat. Commun.* **2017**, *8*, 1654. [CrossRef]
7. Yu, D.; Gu, J.; Chen, Y.; Kang, W.; Wang, X.; Wu, H. Current Strategies to Combat Cisplatin-Induced Ototoxicity. *Front. Pharmacol.* **2020**, *11*, 999. [CrossRef] [PubMed]
8. Hill, G.W.; Morest, D.K.; Parham, K. Cisplatin-Induced Ototoxicity: Effect of intratympanic dexamethoasone injections. *Otol. Neurotol.* **2008**, *29*, 1005–1011. [CrossRef] [PubMed]
9. Schreiber, B.E.; Agrup, C.; Haskard, D.O.; Luxon, L.M. Sudden sensorineural hearing loss. *Lancet* **2010**, *375*, 1203–1211. [CrossRef]
10. Nyberg, S.; Abbott, N.J.; Shi, X.; Steyger, P.S.; Dabdoub, A. Delivery of therapeutics to the inner ear: The challenge of the blood-labyrinth barrier. *Sci. Transl. Med.* **2019**, *11*, 1–12. [CrossRef]
11. Williams, D.M. Clinical pharmacology of corticosteroids. *Respir. Care* **2018**, *63*, 655–670. [CrossRef]
12. Dindelegan, M.G.; Blebea, C.; Perde-Schrepler, M.; Buzoianu, A.D.; Maniu, A.A. Recent Advances and Future Research Directions for Hearing Loss Treatment Based on Nanoparticles. *J. Nanomater.* **2022**, *2022*, 1–15. [CrossRef]
13. Salt, A.N.; Plontke, S.K. Local inner-ear drug delivery and pharmacokinetics. *Drug Discov. Today* **2005**, *10*, 1299–1306. [CrossRef]
14. Dormer, N.H.; Nelson-Brantley, J.; Staecker, H.; Berkland, C.J. Evaluation of a transtympanic delivery system in Mus musculus for extended release steroids. *Eur. J. Pharm. Sci.* **2019**, *126*, 3–10. [CrossRef] [PubMed]

15. Yang, K.-J.; Son, J.; Jung, S.Y.; Yi, G.; Yoo, J.; Kim, D.-K.; Koo, H. Optimized phospholipid-based nanoparticles for inner ear drug delivery and therapy. *Biomaterials* **2018**, *171*, 133–143. [CrossRef] [PubMed]
16. Egli Gallo, D.; Khojasteh, E.; Gloor, M.; Hegemann, S.C. Effectiveness of systemic high-dose dexamethasone therapy for idiopathic sudden sensorineural hearing loss. *Audiol. Neurotol.* **2013**, *18*, 161–170. [CrossRef] [PubMed]
17. Li, X.; Chen, W.-J.; Xu, J.; Yi, H.-J.; Ye, J.-Y. Clinical Analysis of Intratympanic Injection of Dexamethasone for Treating Sudden Deafness. *Int. J. Gen. Med.* **2021**, *14*, 2575–2579. [CrossRef]
18. Czock, D.; Keller, F.; Rasche, F.M.; Häussler, U. Pharmacokinetics and Pharmacodynamics of Systemically Administered Glucocorticoids. *Clin. Pharmacokinet.* **2005**, *44*, 61–98. [CrossRef]
19. Liu, D.; Ahmet, A.; Ward, L.; Krishnamoorthy, P.; Mandelcorn, E.D.; Leigh, R.; Brown, J.P.; Cohen, A.; Kim, H. A practical guide to the monitoring and management of the complications of systemic corticosteroid therapy. *Allergy Asthma Clin. Immunol.* **2013**, *9*, 30. [CrossRef]
20. Toniazzo, T.; Berbel, I.F.; Cho, S.; Fávaro-Trindade, C.S.; Moraes, I.C.; Pinho, S.C. β-carotene-loaded liposome dispersions stabilized with xanthan and guar gums: Physico-chemical stability and feasibility of application in yogurt. *LWT* **2014**, *59*, 1265–1273. [CrossRef]
21. Guldiken, B.; Gibis, M.; Boyacioglu, D.; Capanoglu, E.; Weiss, J. Physical and chemical stability of anthocyanin-rich black carrot extract-loaded liposomes during storage. *Food Res. Int.* **2018**, *108*, 491–497. [CrossRef]
22. Lin, W.; Goldberg, R.; Klein, J. Poly-phosphocholination of liposomes leads to highly-extended retention time in mice joints. *J. Mater. Chem. B* **2022**, *10*, 2820–2827. [CrossRef] [PubMed]
23. Tan, C.; Wang, J.; Sun, B. Biopolymer-liposome hybrid systems for controlled delivery of bioactive compounds: Recent advances. *Biotechnol. Adv.* **2021**, *48*, 107727. [CrossRef] [PubMed]
24. Shah, S.; Famta, P.; Raghuvanshi, R.S.; Singh, S.B.; Srivastava, S. Lipid polymer hybrid nanocarriers: Insights into synthesis aspects, characterization, release mechanisms, surface functionalization and potential implications. *Colloids Interface Sci. Commun.* **2021**, *46*, 100570. [CrossRef]
25. Bhargavi, N.; Dhathathreyan, A.; Sreeram, K. Regulating structural and mechanical properties of pectin reinforced liposomes at fluid/solid interface. *Food Hydrocoll.* **2020**, *111*, 106225. [CrossRef]
26. Zhou, W.; Liu, W.; Zou, L.; Liu, W.; Liu, C.; Liang, R.; Chen, J. Storage stability and skin permeation of vitamin C liposomes improved by pectin coating. *Colloids Surfaces B Biointerfaces* **2014**, *117*, 330–337. [CrossRef]
27. Shao, P.; Wang, P.; Niu, B.; Kang, J. Environmental stress stability of pectin-stabilized resveratrol liposomes with different degree of esterification. *Int. J. Biol. Macromol.* **2018**, *119*, 53–59. [CrossRef] [PubMed]
28. Prajapati, V.D.; Jani, G.K.; Moradiya, N.G.; Randeria, N.P. Pharmaceutical applications of various natural gums, mucilages and their modified forms. *Carbohydr. Polym.* **2013**, *92*, 1685–1699. [CrossRef] [PubMed]
29. Rana, V.; Rai, P.; Tiwary, A.K.; Singh, R.S.; Kennedy, J.F.; Knill, C.J. Modified gums: Approaches and applications in drug delivery. *Carbohydr. Polym.* **2011**, *83*, 1031–1047. [CrossRef]
30. Van Bracht, E.; Raavé, R.; Verdurmen, W.P.R.; Wismans, R.G.; Geutjes, P.J.; Brock, R.E.; Oosterwijk, E.; van Kuppevelt, T.H.; Daamen, W.F. Lyophilisomes as a new generation of drug delivery capsules. *Int. J. Pharm.* **2012**, *439*, 127–135. [CrossRef]
31. Kratz, F. Albumin as a drug carrier: Design of prodrugs, drug conjugates and nanoparticles. *J. Control. Release* **2008**, *132*, 171–183. [CrossRef]
32. Grinberg, O.; Hayun, M.; Sredni, B.; Gedanken, A. Characterization and activity of sonochemically-prepared BSA microspheres containing Taxol—An anticancer drug. *Ultrason. Sonochemistry* **2007**, *14*, 661–666. [CrossRef]
33. Shen, H.J.; Shi, H.; Ma, K.; Xie, M.; Tang, L.L.; Shen, S.; Li, B.; Wang, X.-S.; Jin, Y. Polyelectrolyte capsules packaging BSA gels for pH-controlled drug loading and release and their antitumor activity. *Acta Biomater.* **2013**, *9*, 6123–6133. [CrossRef] [PubMed]
34. Yu, S.; Hu, J.; Pan, X.; Yao, P.; Jiang, M. Stable and pH-sensitive nanogels prepared by self-assembly of chitosan and ovalbumin. *Langmuir* **2006**, *22*, 2754–2759. [CrossRef] [PubMed]
35. Lu, B.; Xiong, S.-B.; Yang, H.; Yin, X.-D.; Zhao, R.-B. Mitoxantrone-loaded BSA nanospheres and chitosan nanospheres for local injection against breast cancer and its lymph node metastases: I: Formulation and in vitro characterization. *Int. J. Pharm.* **2006**, *307*, 168–174. [CrossRef] [PubMed]
36. Shah, S.; Dhawan, V.; Holm, R.; Nagarsenker, M.S.; Perrie, Y. Liposomes: Advancements and innovation in the manufacturing process. *Adv. Drug Deliv. Rev.* **2020**, *154–155*, 102–122. [CrossRef]
37. Jaafar-Maalej, C.; Diab, R.; Andrieu, V.; Elaissari, A.; Fessi, H. Ethanol injection method for hydrophilic and lipophilic drug-loaded liposome preparation. *J. Liposome Res.* **2009**, *20*, 228–243. [CrossRef]
38. Butler, M.F.; Glaser, N.; Weaver, A.C.; Kirkland, A.M.; Heppenstall-Butler, M. Calcium Carbonate Crystallization in the Presence of Biopolymers. *Cryst. Growth Des.* **2006**, *6*, 781–794. [CrossRef]
39. Cipolla, D.; Wu, H.; Salentinig, S.; Boyd, B.; Rades, T.; Vanhecke, D.; Petri-Fink, A.; Rothin-Rutishauser, B.; Eastman, S.; Redelmeier, T.; et al. Formation of drug nanocrystals under nanoconfinement afforded by liposomes. *RSC Adv.* **2016**, *6*, 6223–6233. [CrossRef]
40. Beck, R.; Pohlmann, A.; Hoffmeister, C.; Gallas, M.; Collnot, E.; Schaefer, U.; Guterres, S.; Lehr, C.-M. Dexamethasone-loaded nanoparticle-coated microparticles: Correlation between in vitro drug release and drug transport across Caco-2 cell monolayers. *Eur. J. Pharm. Biopharm.* **2007**, *67*, 18–30. [CrossRef] [PubMed]
41. Gómez-Gaete, C.; Fattal, E.; Silva, L.; Besnard, M.; Tsapis, N. Dexamethasone acetate encapsulation into Trojan particles. *J. Control. Release* **2008**, *128*, 41–49. [CrossRef]

42. Dukovski, B.J.; Plantić, I.; Čunčić, I.; Krtalić, I.; Juretić, M.; Pepić, I.; Lovrić, J.; Hafner, A. Lipid/alginate nanoparticle-loaded in situ gelling system tailored for dexamethasone nasal delivery. *Int. J. Pharm.* **2017**, *533*, 480–487. [CrossRef] [PubMed]
43. Bucatariu, S.; Constantin, M.; Ascenzi, P.; Fundueanu, G. Poly(lactide-co-glycolide)/cyclodextrin (polyethyleneimine) microspheres for controlled delivery of dexamethasone. *React. Funct. Polym.* **2016**, *107*, 46–53. [CrossRef]
44. Hsiao, I.L.; Gramatke, A.M.; Joksimovic, R.; Sokołowski, M.; Gradzielski, M.; Haase, A. Size and Cell Type Dependent Uptake of Silica Nanoparticles. *J. Nanomed. Nanotechnol.* **2014**, *5*, 248–258. [CrossRef]
45. Kroll, A.; Dierker, C.; Rommel, C.; Hahn, D.; Wohlleben, W.; Schulze-Isfort, C.; Göbbert, C.; Voetz, M.; Hardinghaus, F.; Schnekenburger, J. Cytotoxicity screening of 23 engineered nanomaterials using a test matrix of ten cell lines and three different assays. *Part. Fibre Toxicol.* **2011**, *8*, 9. [CrossRef] [PubMed]
46. Rees, K.R. Cells in Culture in Toxicity Testing: A Review. *J. R. Soc. Med.* **1980**, *73*, 261–264. [CrossRef]
47. Perde-Schrepler, M.; Fischer-Fodor, E.; Virag, P.; Brie, I.; Cenariu, M.; Pop, C.; Valcan, A.; Gurzau, E.; Maniu, A. The expression of copper transporters associated with the ototoxicity induced by platinum-based chemotherapeutic agents. *Hear. Res.* **2020**, *388*, 107893. [CrossRef]
48. Kalinec, G.; Thein, P.; Park, C.; Kalinec, F. HEI-OC1 cells as a model for investigating drug cytotoxicity. *Hear. Res.* **2016**, *335*, 105–117. [CrossRef]
49. Pons, M.; Foradada, M.; Estelrich, J. Liposomes obtained by the ethanol injection method. *Int. J. Pharm.* **1993**, *95*, 51–56. [CrossRef]
50. Charcosset, C.; Juban, A.; Valour, J.-P.; Urbaniak, S.; Fessi, H. Preparation of liposomes at large scale using the ethanol injection method: Effect of scale-up and injection devices. *Chem. Eng. Res. Des.* **2015**, *94*, 508–515. [CrossRef]
51. Anitha, A.; Maya, S.; Deepa, N.; Chennazhi, K.; Nair, S.; Tamura, H.; Jayakumar, R. Efficient water soluble O-carboxymethyl chitosan nanocarrier for the delivery of curcumin to cancer cells. *Carbohydr. Polym.* **2011**, *83*, 452–461. [CrossRef]
52. Paşcalău, V.; Tertis, M.; Pall, E.; Suciu, M.; Marinca, T.; Pustan, M.; Merie, V.; Rus, I.A.; Moldovan, C.; Topala, T.; et al. Bovine serum albumin gel/polyelectrolyte complex of hyaluronic acid and chitosan based microcarriers for Sorafenib targeted delivery. *J. Appl. Polym. Sci.* **2020**, *137*, 1–16. [CrossRef]
53. Kalinec, G.M.; Webster, P.; Lim, D.J.; Kalinec, F. A Cochlear Cell Line as an in vitro System for Drug Ototoxicity Screening. *Audiol. Neurotol.* **2003**, *8*, 177–189. [CrossRef]
54. Devarajan, P.; Savoca, M.; Castaneda, M.; Park, M.S.; Esteban-Cruciani, N.; Kalinec, G.; Kalinec, F. Cisplatin-induced apoptosis in auditory cells: Role of death receptor and mitochondrial pathways. *Hear. Res.* **2002**, *174*, 45–54. [CrossRef]
55. Kalinec, G.M.; Park, C.; Thein, P.; Kalinec, F. Working with Auditory HEI-OC1 Cells. *J. Vis. Exp.* **2016**, *2016*, e54425. [CrossRef] [PubMed]

 *gels*

Article

Formulation and In Vivo Pain Assessment of a Novel Niosomal Lidocaine and Prilocaine in an Emulsion Gel (Emulgel) of Semisolid Palm Oil Base for Topical Drug Delivery

Aidawati Mohamed Shabery [1], Riyanto Teguh Widodo [2,*] and Zamri Chik [1,3,*]

[1]   Department of Pharmacology, Faculty of Medicine, Universiti Malaya, Kuala Lumpur 50603, Malaysia
[2]   Department of Pharmaceutical Technology, Faculty of Pharmacy, Universiti Malaya, Kuala Lumpur 50603, Malaysia
[3]   Universiti Malaya Bioequivalence Testing Centre (UBAT), Faculty of Medicine, Universiti Malaya, Kuala Lumpur 50603, Malaysia
*   Correspondence: riyanto@um.edu.my (R.T.W.); zamrichik@ummc.edu.my (Z.C.)

Abstract: This study aimed to formulate semisolid niosomal encapsulated lidocaine and prilocaine using the patented palm oil base Hamin-C® for further characterization and in vivo pain assessment. Seven formulations were initially studied with various chemical compositions. A thin-layer film hydration method was used to produce niosome using a mixture of surfactant (Span® 40 or Span® 60) and cholesterol (CHOL) at a 1:1 ratio, with/without a charge-inducing agent (diacetyl phosphate) (DCP) and with/without labrasol®. Niosome F1 formulation had been identified as the highest %EE achieved, at 53.74 and 55.63% for prilocaine and lidocaine, respectively. Furthermore, NIO-HAMIN F1 emulgel indicated the best formulation with higher permeability of prilocaine and lidocaine compared to the rest of the formulations. The reformulation of optimization of NIO-HAMIN F1 emulgel using a cold process to NIO-HAMIN F1-C emulgel to improve the viscosity resulted in higher diffusion of prilocaine and lidocaine by 5.71 and 33.38%, respectively. In vivo pain perception studies by verbal rating score (VRS) and visual analogue score (VAS) on healthy subjects show a comparable local anesthetic effect between NIO-HAMIN F1-C emulgel and EMLA® cream.

Keywords: niosome; lidocaine; prilocaine; topical drug delivery; local anesthetic; nanoparticles

Citation: Shabery, A.M.; Widodo, R.T.; Chik, Z. Formulation and In Vivo Pain Assessment of a Novel Niosomal Lidocaine and Prilocaine in an Emulsion Gel (Emulgel) of Semisolid Palm Oil Base for Topical Drug Delivery. *Gels* **2023**, *9*, 96. https://doi.org/10.3390/gels9020096

Academic Editors: Maddalena Sguizzato, Rita Cortesi and Rachel Yoon Chang

Received: 20 December 2022
Revised: 13 January 2023
Accepted: 18 January 2023
Published: 22 January 2023

1. Introduction

Local anesthetics are used clinically to reduce or eliminate pain whether for surgery or the management of other acute and chronic pain conditions [1]. Local anesthesia can be in the form of an injection or topical application. However, injectable anesthesia can be painful and hard to use, especially for children and patients with needle phobia [2]. Therefore, the topical application of local anesthetics is preferable in several medical procedures such as minor surgery because it is convenient, effective, and easy to apply [3].

The prototypical amide-type local anesthetic lidocaine (Figure 1a), which was introduced by Lofgren in 1943, has a rapid onset of action, inherent potency, and moderate duration of action [4]. Many over-the-counter topical anesthetics contain lidocaine whether in a eutectic mixture or as monotherapy in the form of a gel, ointment, or patches [5]. Prilocaine (Figure 1b) is also an amide-type local anesthetic that is frequently used for regional pain relief through the injectable and transdermal route. Transdermal administration is preferable to injection due to it being self-administered and painless [6]. Prilocaine is mostly used in dentistry as an injectable form in combination with lidocaine as a topical preparation; it is used for dermal anesthesia and paresthesia. It shows low cardiac toxicity and is very commonly used in intravenous regional anesthesia. Prilocaine is slightly less potent than lidocaine but is considerably less toxic, produces less tissue vasodilation, and can be used reliably in plain solution form for short-duration procedures [7].

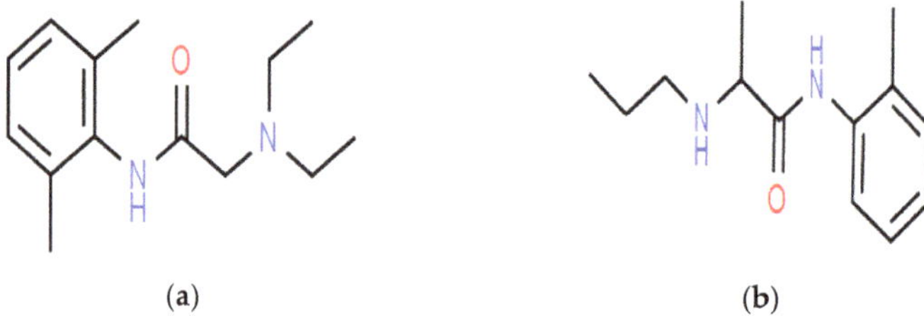

Figure 1. Chemical structures of (**a**) lidocaine and (**b**) prilocaine.

Astra Pharmaceuticals, USA, produced EMLA®, cream which is considered the gold standard for other topical anesthetics and has proven to have anesthetic efficacy in several clinical trials [2]. EMLA® cream is formulated in an emulsion form, containing 2.5% lidocaine and 2.5% as a eutectic mixture in its oil phase [3]. Alireza et al. [8] have shown that the eutectic system of EMLA® cream is efficient in providing an anesthetic effect for topical application.

Topical anesthetics required 30–60 minutes of dwell time after application before anesthetic effects took place [9]. Including EMLA® cream, dwell times of 60 min are not very efficient, especially in emergency cases, and this will limit its clinical usage. Therefore, new topical anesthetics are commercially available to improve the onset of action, including ELA-Max, Amethocaine, and Topicaine. ELA-Max provides effective analgesia after 30 min of application, but in general, recommended application time for ELA-Max is 60 min, while Amethocaine has been shown to provide effective analgesia within 30–45 min and Topicaine's recommended application time from the manufacturer is 30–60 min before venipuncture [10]. Three topical anesthetics can be alternatives to EMLA® cream in terms of effectiveness, namely, tetracaine, liposome-encapsulated tetracaine, and liposome-encapsulated lidocaine [2].

The effectiveness of a topical local anesthetic can be affected by its delivery system. A topical delivery system of local anesthetics may be different in formulations and presentations, whether solid (patches), semisolid, or liquid [1]. Good local anesthetic topical delivery systems are characterized by less concentration of the drugs, better permeability and absorption, keeping the drugs at the target site longer, decreasing the clearance, and limiting local and systemic toxicity [11]. A novel means of improving drug delivery is with vesicular systems. A vesicular system is able to enhance the bioavailability of encapsulated drugs and provides therapeutic activity in a controlled manner for a prolonged period [12].

Vesicular systems consist of vesicular carriers such as soft lipid vesicles. Niosomes are one type of soft lipid vesicles that are elastic and deformable, which are used to encapsulate local anesthetics for dermal delivery efficiency and hence are able to slowly release large doses of local anesthetic agents to provide a prolonged period of analgesia without toxicity [1]. Moreover, niosomes can solve the issues related to insolubility, instability, low bioavailability, and rapid degradation of drugs [13]. They also can be easily and reliably produced in the laboratory compared to liposomes [14]. Niosomes are more stable and may provide faster penetration to the stratum corneum than liposomes [10].

Hamin-C® palm oil base is a self-emulsifying base; it is a mixture of hydrogenated palm kernel oil and hydrogenated palm kernel stearin [15], making it more thermally stable and robust to the temperature changes. Therefore, Hamin-C® palm oil base becomes a new revolution in the drug delivery system for semisolid and suppositories. In previous study, Hamin-C® palm oil base has proved to be successful in delivering aspirin systemically via suppository formulation, in the testosterone transdermal delivery system (TDDS) in an animal model, producing the hypoglycemic effect for glucose control in rabbits and delivering lidocaine, which is comparable to EMLA® cream [2]. Hamin-C® palm oil base, which contains 5% lidocaine, demonstrates the cumulative amount of lidocaine release and is slightly higher than EMLA® cream. Clinical studies also showed that Hamin-C® palm oil base produces adequate numbness and is comparable to EMLA® cream [2].

This study aimed to prepare niosome-encapsulated prilocaine and lidocaine loaded in Hamin-C® palm oil base for topical anesthetic application. To increase the diffusion of lidocaine and prilocaine through the membrane, a smaller size vesicle in the form of a niosome was produced using the thin-film hydration method with some modifications.

2. Results and Discussion

2.1. Niosome Encapsulation Efficiency (%EE)

All formulated niosome-encapsulated drugs were subjected to %EE to evaluate the amount of drug encapsulated in the niosome. Table 1 shows the %EE for all niosome. The highest %EE was achieved for formulations F1 and F6. Formulation F1, which was formulated using Span® 40, showed greater encapsulation efficiency, with 53.74% and 55.63% for lidocaine and prilocaine, respectively. Formulation F6, which contains Span® 60, gives a slightly lower %EE because its hydrophobic alkyl chain is shorter than in Span® 40 [16].

Table 1. Niosome encapsulated lidocaine and prilocaine percentage of entrapment efficiency (%EE), particle size (nm), zeta potential (mv), and polydispersity index (PDI).

Formulation Code	Vesicular Size (nm)	PDI	Zeta-Potential	%EE	
				LDC	PLC
F1	748.6	0.732	-28.4 ± 18.2	53.74 ± 0.03	55.63 ± 0.03
F2	362.2	0.697	-54.2 ± 12.3	34.36 ± 0.03	32.63 ± 0.03
F3	1206	1.000	-66.3 ± 13.4	27.60 ± 0.04	29.51 ± 0.04
F4	1035	0.945	-62.8 ± 12.4	24.36 ± 0.03	26.83 ± 0.03
F5	1902	0.956	-50.3 ± 6.33	26.45 ± 0.08	28.23 ± 0.08
F6	4091	0.533	-46.8 ± 8.29	50.26 ± 0.09	47.73 ± 0.08
F7	623.7	0.885	-72.9 ± 24.9	8.94 ± 0.03	12.27 ± 0.03

Additional negatively charged DCP in formulations F1 and F6 resulted in higher %EE and lower %EE in F3, F4, and F5. This result showed that additional DCP alters encapsulation efficiency but also depends on other factors such as surfactant alkyl-side chain [17].

2.2. Niosome Size, Zeta Potential, and Polydispersity Index (PDI)

Size of vesicles or nanoparticles is very crucial in the transdermal drug delivery system. The size must be below 300 nm in order to be able to pass through the deepest skin layer. PDI values that are close to zero show less degree of heterogeneity of the particle size [18]. Zeta potential is important in determining the stability of the niosome, and higher values indicate a more stable niosome [19,20].

The lowest niosome size was achieved in F2, at 362.2 nm, and the highest was 4091 nm for F6. Niosome F2 formulation used Span® 40 as a surfactant, while niosome F6 used Span® 60. F2 does not contain DCP, while F6 contains 0.01 M DCP (Table 1). Span® 60 produces a larger size niosome due to the longer alkyl group that produces a wider bilayer, resulting in a larger vesicle [21].

Niosome formulated using Span® 40 seems to have a lower particle size if compared to Span® 60 without labrasol, except for F7. Niosome F7 was sonicated at higher amplitude, thus decreasing its size. Niosome formulated using Span® 60 was not added with labrasol because it will increase the vesicle size [22]. Since the method used to prepare niosome is film hydration, multilamellar vesicles (MLV) should be produced with a size in the range of 0.5–10 μm [23]. However, the size of the niosome produced in this study was in the range of 0.3–4.0 μm due to the use of probe sonication [24].

Niosome formulations F2, F5, and F6 were within negative zeta potential values ranging between -41.7 and -58.4 mv, which are sufficiently high for the electrostatic stabilization of niosomes [23]. F1 has a lower zeta potential, which was -28.4 mv, and is considered stable but with incipient instability. While other formulations have very high zeta potential, this shows very good stability [25]. The polydispersity index (PDI) of all formulations was more than 0.3 and did not correspond to a homogeneous population of colloidal systems [23]. All niosome formulation was characterized by a polydispersed population of particles, with the lowest value of PDI shown by F6 at 0.533, while F3, F4, and F5 have the highest PDI value, between 0.9–1.0, which is considered highly polydisperse with multiple particle size populations [26–28].

2.3. In Vitro Permeability Test

The cumulative prilocaine and lidocaine drug permeability per unit area from NIO-HAMIN emulgels across Strat M^{TM} membrane over 120 min were plotted on two separate scales, as shown in Figures 2 and 3.

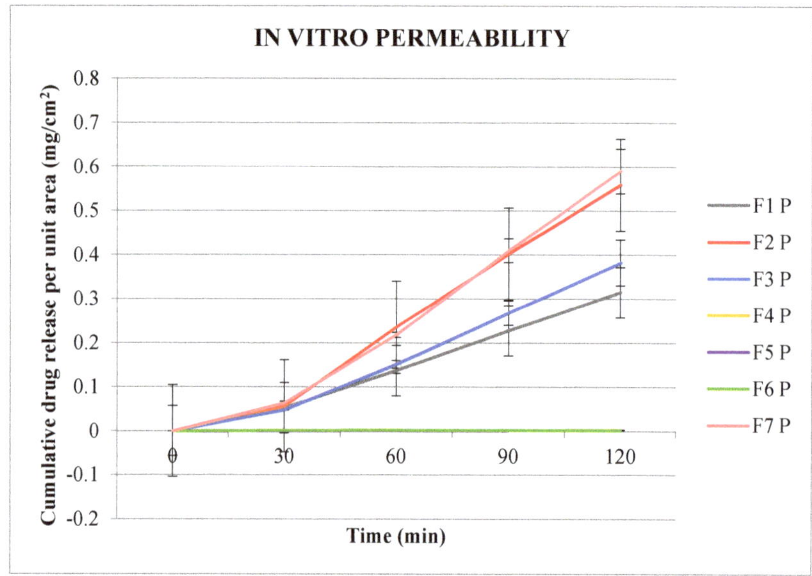

Figure 2. In vitro permeability of prilocaine for the various formulation of NIO–HAMIN emulgels. (P: prilocaine; L: lidocaine).

Figure 2 plots show that after 30 min, the cumulative amount of prilocaine that permeated the membrane was higher for NIO-HAMIN F1, F2, F3, and F7 as compared to NIO-HAMIN F4, F5, and F6. NIO-HAMIN F2 and F7 show consistently higher amounts of prilocaine compared to NIO-HAMIN F1 and F3 over a 2 h period (Figure 2). With a vesicle size of less than one micron in NIO-HAMIN F1, F2, and F7, a higher amount of prilocaine was diffused through the membrane. Additional labrasol® in NIO-HAMIN F3 does enhance the permeability of prilocaine, although the size is larger than other formulations. This was attributed to the fact that labrasol® is a nonionic surfactant capable of improving the solubility of prilocaine, while NIO-HAMIN F4, F5, and F6 give lower prilocaine permeability because the amount of niosome added in Hamin-C® was only 2.5%. Additionally, the size of the vesicle is also larger if compared to other formulations [29].

Figure 3 shows that the cumulative amount of lidocaine at 30 min achieved for NIO-HAMIN F1 and F3 was higher compared to NIO-HAMIN F2, F4, F6, and F7. However, NIO-HAMIN F2 shows higher permeability at 60, 90, and 120 min compared to NIO-HAMIN F4, F6, and F7 but is still lower compared to NIO-HAMIN F1 and F3.

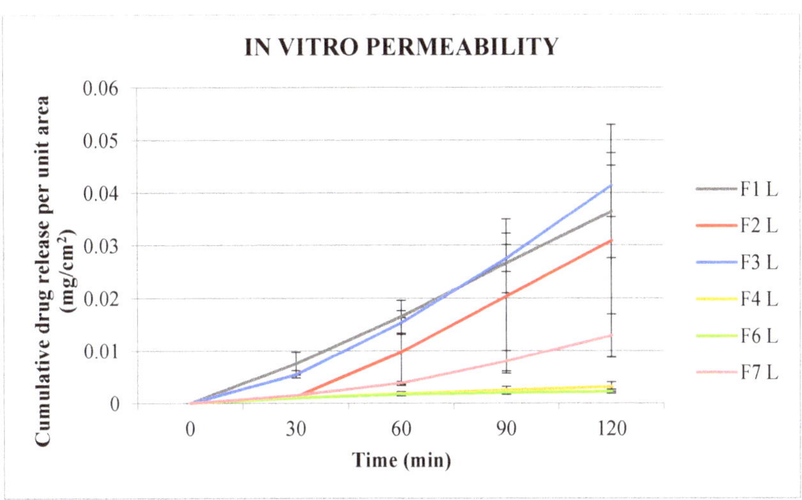

Figure 3. In vitro permeability of lidocaine for the various formulation of NIO-HAMIN emulgels. (P: prilocaine; L: lidocaine).

All formulas have a lower amount of lidocaine permeability as compared to prilocaine, although the amount added into the donor compartment of the Franz diffusion cell was the same as prilocaine. This is probably due to the ability of lidocaine to permeate through the membrane since it is lower than the prilocaine, or because the release of lidocaine from the niosome is lesser compared to the prilocaine. In other words, lidocaine may have a stronger affinity with the niosome vesicles. This will be further investigated for the continuous development of NIO-HAMIN.

Since NIO-HAMIN F1 has the highest %EE and high prilocaine and lidocaine permeability compared to the other formulations, NIO-HAMIN F1 was selected for this study for further optimization. A high %EE leads to a high drug load in the emulgel; therefore, less emulgel needs to be applied onto the skin to obtain a high anesthetic effect.

To improve NIO-HAMIN F1 permeability, it was reformulated and optimized to become NIO-HAMIN F1-C using a cold process to adjust the pH of the emulgel to 9 since local anesthetics give a more rapid onset of action and are longer-acting in the nonionized-form at more basic pH [30]. This is because NIO-HAMIN emulgel from the hot process method will become watery after adjusting to pH 9 due to the decreasing viscosity of carbopol polymer in the presence of excess electrolytes [31]. As the results demonstrated, prilocaine in vitro permeability of NIO-HAMIN F1-C shows 5.71% higher permeability, while lidocaine permeability was 33.38% higher after 30 min (Figure 4) when comparing to NIO-HAMIN F1 (hot process). With a high permeability rate in NIO-HAMIN F1-C, this may lead to a high effective rate to gain an anesthetic effect at the site of application.

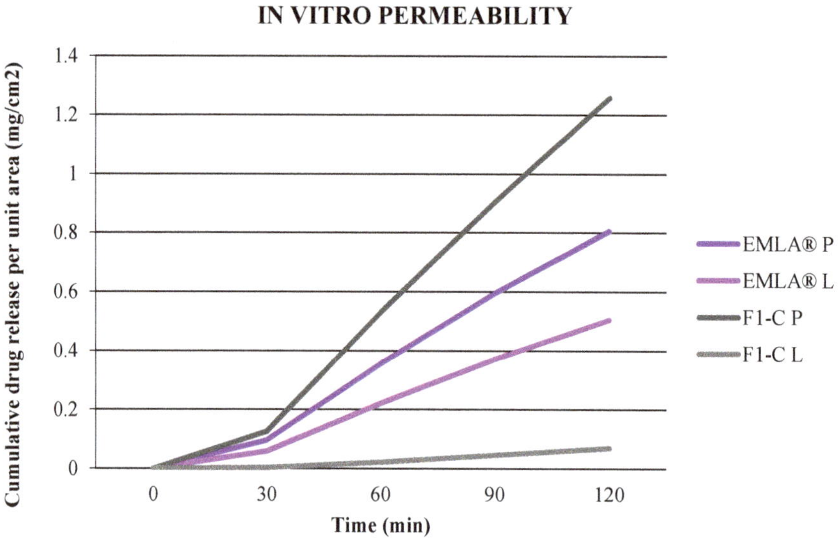

IN VITRO PERMEABILITY

Figure 4. In vitro permeability of prilocaine and lidocaine of NIO-HAMIN F1-C emulgel after reformulated using a cold process method VS EMLA® cream. (P: prilocaine; L: lidocaine).

2.4. NIO-HAMIN F1-C Emulgel Characterization

Physical characterizations of NIO-HAMIN F1-C emulgel are summarized in Table 2. The mean pH of the emulgel is 8.86 ± 0.44, which is alkaline. The emulgel color was measured using CIE LAB color space. The L* value indicates the lightness, the a* value indicates the color between red to green, and the b* value the color between yellow to blue [32]. Using CIE LAB color calculator (http://colorizer.org/, accessed on 27 November 2020), one can conclude that the cream is slightly whitish grey.

Table 2. NIO-HAMIN F1-C emulgel formulation characterization.

Parameter	NIO-HAMIN F1-C Emulgel			Mean	±SD
	Day 0	**Month 1**	**Month 2**		
pH	9.36	8.53	8.67	8.86	0.44
Viscosity (P)					
2.5 rpm	406.67	1244.67	1529.33	1060.22	583.62
5 rpm	242.67	675.33	833.33	583.78	305.79
10 rpm	147.67	376.67	440.00	321.44	153.79
20 rpm	91.83	217.00	244.18	184.34	81.26
50 rpm	51.87	106.80	138.80	99.16	43.97
Color					
L*	85.52	86.00	85.49	85.67	0.29
a*	−0.70	−0.95	−1.24	−0.96	0.27
b*	5.02	5.54	4.62	5.06	0.46

The viscosity and rheogram of the NIO-HAMIN F1-C emulgel are shown in Figures 5 and 6. The shear stress of the emulgel increased as the shear rate increased. The viscosity of the emulgel decreased as the shear rate increased, which shows that the emulgel is a pseudoplastic exhibiting non-Newtonian flow behavior [33]. This shear-thinning behavior makes the emulgel easy to apply and stays static when rubbed on the skin.

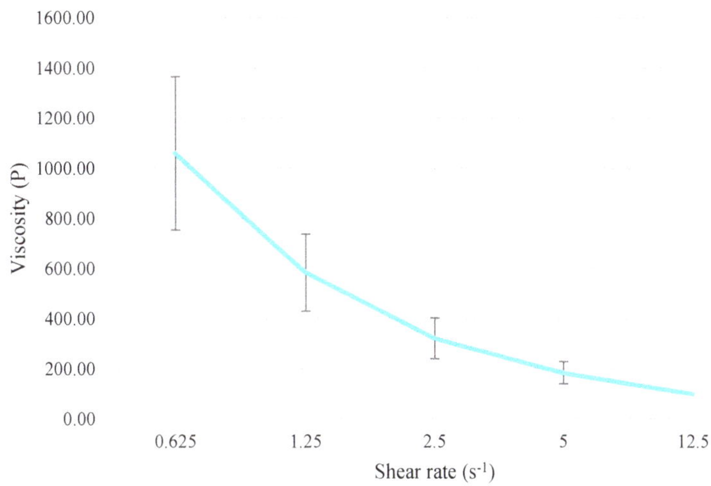

Figure 5. Influence of shear rate on viscosity of NIO-HAMIN F1−C emulgel. Data presented as mean ± SD ($n = 3$).

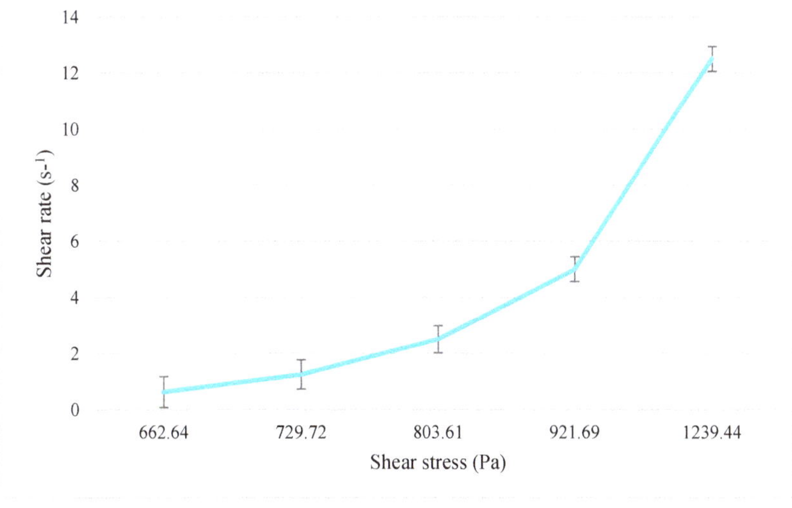

Figure 6. Rheogram of NIO-HAMIN F1−C emulgel: shear stress as a function of shear rate. Data presented as mean ± SD ($n = 3$).

Drug content for NIO-HAMIN F1-C emulgel was analyzed using a developed HPLC to study the consistency of the content, and the results were consistent at 25.21 ± 0.53 mg/g emulgel ($n = 10$).

2.5. In Vivo Pain Assessment

Verbal rating score (VRS) and visual analogue score (VAS) [34,35] were adopted as the methods to assess the efficacy of NIO-HAMIN F1-C emulgel as a topical local anesthetic preparation in comparison against EMLA® cream, with the results as discussed below.

Study Design and Subject Admission

The clinical study was conducted in two phases:

(1) Study 1

Figures 7 and 8 show the VRS and VAS scores for the 20 subjects who participated in the study. Statistical analysis showed that there was no significant difference between NIO-HAMIN and placebo for both application times applied ($p < 0.05$).

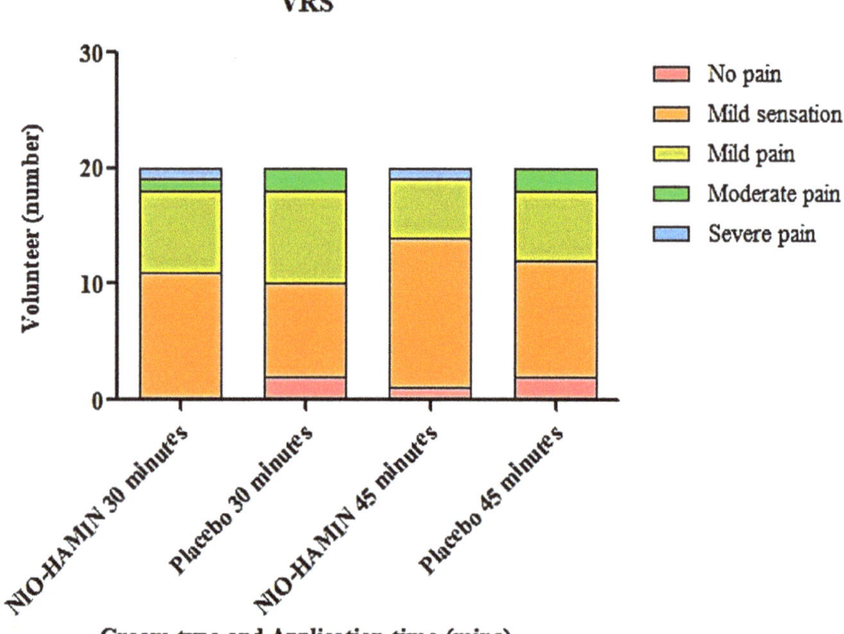

Figure 7. Verbal rating score (VRS) for NIO-HAMIN F1-C emulgel compared to placebo. Values are presented as number of subjects vs. treatment ($n = 20$, $p < 0.05$).

However, the distribution of pain scores was reduced by an increase in the application time of NIO-HAMIN such as from 30 min to 45 min. On the contrary, placebo pain score distribution increased with the increased application time (Figure 9). Since there was no significant difference between the pain score of NIO-HAMIN and placebo, for study 2, application time was decided at 60 min, which is the recommended application time for EMLA® cream [36].

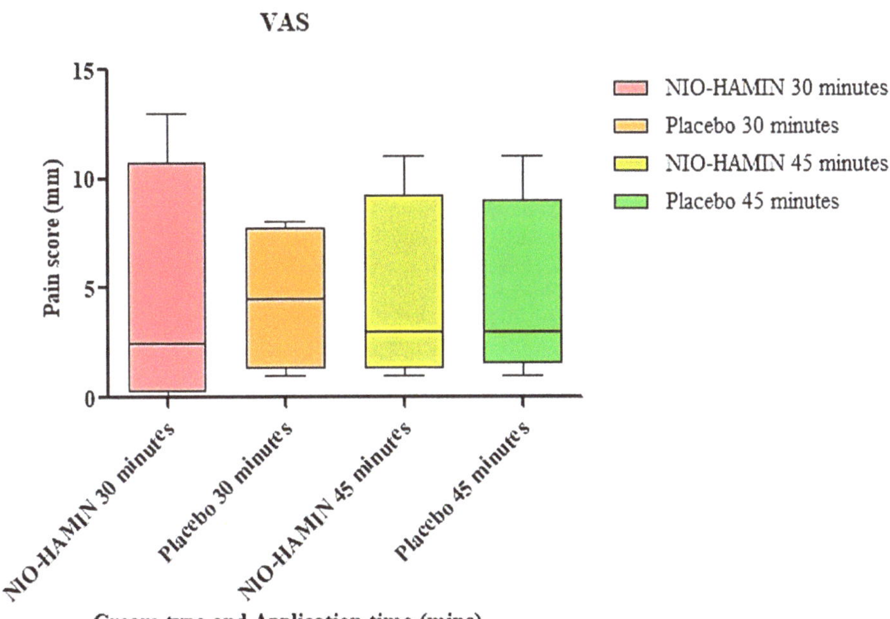

Figure 8. Visual analogue score (VAS) for NIO-HAMIN F1-C emulgel compared to placebo. Values are presented as pain score vs. treatment ($n = 20$, $p < 0.05$).

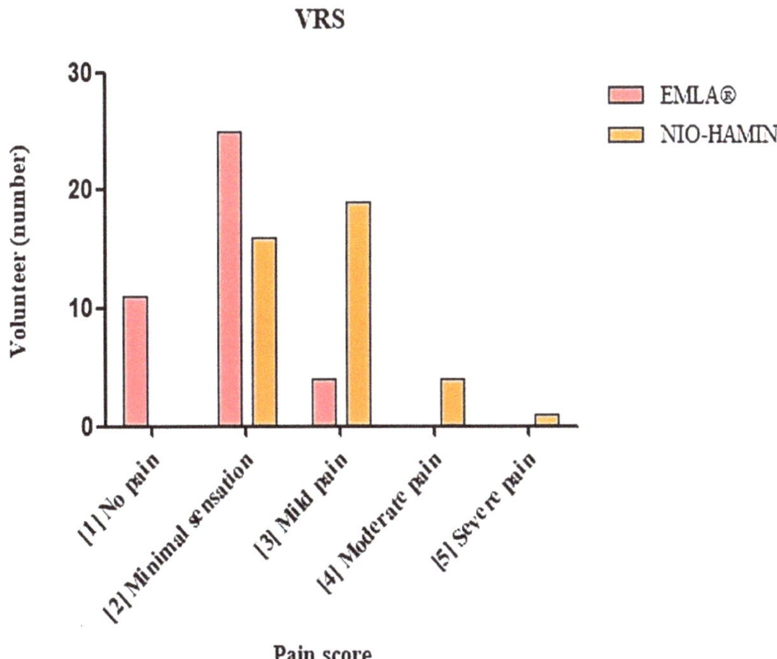

Figure 9. Verbal rating score (VRS) for NIO-HAMIN F1-C emulgel compare to EMLA® cream. Values are presented as the number of subjects vs. pain score ($n = 40$, $p < 0.05$).

(2)   Study 2

In total, 40 healthy subjects were involved in the study. The VRS result showed that 11 subjects scored no pain for EMLA® cream, while none did so for NIO-HAMIN F1-C emulgel. A total of 25 subjects scored minimal sensation for EMLA® cream and 16 subjects did so for NIO-HAMIN F1-C emulgel. Most of the subjects scored mild pain for NIO-HAMIN F1-C emulgel ($n = 19$). There were four subjects who scored moderate pain and one subject who scored severe pain for NIO-HAMIN F1-C emulgel but none for EMLA® cream. In reference to VRS, the statistical analysis of both pain scores (minimal sensation and mild pain) regarding NIO-HAMIN F1-C emulgel and EMLA® were found significantly effective ($p < 0.05$). However, the result showed a significant difference for "No Pain" since none of the subjects scored no pain for NIO-HAMIN F1-C emulgel ($p < 0.05$) (Figure 10). VAS analysis for study 2 showed no significant difference or was significantly effective between EMLA® cream and NIO-HAMIN F1-C emulgel with median differences of 4.35% ($p < 0.05$, p = ns) (Figure 10).

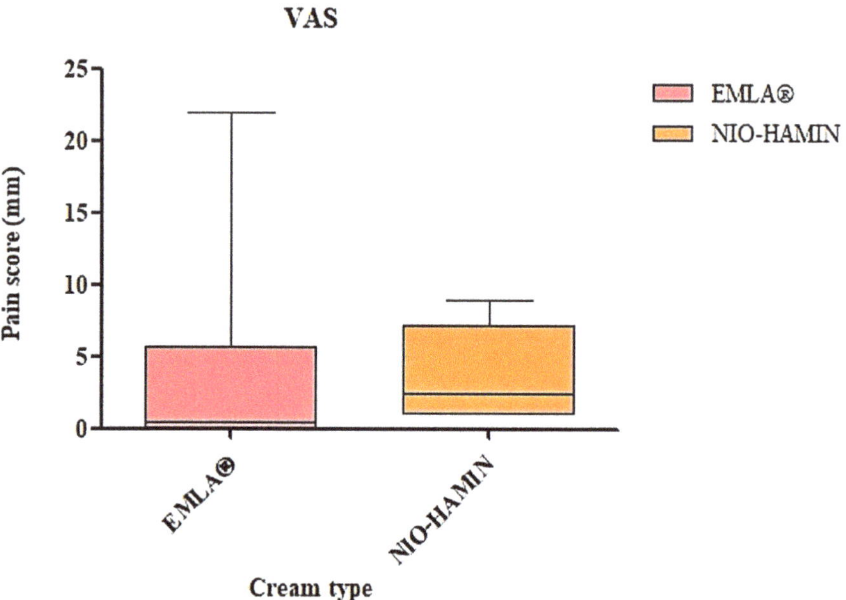

Figure 10. Visual analogue score (VAS) for NIO-HAMIN F1-C emulgel compared to EMLA® cream. Values are presented as pain score vs. treatment ($n = 40$, $p < 0.05$).

3. Conclusions

In conclusion, we successfully formulated and characterized a new semisolid niosomal local anesthetic emulgel using a patented palm oil base, Hamin-C®, which contains lidocaine and prilocaine. This new formulation was tested and characterized by having good properties of topical local anesthetic and enhanced release of local anesthetic agents, in vitro. In vivo study shows that the NIO-HAMIN F1-C emulgel local anesthetic effect is comparable but slightly less effective as compared to EMLA® cream. This is probably because the lidocaine release from NIO-HAMIN F1-C is less as compared to EMLA® cream. However, the prilocaine released was higher for NIO-HAMIN F1-C compared to EMLA® cream. More investigation and modification in the formulation are required to improve the release of lidocaine from the niosome. In this case, to improve the release we should have a smaller size with a lower polydispersity index of the niosome. These changes would probably produce a better in vivo effect.

4. Materials and Methods

4.1. Materials

Lidocaine hydrochloride, Span® 60, SPAN® 40, and Diacetyl phosphate (DCP) were purchased from Sigma-Aldrich Co (St. Louis, MO, USA). Prilocaine hydrochloride was purchased from UTRC (Toronto, ON, Canada). Disodium hydrogen phosphate was purchased from R&M Chemicals (Essex, UK). Sodium dihydrogen phosphate anhydrous was purchased from QReC (Selangor, Malaysia). Sodium hydroxide was purchased from Sharlau Chemie (S.A, Barcelona, Spain). Cholesterol, 95% stabilized, was purchased from Acros Organics (Fair Lawn, NJ, USA). Acetonitrile and chloroform were purchased from Fisher Scientific (Fair Lawn, NJ, USA). Labrasol was purchased from Gattefosse (Cedex, France). Propylene glycol was purchased from Dow Chemical (Thailand). Hamin-C® palm oil base was purchased from Oleopharma Sdn. Bhd., (Malaysia). Acrylates/C10–30 Alkyl Acrylate Crosspolymer was purchased from Lubrizol (Clifton, NJ, USA). Strat M™ membrane was purchased from Merck (Darmstadt, Germany).

4.2. Methods

4.2.1. Preparation and Evaluation of Niosome-Encapsulated Prilocaine and Lidocaine

Niosome was prepared using a thin-film hydration technique. Surfactant (Span® 40 or 60) and cholesterol were composed in a molar ratio of 1:1. The composition of surfactant, cholesterol, labrasol, Diacetyl phosphate, lidocaine, and prilocaine was dissolved in chloroform according to the formulation code as summarized in Table 3. Thin-layer film was achieved by rotary evaporation (EYELA, Rikakika, Tokyo, Japan) of the mixture at 60 °C. Phosphate buffer (PBS) pH 6 was added to the thin-layer film to hydrate it. The niosome was ultrasonicated at 40/60 amplitudes with 2000 J energy and 30 W powers for 60 min or 120 min using an ultrasonic homogenizer (Biologics, Inc., Bristow, VA, USA) with a titanium probe to agitate particles. Niosome then was centrifuged (Sartorius, Goettingen, Germany) at 11,000 rpm for 60 min to remove the excess drug. The supernatant was then removed, and distilled water was added and vortex mixed (Snijders Scientific, Tilburg, Holland) and ultrasonicated (Kudos, Shanghai, China) to resuspend the niosome. The finished niosome was stored at a refrigerator temperature between 4 °C to 8 °C before further use. Table 1 summarizes the composition of niosomes for various formulations prepared.

Table 3. Composition of the niosome in various formulations.

Formulation Code	Lidocaine (g)	Prilocaine (g)	SURFACTANT (M)		Cholesterol (M)	Labrasol (%)	DCP (M)
			Span® 40	Span® 60			
F1	0.25	0.25	1	-	1	2	0.01
F2	0.25	0.25	1	-	1	2	-
F3	0.50	0.50	1	-	1	2	0.01
F4	0.25	0.25	1	-	1	-	0.01
F5	0.25	0.25	-	1	1	-	0.01
F6	0.25	0.25	-	1	1	-	0.01
F7	0.25	0.25	-	1	1	-	-

4.2.2. Encapsulation Efficiency

Niosome was mixed with Isopropyl alcohol (IPA) at a 1:10 ratio and vortex mixed until a clear solution was obtained. The clear solution was filtered, and the quantification of drug content was analyzed using a developed high-performance liquid chromatography (HPLC) as described below. Encapsulation efficiency (EE) percentage was calculated using the following formula:

$$\%EE = (\text{Total drug content}/\text{Total drug added}) \times 100 \times df$$

where

1. Total drug content = amount of drug obtained from HPLC analysis;
2. Total drug added = amount of drug added into the formulation;
3. df = dilution factor.

4.2.3. Development of High-Performance Liquid Chromatography (HLPC) for Lidocaine and Prilocaine Analysis

HPLC (Shimadzu, Kyoto, Japan) series LC 20AD was used in this study. Phenomenex Synergi^{TM} 4 μm Fusion-RP 80 Å, with dimensions of 4 μm, 150 × 4.6 mm C18 column, was used as the stationary phase, while 0.01 M phosphate buffer pH 6 and acetonitrile at 55:45 ratio was used as the mobile phase. The injection volume was set to 10 μL, and the analysis was run at a flow rate of 1 mL/min for 8 min. The temperature of the HPLC system was maintained at 25 °C. Lidocaine and prilocaine were detected at 210 nm using a UV detector [4]. The standard curve showed good linearity with regression of coefficient (r^2) greater than 0.998. The retention times were 3.10 and 4.10 min for prilocaine and lidocaine, respectively. The limit of detection (LOD) for prilocaine and lidocaine were 0.3 mg/L and 0.4 mg/L, respectively, while the limit of quantification (LOQ) was 1.0 mg/L and 1.2 mg/L, respectively.

4.2.4. Niosome Particle Size, Zeta Potential, and Polydispersity Index (PDI)

The size and zeta potential of niosome-encapsulated lidocaine and prilocaine were measured using a Malvern Zetasizer (Malvern Panalytical, Malvern, UK). A total of 100 μL of niosomes were pipetted and diluted in 900 μL of distilled water. All measurements were taken at 25 °C.

4.2.5. Preparation of Niosome Encapsulated Lidocaine and Prilocaine NIO-HAMIN Emulgel
Preparation of NIO-HAMIN Emulgel by Hot Process

Hamin-C$^{®}$-based cream was prepared according to Khamdiah Khodari et al. [4]. The water phase consists of distilled water, propylene glycol (PG), and sodium hydroxide, 10% (NaOH), while the oil phase containing Hamin-C$^{®}$ base oil and carbomer was heated separately to 40–50 °C. Once both phases' temperatures were achieved, the oil phase was added to the water phase while stirring using an overhead stirrer (IKA, Germany) at gradual speed from 400 to 800 rpm. Once the emulgel formed, the speed was stopped gradually until it cooled down. Lastly, the base emulgel was transferred into a container and left at room temperature for 24 h to stabilize the whole texture. All the ingredients involved were weighed in % *w/w* [4].

Hamin-C$^{®}$-based emulgel from the above preparation was weighed and stirred while mixed with niosome at different percentages (97.5% for F1 and F7, 95% for F4–F6 and 50% for F2 and F3) to form NIO-HAMIN emulgels. Different percentages of niosome were used to find the right percentage of drug content to achieve the highest diffusion amount using the Franz diffusion test.

Preparation of NIO-HAMIN Emulgel by Cold Process

Niosome, propylene glycol, and carbomer were weighed separately and mixed by stirring slowly until the carbomer was wetted. Sodium hydroxide (10%) was added to form a gel. Hamin-C$^{®}$ palm oil base was added to the gel and the mixture was stirred until all the Hamin-C$^{®}$ bases were thoroughly mixed. The preservative was added and stir-mixed before the pH of the emulgel was adjusted to 9 with 10% NaOH.

4.2.6. In Vitro Skin Permeation Test

A Franz diffusion cell (Hanson, California, USA) was used to evaluate in vitro skin permeation of the NIO-HAMIN local anesthetic emulgel preparations, with EMLA$^{®}$ cream used as a reference. Strat M^{TM} membrane (Merck, Darmstadt, Germany) was selected in this study because it correlates more closely to human skin than any other synthetic membrane. The surface area of the membrane was fixed at 1.767 cm^2 and phosphate buffer at pH 6 as the medium. About 1 g of sample was loaded into the donor compartment and left for 2 h. The period of sample collection was every 30 min. The system was performed under a controlled temperature of 37 °C. The collected samples were analyzed using a developed HPLC method as described in 2.2.3. The permeability profile was demonstrated using the following calculation [2]:

$$Q = Cn \cdot V / A$$

where

1. Q = cumulative amount of drug release;
2. Cn = concentration of lidocaine/prilocaine in the receiver compartment (μg/mL);
3. V = volume of the receiver compartment;
4. A = surface area of the membrane in cm^2.

4.2.7. Physicochemical Characterizations

The physicochemical characterization of an optimized NIO HAMIN emulgel covered the determination of color, pH, rheology, and viscosity properties. The color was determined using a color spectrophotometer (Minolta, Japan) with parameters a* = green-to-red ratio, b* = yellow-to-blue ratio, and L* = light-to-dark ratio measured. The pH was measured using a pH meter (Eutech Instrument, Singapore). Rheology/viscosity properties were evaluated at room temperature using a viscometer (Brookfield, Stoughton, WI, USA) with spindle no. S29 at 2.5–50 RPM for 30 s.

4.2.8. Drug Uniformity Content

A total of 1 g of an optimized NIO-HAMIN emulgel was weighed and extracted using 15 mL methanol with the aid of ultrasonication for 15 min. The solutions were filtered into a 25 mL volumetric flask; Isopropyl alcohol was used to top up to the mark [37]. The prepared sample was analyzed using a developed HPLC, as described in 4.2.3.

4.2.9. In Vivo Pain Assessment

The efficacy of an optimized NIO-HAMIN emulgel as a topical local anesthetic through human skin and to produce numbness at the applied area was assessed in 60 healthy adult subjects. Two methods of pain assessment were adopted, namely, the verbal rating score (VRS) and the visual analogue score (VAS) [34]. VRS assessment was conducted according to the severity of the pain. Subjects were provided with a choice of five answers: no pain, minimal sensation, mild pain, moderate pain, and severe pain [35]. On the other hand, VAS was performed by designing a 100 mm horizontal line and marks of "no pain" at the end line and "severe pain" at the other end of the line were given. Subjects were requested to make a vertical cross on the line which relates to the intensity of pain experienced during the procedure.

4.2.10. Study Design and Subject Admission
The First Phase of the Clinical Study (Study 1)

Study 1 was conducted to determine the onset of action of an optimized NIO-HAMIN emulgel in 20 healthy adult subjects with normal skin conditions. The study was designed as one that is single-blinded and placebo-controlled. The ventral aspect of the forearm of the right and left upper limbs was marked with an area of 10 cm^2 at two sites. Approximately 2 g of an optimized NIO-HAMIN emulgel and 1 g of EMLA® cream were applied for 30 and 45 min. Upon 5 min removal of the two applied preparations, 20 pinpricks were performed dispersedly on the mark site [4]. Subjects rated the pain they experienced using VAS and VRS pain assessment as described above.

The Second Phase of the Clinical Study (Study 2)

Study 2 was conducted to evaluate the effectiveness of an optimized NIO-HAMIN emulgel as a local anesthetic preparation in comparison against EMLA® cream, after the application period as decided in the study 1, using the same pain assessment methods as above. The study design was a single-dose, blinded, crossover, randomized, and balanced study on 40 healthy adult subjects with normal skin condition. The study was carried out after receiving approval from the Universiti Malaya Medical Ethics Committee (MREC ID No.: 202032-8335).

4.2.11. Statistical Analysis

A Student's *t*-test and Wilcoxon signed ranked test were used to analyze the differences in means and clinical study data, respectively. Test results with $p < 0.05$ were considered to be significant.

Author Contributions: Conceptualization, R.T.W. and Z.C.; methodology, R.T.W. and Z.C.; software, A.M.S.; validation, A.M.S.; formal analysis, A.M.S.; investigation, A.M.S.; resources, Z.C.; data curation, A.M.S.; writing—original draft preparation, A.M.S.; writing—review and editing, R.T.W. and Z.C.; visualization, A.M.S.; supervision, R.T.W. and Z.C.; project administration, Z.C.; funding acquisition, Z.C. All authors have read and agreed to the published version of the manuscript.

Funding: This work was supported by Mahmood Merican Medical Research Grant (UM-71-KWG-DM).

Institutional Review Board Statement: The study was carried out after receiving approval from the Universiti Malaya Medical Ethics Committee (MREC ID No. 202032-8335).

Informed Consent Statement: Informed consent was obtained from all subjects involved in the study.

Data Availability Statement: data is unavailable due to privacy or ethical restrictions.

Conflicts of Interest: The authors declare no conflict of interest.

References

1. Shipton, E.A. New Formulations of Local Anaesthetics—Part 1. *Anesthesiol. Res. Pract.* **2012**, *2012*, 546409.
2. Scott, M.B.; Valerie, R.; Jenny, B.; Kevin, W.; Stephen, W.D.; Joanna, B. A randomized, controlled trial to evaluate topical anesthetic for 15 min before venipuncture in pediatrics. *Am. J. Emerg. Med.* **2013**, *31*, 20–25.
3. Pratik, G.; Nitin, M.; Sandhya, C.; Madhur, K.R. Comparison of topical anesthetics for radiofrequency ablation of achracordons: Eutectic mixture of lignocaine/prilocaine versus lidoacine/tetracaine. *Int. Sch. Res. Not.* **2014**, *2014*, 743027.
4. Khamdiah Khodari, S.N.; Zamri, C.; Mohamed Ibrahim, N.; Lucy, C. In vitro and in vivo evaluation of new topical anaesthetic cream formulated with with palm-oil base. *Curr. Drug Deliv.* **2016**, *14*, 690–695.
5. Tina, A. Review of lidoacine/tetracaine cream as a topical anesthetic for dermatologic laser procedures. *Pain Ther.* **2013**, *2*, 11–19.
6. Kang, C.; Shin, S.C. Development of prilocaine gels for enhanced local anesthetic action. *Arch. Pharm. Res.* **2012**, *35*, 1197–1204.
7. Alsharif, A.; Omar, E.; Badr Alolayan, A.; Bahabri, R.; Ghazal, G. 2% lidocaine versus 3% prilocaine for oral and maxillofacial surgery. *Saudi J. Anaesth.* **2018**, *12*, 571–577.
8. Alireza, D.; Syyed, M.A.; Pedram, D.; Amin, D. The efficacy of eutectic mixture of local anesthetics as a topical anesthetic agent used for dental procedures: A brief review. *Anesth. Essays Res.* **2016**, *10*, 383–387.
9. Mritunjay, K.; Ranjiv, C.; Manish, G. Topical anesthesia. *J. Anaesthesiol. Clin. Pharmacol.* **2015**, *31*, 450–456.
10. Eneida, D.P.; Cintia, M.S.C.; Giovana, R.T.; Michelle, F.M.; Leonardo, F.F.; Daniele, R.D.A. Drug delivery systems for local anesthetics. *Recent Pat. Drug Deliv. Formul.* **2010**, *4*, 23–34.
11. Ali, N.; Karikumar, S.L.; Kaur, A. Niosomes: An excellent tool for drug delivery. *Int. J. Res. Pharm. Chem.* **2012**, *2*, 479–487.
12. Ali, B.; Boon-Seang, C.; Harisun, Y. Niosomal drug delivery systems: Formulations, preparation and applications. *World Appl. Sci. J.* **2014**, *32*, 1671–1685.
13. Gannu, P.K.; Pogaku, R. Nonionic surfactant vesicular systems for effective drug delivery—An overview. *Acta Pharm. Sin. B* **2011**, *1*, 208–219.
14. Sritharan, N.; Zamri, C.; Mohamed Ibrahim, N. In vitro characteristic of an insulin suppository developed using palm oil base (Hamin) and its hypoglycaemic effect on rabbits. *Front. Life Sci.* **2015**, *8*, 256–263.
15. Avinash, S.; Gowda, D.V.; Suresh, J.; Avarind, R.A.S.; Atul, S.; Riyaz, A.M.O. Formulation and evaluation of topical gel using Eupatorium glandulosum michx. for wound healing activity. *Der. Pharm. Lett.* **2016**, *8*, 52–63.
16. Rajalakhsmi, S.V.; Vinaya, O.G. Formulation development, evaluation and optimization of medicated lip rouge containing niosomal acyclovir for the management of recurrent herpes labialis. *Int. J. Appl. Pharm.* **2017**, *9*, 21–27.
17. Kandasamy, R.; Veinthramuthu, S. Formulation and optimization of zidovudine niosomes. *AAPS Pharm. Sci. Tech.* **2010**, *11*, 1119–1127.
18. Delly, R.; Glodie, A.W.; Effionora, A. Novel transdermal ethosomal gel containing green tea (camellia sinensis L. kuntze) leaves extract: Formulation and in-vitro penetration study. *J. Young Pharma.* **2017**, *9*, 336–340.

19. Xuemei, G.; Minyan, W.; Suna, H.; Wei-En, Y. Advances of non-ionic surfactant vesicles (niosomes) and their application in drug delivery. *Pharmaceutics* **2019**, *11*, 55.
20. Khan, D.; Bashir, S.; Figueiredo, P.; Santos, H.; Khan, M.; Peltonen, L. Process optimization of ecological probe sonication technique for production of rifampicin loaded niosomes. *J. Drug Deliv. Sci. Technol.* **2019**, *50*, 27–33.
21. Ameerah, A.D. Benazepril hydrochloride loaded niosomal formulation for oral delivery: Formulation and characterization. *Int. J. Appl. Pharm.* **2018**, *10*, 66–70.
22. Ya'akob, H.; Siew Chin, C.; Abd Aziz, A.; Ware, I.; Fauzi, M.A.J.; Rashidah, N.A.; Sabtu, R. Effect of span 60, labrasol, and cholestrol on labisia pumila loaded niosomes quality. *Int. Sch. Sci. Res. Innov.* **2017**, *10*, 521–524.
23. Seleci, D.A.; Seleci, M.; Walter, J.G.; Stahl, F.; Scheper, T. Niosomes as nanoparticle drug carriers: Fundamentals and recent applications. *J. Nanomater.* **2016**, *2016*, 7372306.
24. Sezgin-Bayindir, Z.; Antep, N.M.; Yuksel, N. Development and characterization of mixed niosomes for oral delivery using candesartan cilexetil as a model poorly water-soluble drug. *Am. Assoc. Pharm. Sci.* **2014**, *16*, 108–117.
25. Kumar, A.; Chandra, K.D. Methods for characterization of nanoparticles [editorial]. *Adv. Nanomed. Deliv. Ther. Nucleic* **2017**, 43–58. [CrossRef]
26. Danaei, M.; Dehghankhold, M.; Ataei, S.; Hasanzedah, D.F.; Javanmard, R.; Dokhani, A.; Khorasani, A.; Mozafari, M.R. Impact of praticle size and polydispersity index on the clinical application of lipidic nanocarrier systems. *Pharmaceutics* **2018**, *10*, 57.
27. Katharotiya, K.; Shinde, G.; Katharotiya, D.; Shelke, S.; Patel, R.; Kulkarni, D. Development, evaluation and biodistribution of stealth liposomes of 5-fluorouracil for effective treatment of breast cancer. *J. Liposome Res.* **2022**, *32*, 146–158.
28. De Silva, L.; Fu, J.Y.; Htar, T.T.; Muniyandy, S.; Kasbollah, A.; Wan Kamal, W.H.B.; Chuah, L.H. Characterization, optimization, and in vitro evaluation of Technetium-99m-labeled niosomes. *Int. J. Nanomed.* **2019**, *14*, 1101–1117.
29. Yusuf, M.; Sharma, V.; Pathak, K. Nanovesicles for transdermal delivery of felodipine. Development, characterization, and pharmacokinetics. *Int. J. Pharm. Investig.* **2014**, *4*, 119–130.
30. Georgette, O.; Spencer, B.; Jeffrey, K. Comparison of five commonly-available, lidocaine-containing topical anesthetics and their effct on serum levels of lidocaine and its metabolite monoethylglycinexylidide (MEGX). *Aesthet. Surg. J.* **2011**, *32*, 495–503.
31. Lubrizol. *Viscosity of Carbopol Polymers in Aqueous Systems*; Lubrizol Advanced Materials, Inc.: Cleveland, OH, USA, 2010.
32. Mokrzycki, W.S.; Tatol, M. Perceptual difference in L* a* b* color space as the bae for object colour identification. In Proceedings of the 1st International Conference on Image Processing & Communications, Athens, Greece, 25 September 2009; ResearchGate: Olsztyn, Poland, 2009; pp. 1–8.
33. Kasuhita, B.; Thiagarajan, N.; Padma, T. Formulation optimization, rheological characterization and suitability studies of polyglucoside-based azadirachta indica a. juss emollient cream as a dermal base for sun protection application. *Indian J. Pharm. Sci.* **2017**, *79*, 914–922.
34. Gupta, P.; Maqbool, T.; Sleemuddin, M. Oriented immobolization of stem bromelin via the lone histidine on metal affinity support. *J. Mol. Cat B Enzym.* **2007**, *45*, 78–83.
35. Torchilin, V.P. Fluorescence microscopy to follow the targeting of liposomes and micelles to cells and their fate. *Adv. Drug Deliv. Rev.* **2005**, *57*, 95–109.
36. Sawyer, J.; Febbraro, S.; Masud, S.; Ashburn, M.A.; Campbell, J.C. Heated lidocaine/tetracaine patch (Synergy TM, Rapydan TM) compared with lidocaine/prilocaine cream (EMLA (R)) for topical anaesthesia before vascular access. *BJA* **2009**, *102*, 210–215.
37. Gowekar, N.M.; Wadher, S.J. Development and validation of HPLC method for simultaneous determination of lidocaine and prilocaine in topical formulation. *Asian J. Pharm. Clin. Res.* **2017**, *10*, 179–182.

 *gels*

Article

Ginger Extract-Loaded Sesame Oil-Based Niosomal Emulgel: Quality by Design to Ameliorate Anti-Inflammatory Activity

Marwa H. Abdallah [1,2,*], Hanaa A. Elghamry [2], Nasrin E. Khalifa [1,3], Weam M. A. Khojali [4,5], El-Sayed Khafagy [6,7], Amr S. Abu Lila [1,2], Hemat El-Sayed El-Horany [8,9] and Shaimaa El-Housiny [10]

[1] Department of Pharmaceutics, College of Pharmacy, University of Ha'il, Ha'il 81442, Saudi Arabia
[2] Department of Pharmaceutics and Industrial Pharmacy, Faculty of Pharmacy, Zagazig University, Zagazig 44519, Egypt
[3] Department of Pharmaceutics, Faculty of Pharmacy, University of Khartoum, Khartoum 11115, Sudan
[4] Department of Pharmaceutical Chemistry, College of Pharmacy, University of Ha'il, Ha'il 81442, Saudi Arabia
[5] Department of Pharmaceutical Chemistry, Faculty of Pharmacy, Omdurman Islamic University, Omdurman 14415, Sudan
[6] Department of Pharmaceutics, College of Pharmacy, Prince Sattam Bin Abdulaziz University, Al-kharj 11942, Saudi Arabia
[7] Department of Pharmaceutics and Industrial Pharmacy, Faculty of Pharmacy, Suez Canal University, Ismailia 41552, Egypt
[8] Department of Biochemistry, College of Medicine, University of Ha'il, Ha'il 81442, Saudi Arabia
[9] Department of Medical Biochemistry, Faculty of Medicine, Tanta University, Tanta 31511, Egypt
[10] Department of Pharmaceutics and Industrial pharmacy, Faculty of Pharmacy, Modern University for Technology and Information, Cairo 4410240, Egypt
* Correspondence: mh.abdallah@uoh.edu.sa

Citation: Abdallah, M.H.; Elghamry, H.A.; Khalifa, N.E.; Khojali, W.M.A.; Khafagy, E.-S.; Abu Lila, A.S.; El-Horany, H.E.-S.; El-Housiny, S. Ginger Extract-Loaded Sesame Oil-Based Niosomal Emulgel: Quality by Design to Ameliorate Anti-Inflammatory Activity. *Gels* 2022, *8*, 737. https://doi.org/10.3390/gels8110737

Academic Editors: Maddalena Sguizzato, Rita Cortesi and Rachel Yoon Chang

Received: 26 October 2022
Accepted: 10 November 2022
Published: 14 November 2022

Publisher's Note: MDPI stays neutral with regard to jurisdictional claims in published maps and institutional affiliations.

Abstract: Ginger, a natural plant belonging to the Zingeberaceae family, has been reported to have reasonable anti-inflammatory effects. The current study aimed to examine ginger extract transdermal delivery by generating niosomal vesicles as a promising nano-carrier incorporated into emulgel prepared with sesame oil. Particle size, viscosity, in vitro release, and ex vivo drug penetration experiments were performed on the produced formulations (ginger extract loaded gel, ginger extract loaded emulgel, ginger extract niosomal gel, and ginger extract niosomal emulgel). Carrageenan-induced edema in rat hind paw was employed to estimate the in vivo anti-inflammatory activity. The generated ginger extract formulations showed good viscosity and particle size. The in vitro release of ginger extract from niosomal formulation surpassed other formulations. In addition, the niosomal emulgel formulation showed improved transdermal flux and increased drug permeability through rabbit skin compared to other preparations. Most importantly, carrageenan-induced rat hind paw edema test confirmed the potential anti-inflammatory efficacy of ginger extract niosomal emulgel, compared to other formulations, as manifested by a significant decrease in paw edema with a superior edema inhibition potency. Overall, our findings suggest that incorporating a niosomal formulation within sesame oil-based emulgel might represent a plausible strategy for effective transdermal delivery of anti-inflammatory drugs like ginger extract.

Keywords: ginger extract; niosomes; emulgel; sesame oil; quality by design; anti-inflammatory effect

1. Introduction

Drug delivery via the transdermal route has enabled delivering a wide variety of drugs for many years [1]. To access the systemic circulation, the drug needs first to travel through skin layers. Afterwards, the bioactive agent is subsequently carried throughout the body via the bloodstream to produce its therapeutic effect. In comparison to the other administration routes, transdermal drug delivery shows potential advantages including avoiding first pass hepatic metabolism, prolonging the duration of action of the drug, reducing fluctuation in drug levels, augmenting pharmacological action, reducing side effects, and improving patient compliance [2]. Most importantly, transdermal drug delivery systems (TDDSs) can be properly used for long-term or chronic use. Consequently, it is a valid choice to develop TDDS for the treatment of a number of pathological illnesses, including inflammation.

Nevertheless, transdermal treatment is only effective with certain kinds of bioactive compounds because of the stratum corneum, which acts as a barrier for the penetration of substances through the skin [3]. As a result, numerous strategies have been implemented to modify stratum corneum permeability. Among them, the implementation of nano-formulations has been acknowledged for the efficacy to bypass transdermal therapy's limitations [4]. Nano-formulations have been deemed effective TDDSs owing to their inherent characteristics of small particle size, high drug retention, and efficient targeting capabilities. Lipid-based vesicles, such as transfersomes, niosomes, liposomes, and ethosomes, as well as nanoparticles and nano-emulsions, can all be classified as nano-formulations.

Niosomes, also known as non-ionic surfactant-based vesicles, have attracted much attention in pharmaceutical fields as a potential substitute for liposomes as a drug carrier. Despite their comparable features, niosomes offer potential advantages over liposomes including higher stability, lower cost, ease of formulation, and scalability. Furthermore, niosomes have gained popularity as potential transdermal drug delivery systems owing to their ability to alter the stratum corneum's membrane characteristics, which augment drug penetration through skin layers, and eventually, promote efficient drug delivery to systemic circulation [5]. In topical drug delivery systems, niosomes have been reported to act as penetration enhancers, local drug storage for prolonged drug release, solubility enhancers for drugs with weak solubility, and rate-limiting membrane barriers for controlled delivery systems [6].

Nevertheless, due to the low viscosity of niosomal vesicles, which results in improper skin application, niosomal formulation may provide certain challenges when applied topically [7]. Instead, incorporating niosomes into emulgel preparation may theoretically lead to viscosity increasing of the preparation, encouraging its topical administration [8]. Emulgel is an effective drug delivery vehicle that combines the characteristics of both the emulsion and the gel base [9]. Emulgel possesses the advantages of being easily applied topically and could be intended for transdermal application [9]. Most importantly, emulgel has the capability to improve skin permeability, which in turn, increases therapeutic effectiveness [10]. Therefore, the application of niosomal emulgel is supposed to enhance the penetration of drugs through skin.

Natural products are compounds that can be found naturally or extracted from various forms of medicinal plants. Natural products have historically been used since ancient times as a source of many therapeutic agents. Nowadays, there is a renewed interest for the utilization of natural products for treating a wide variety of disorders, including inflammation. The herbaceous plant *Zingeber officinale*, also known as ginger, a member of the *Zingeberaceae* family, is frequently used as a spice, condiment, and herb [11]. It has been utilized as traditional medicine to treat a variety of diseases, including inflammatory disorders. Gingerols, shagaols, and paradol, the active ingredients in ginger, have been repeatedly reported to exert anti-inflammatory, antioxidant, anti-cancer, and anti-atherosclerotic activities [12].

Additionally, sesame oil, which is made from sesame seeds (Sesamum indicum), and is one of the plant oils, is made of protein, unsaturated and saturated fatty acids, as well as trace amounts of other nutrients such as sesamolin, sesamol, and sesamin in addition to tocopherol [13]. Sesame oil has anti-cancer [14], anti-hypertensive [15], antiaging, antioxidant, anti-inflammatory [16], and immunoregulatory characteristics. Sesame oil can be used to create emulgel [17] and microemulsion [18] by combining it with a variety of carriers. The drug's delivery and efficacy could be enhanced by combining the sesame oil-based emulgel with a niosomal formulation of ginger extract.

The target of this investigation, therefore, was to formulate ginger extract-loaded sesame oil-based niosomal emulgel. The central composite design was used to optimize the developed niosomal emulgel formulations. Then, the developed formulations were analyzed for the physical properties. In addition, a carrageenan-induced rat hind paw edema test was employed to determine the anti-inflammatory efficacy of the optimized niosomal emulgel formulation.

2. Results and Discussion

2.1. Characterization of Niosomes Loaded with Ginger Extract

Based on our previous investigations, niosomal dispersion with 1:1 molar ratio of cholesterol and nonionic surfactant was developed. Entrapment efficiency of the developed ginger extract loaded niosomes was investigated using centrifugation method. The entrapment efficiency of niosomes loaded with ginger extract was $63.21 \pm 2.45\%$. Particle size and PDI measurement of ginger extract-loaded niosomes were determined, and the results are shown in Figure 1. Ginger extract-loaded niosomes exhibited a particle size of 232.3 ± 3.01 nm, with good size distribution (PDI 0.310).

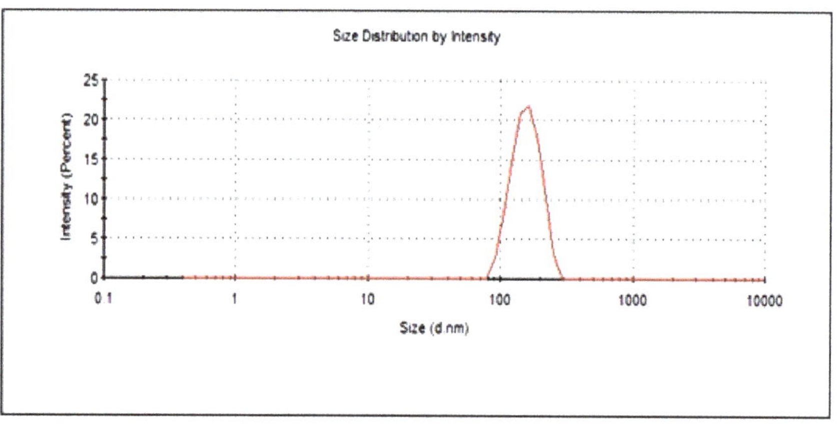

Figure 1. Vesicular size and size distribution curve of ginger extract-loaded niosomes.

The stability of formulated ginger extract-loaded niosomes, in terms of particle size and entrapment efficiency, was evaluated over one and three months at 25 °C and 4 °C. As depicted in Figure 2, particle size, PDI, and entrapment efficiency (EE) of freshly prepared niosomes did not vary significantly upon storage for one or three months at 25 °C and 4 °C ($p > 0.05$). These results confirmed the stability and capacity of niosomes to transport drugs.

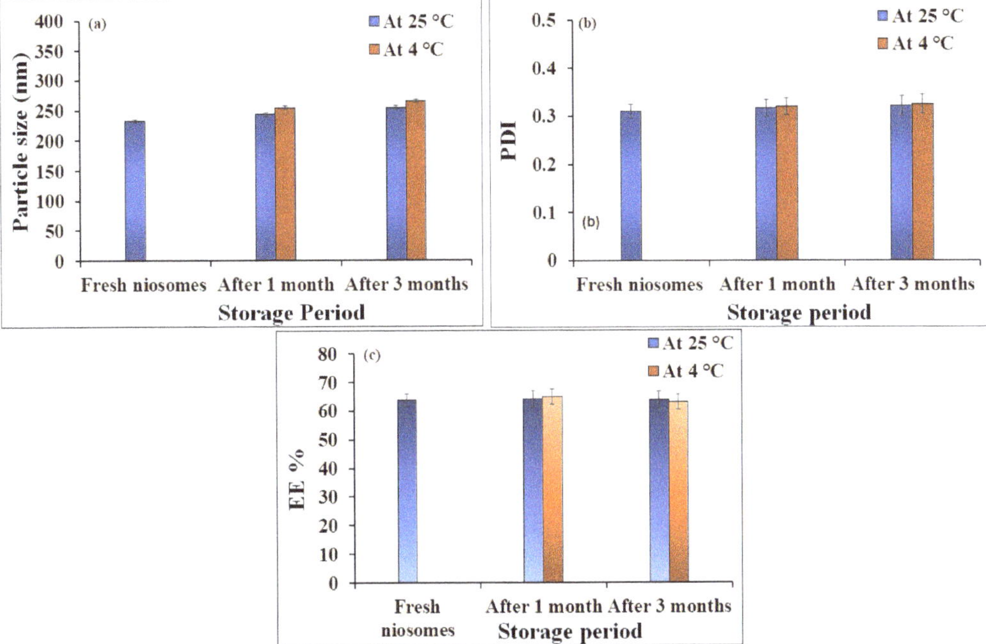

Figure 2. Outline of stability study for ginger extract-loaded niosomal formulation for 1 and 3 months at 4 °C and 25 °C in terms of (**a**) Particle size (nm); (**b**) PDI; (**c**) EE% in comparison to freshly prepared niosomes.

2.2. Development of Ginger Extract Niosomal Emulgel

2.2.1. Solubility Studies

Solubility investigations of ginger extract in various surfactants (Tween 20, Tween 80, Span 20, and Span 80,) and co-surfactants (Propylene glycol and PEG 400) were conducted in order to select the suitable (S_{mix}) surfactant/co-surfactant mixture for the emulsification process. Tween 80 and PEG 400 had the highest solubility of ginger extract when compared to other systems comprising Tween 20, Span 20, Span 80, or Propylene glycol (PG) as illustrated in Figure 3. Tween 80 has the best solubilizing capacity for ginger extract (53.67 mg/mL). Tween 80 is a non-ionic surfactant with high HLB value (14.9) that can be mixed with lipid vehicles to enhance self-emulsification [19]. As a consequence, for subsequent emulsification processes, a surfactant/co-surfactant (S_{mix}) comprising Tween 80 and PEG 400 was used.

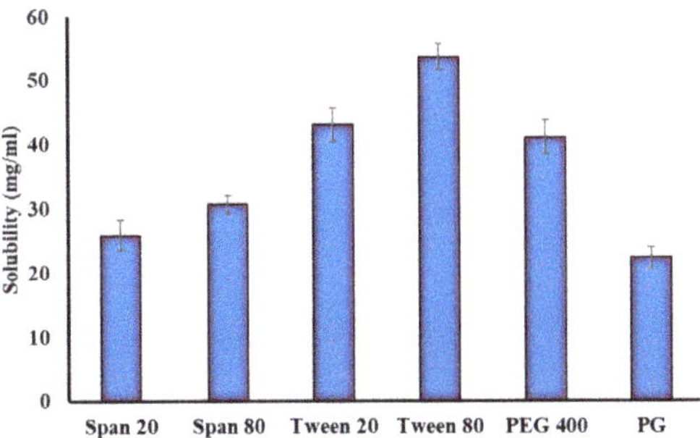

Figure 3. Solubility of ginger extract in different surfactants and co-surfactants.

2.2.2. Construction of Phase Diagrams

The pseudoternary phase diagrams with various volume ratios of Tween 80 to PEG400 (1:1, 1:2, 1:3, 2:1, and 3:1) are depicted in Figure 4. It was evident that the nanoemulsion (ME) area in the phase diagram remarkably increased from 29% to 31% with increasing surfactant to co-surfactant (S_{mix}) ratio 1:1 to 2:1 (Figure 4a,d). However, further increase in the S_{mix} to 3:1 resulted in a reduction of the ME area (22%) (Figure 4e). At an S_{mix} ratio of 1:1, the surfactant concentration could not be sufficient to form a closely packed barrier film, whilst a higher surfactant concentration (S_{mix} ratio 3:1) might result in the development of high concentration of micelles, which contributed to the turbidity of the prepared dispersion resulting in the reduction in the ME region [20]. In the same context, at low co-surfactant (PEG400) concentration with respect to Tween 80 (S_{mix} 3:1), the nanoemulsion area was decreased (Figure 4e). The decrease in the nanoemulsion area might be due to the reduced ability of water incorporation to the ME system [19]. On the other hand, increasing co-surfactant concentration in the surfactant/co-surfactant mixture (1:2 and 1:3) (Figure 4b,c) resulted in a remarkable decrease in the ME area (24% and 18%, respectively). The reduced ME areas could be ascribed to low surfactant concentration, which reduced the amount of micelles and decreased the solubilization capacity of ME [19]. Consequently, it was concluded that the maximum ME area was obtained at an S_{mix} ratio of 2:1 and this ratio was chosen for the development of emulgel loaded with the drug.

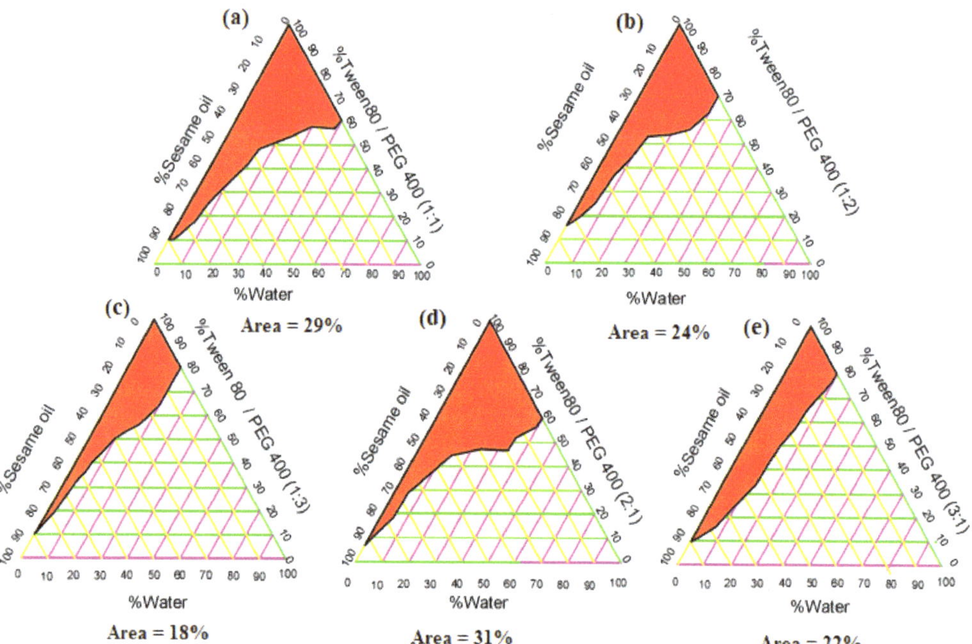

Figure 4. Study of pseudoternary phase diagrams using sesame oil, Tween 80, and PEG 400 at different S_{mix} ratios of (**a**) 1:1, (**b**) 1:2, (**c**) 1:3, (**d**) 2:1, (**e**) 3:1.

2.3. Central Composite (CC) Design for Emulgel Optimization

Conducting central composite (CC) design software resulted in the generation of 18 runs with four center points. Table 1 clearly shows the impact of different formulation variables on the investigated dependent responses of the prepared niosomal emulgel formulations.

Table 1. The independent variables used for optimizing different formulations and the detected results of dependent variables.

Formulation	Independent Variables			Dependent Variables		
	A	B	C	Y_1 (cP)	Y_2 (%)	Y_3 (%)
F1	5	15	7.5	9301.23 ± 114.31	53.45 ± 0.69	82.80 ± 1.15
F2	4	10	10	9899.34 ± 127.23	68.86 ± 1.09	91.37 ± 0.24
F3	6	10	5	8388.45 ± 175.22	53.43 ± 1.79	75.62 ± 0.88
F4	5	15	7.5	9378.47 ± 114.10	52.12 ± 0.81	82.27 ± 0.63
F5	6	20	10	9845.65 ± 234.52	62.76 ± 0.74	92.03 ± 1.36
F6	4	10	5	7685.87 ± 128.51	76.40 ± 1.87	70.96 ± 0.46
F7	6	10	10	9287.12 ± 165.32	69.75 ± 1.65	93.81 ± 0.90
F8	5	15	7.5	9465.47 ± 188.24	52.91 ± 1.17	84.38 ± 0.62
F9	5	15	7.5	9345.65 ± 129.65	53.34 ± 0.75	86.04 ± 0.38
F10	4	20	5	7678.57 ± 137.21	80.25 ± 1.92	73.18 ± 0.53
F11	6	20	5	9578.33 ± 145.45	67.28 ± 1.67	72.41 ± 2.32
F12	5	23.41	7.5	9198.10 ± 152.32	64.15 ± 0.83	81.06 ± 1.40
F13	5	15	3.29	7885.34 ± 134.01	78.25 ± 0.97	72.25 ± 1.53
F14	6.68	15	7.5	9387.38 ± 117.08	52.14 ± 0.92	86.82 ± 0.57
F15	5	6.59	7.5	8752.28 ± 115.32	61.82 ± 1.11	80.50 ± 0.73
F16	4	20	10	9645.87 ± 128.24	58.07 ± 1.10	89.37 ± 1.00
F17	3.32	15	7.5	8545.06 ± 136.36	68.05 ± 1.22	78.95 ± 1.53
F18	5	15	11.70	10,698.20 ± 201.18	72.07 ± 1.05	97.58 ± 2.88

A: Polymer concentration (% w/w); B: oil concentration (% v/v); C: S_{mix} concentration (% v/v); Y_1: viscosity (cP); Y_2: in vitro drug release (%); Y_3: drug content (%).

2.3.1. Design Statistical Analysis

Central composite (CC) design was used to conduct the analysis of variance for the investigated dependent responses, and ANOVA was used to acquire specific parameters such as p-value, F-value, and model F-value [21]. The quadratic model had the best fit for all responses comparing to all other models. Table 2 shows that most responses had a p-value < 0.0001, indicating a significant impact of the independent factors on the tested dependent variables. With regards to the F-value of the responses, it has been proved that higher values are recommended for a model with less error. The F-values for Y_1, Y_2, and Y_3 in the model are 65.72, 122.50, and 31.10, respectively, indicating the significance of the model. Moreover, the lack of fit was 3.73, 5.60, and 1.52, respectively, which were not significant in comparison to the pure error, as well as their corresponding p-values, were 0.0943, 0.0935, and 0.3871, for Y_1, Y_2, and Y_3, respectively.

Table 2. Statistical and regression analysis results for all responses.

Source	Y_1		Y_2		Y_3	
	F-Value	p-Value	F-Value	p-Value	F-Value	p-Value
Model	65.72	<0.0001 *	122.50	<0.0001 *	31.10	< 0.0001 *
A	51.23	<0.0001 *	171.32	<0.0001 *	9.49	0.0151 *
B	19.72	0.0014 *	0.7694	0.4060	0.2827	0.6094
C	400.03	<0.0001 *	42.02	0.0002 *	263.00	< 0.0001 *
AB	27.15	0.0006 *	17.03	0.0033 *	0.8908	0.3729
AC	61.12	<0.0001 *	154.35	<0.0001 *	0.0488	0.8307
BC	5.18	0.1026	112.76	<0.0001 *	0.2520	0.6292
A^2	16.65	0.0040 *	59.78	<0.0001 *	0.6749	0.4351
B^2	15.98	0.0040 *	117.11	<0.0001 *	4.75	0.0610
C^2	1.16	0.2681	565.32	<0.000 *1	0.2354	0.6406
Lack of Fit	5.56	0.0943	5.60	0.0935	1.52	0.3871

Table 2. *Cont.*

Source	Y_1		Y_2		Y_3	
	F-Value	***p*-Value**	**F-Value**	***p*-Value**	**F-Value**	***p*-Value**
			R^2 analysis			
R^2	0.9867		0.9928		0.9722	
Adjusted R^2	0.9716		0.9847		0.9410	
Predicted R^2	0.9036		0.9444		0.8289	
Adequate precision	28.3038		32.3642		19.8035	
Model	Quadratic		Quadratic		Quadratic	

A: Polymer concentration (% w/w); B: oil concentration (% v/v); C: S_{mix} concentration (% v/v); Y_1: viscosity (cP); Y_2: in vitro drug release (%); Y_3: drug content (%); *: significant.

2.3.2. Effect of Independent Variables on Viscosity (Y_1)

The viscosity of an emulgel formulation is an important criterion since it can change the rate at which the drug diffuses from the vehicle, and thereby could exert a considerable impact on the in vitro drug release [22]. Table 1 summarizes the findings on the viscosity of several prepared ginger extract niosomal emulgels. The viscosity of the formulations fluctuates from 7678.57 ± 137.21 to 10698.20 ± 201.18 cP. It was clear that raising A (gelling agent concentration), B (oil concentration), and C (S_{mix} concentration) resulted in a marked increase in the formulation viscosity, which was most likely related to the formulation compositions [23,24]. The following mathematical equation further demonstrates the noticeable effect of the three independent factors on viscosity:

$$Y_1 = -3184.4 + 2205.4A + 31.3B + 1279.65C + 50.2AB - 150.7AC - 8.8BC - 156.4A^2 - 6.1B^2 - 6.6C^2 \tag{1}$$

The positive sign in the equation confirmed that A, B, and C had a synergistic effect on Y1 response. In addition, as shown in Figure 5, the positive effect of different formulation variables on formulation viscosity is represented by 3D-response surface plots and 2D contour.

2.3.3. Effect of Independent Factors on the in Vitro Drug Release (Y_2)

The studies were conducted for 6 h, and the cumulative drug release (%) was calculated. The cumulative drug release ranged from $52.12 \pm 0.81\%$ to $80.25 \pm 1.92\%$. It was obvious that the in vitro drug released from all formulations was significantly influenced by the independent variables studied; there was a negative relationship between percent of in vitro drug release and the independent factors. Increasing the polymer (A), oil (B), and S_{mix} (C) concentrations slow down the drug release process from the examined formulations, presumably owing to the higher viscosity which causes resistance to drug dispersion and movement [25]. The following regression equation explains how the independent factors A, B, and C influence the in vitro release response Y_2:

$$Y_2 = 310.72 - 50.62A - 3.32B - 24.85C + 0.34AB + 2.08AC - 0.35BC + 2.57A^2 + 0.14B^2 + 1.26C^2 \tag{2}$$

The impact of independent variables on in vitro drug release (Y_2) was further studied by the model graphs, as depicted in Figure 6.

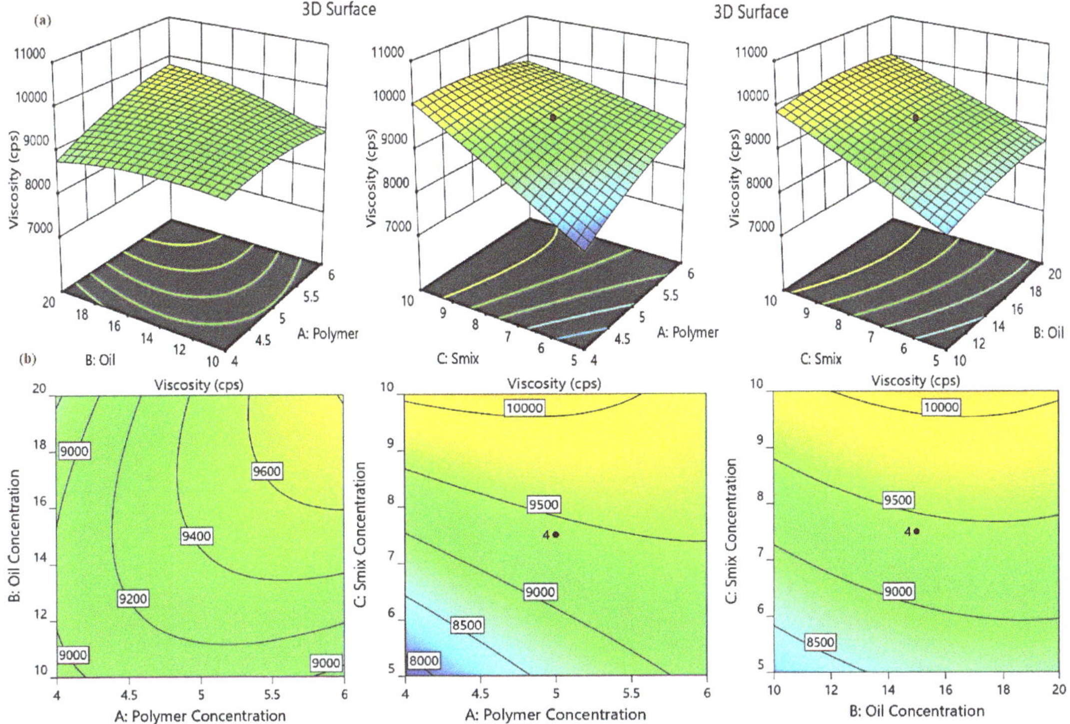

Figure 5. 3D response surface plots (**a**) and corresponding contour plots (**b**) showing the effects of the independent variables on viscosity (Y_1). Two independent variables are considered at a time, while the third one remains constant.*Cont.*

2.3.4. Effect of Independent Variables on the % of Drug Content (Y_3)

The drug content in the developed emulgel formulations ranged from $70.96 \pm 0.46\%$ to $97.58 \pm 2.88\%$, indicating the efficient drug loading within the emulgel formulation. The F-values for drug content (Y_3) in the model were 1.52 and the p-values of the model were less than 0.05, indicating the significance of the model. Equation (3), when expressed in terms of actual factors, can be used to predict the response using different levels of each independent variable. The three independent variables, polymer, oil, and S_{mix} concentrations, had a positive impact on the % drug content, as illustrated in Figure 7.

$$Y_3 = 20.67 + 7.63A + 2.24B + 2.89C - 0.13AB + 0.61AC - 0.03BC - 0.45A^2 - 0.05B^2 + 0.04C^2 \tag{3}$$

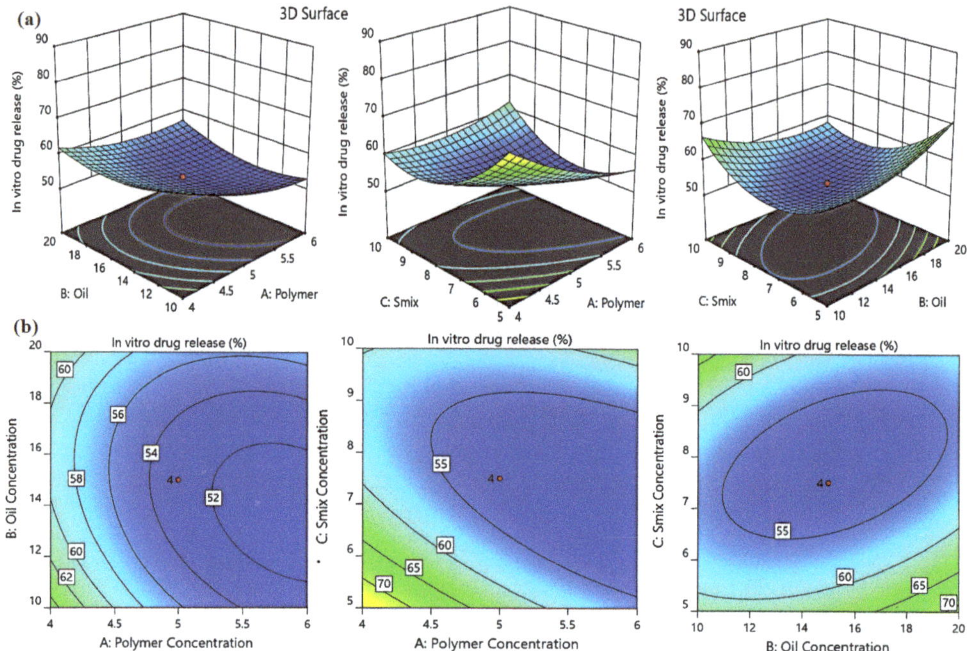

Figure 6. 3D response surface plots (**a**) and corresponding contour plots (**b**) showing the effects of the independent variables on the percent of in vitro drug release (Y_2). Two independent variables are considered at a time, while the third one remains constant.

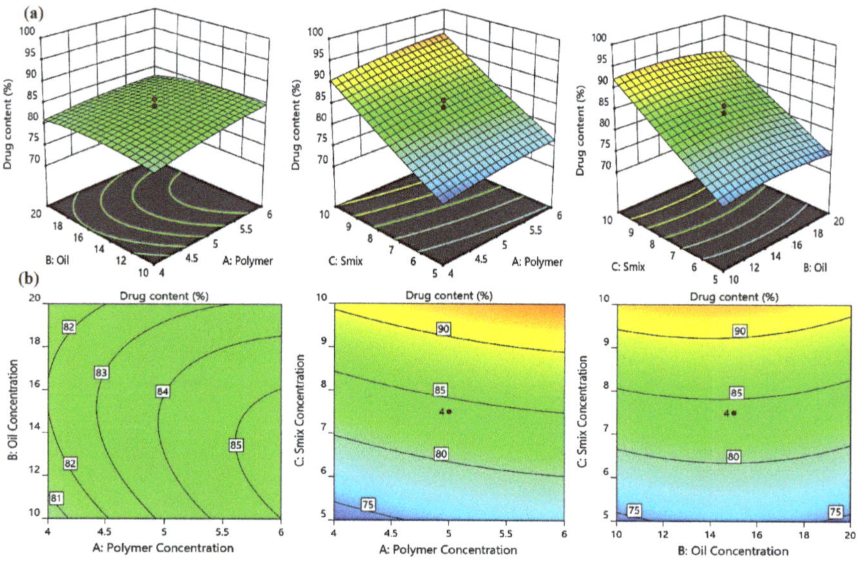

Figure 7. 3D response surface plots (**a**) and corresponding contour plots (**b**) showing the effects of the independent variables on the percent of drug content (Y_3). Two independent variables are considered at a time, while the third one remains constant.

2.3.5. Optimized Formulation of Niosomal Emulgel

The analysis of the polynomial model generated by the CC design was utilized to optimize the emulgel development process with fewer formulations. The numerical optimization was applied by the Design Expert® software (version 12.0, Stat-Ease, Minneapolis, MN, USA). The optimized formulation was obtained by orienting the results towards specific characters that expected to accomplish the desired preparation using point prediction technique. The recommended concentrations of the independent variables acquired by the design of experiment to reach to the desired criteria were detected depending on the maximum desirability (0.822). The optimized formula was obtained at a polymer concentration of 4.75%, oil concentration of 10.07%, and S_{mix} concentration of 9.90%. The actual viscosity, in vitro drug release, and drug content were 9510.33 ± 162.79 cP, 64.84 ± 2.35%, and 93.15 ±1.73%, respectively, for the optimized formulation, which were closed to the predicted values (9915.71 ± 136.34 cP, 66.04 ± 1.18%, 91.18 ± 1.95%, respectively).

2.4. Characterization of Optimized Niosomal Emulgel Formulation

The parameters of the optimized ginger extract niosomal emulgel were investigated and compared to those of the conventional ginger extract niosomal gel, as shown in Table 3. All the formulations had an appealing appearance with a smooth texture. The pH was in the satisfactory range to avoid possible sensitivity and irritation. In addition, the viscosity of the formulations was investigated, and it was revealed that incorporating sesame oil into niosomal emulgel resulted in significantly greater viscosity than niosomal gel ($p = 0.006$); nonetheless, the viscosity of both preparations was within a good range, supporting topical application [26]. The spreadability of the formulations was tested, and the results proved that they could spread uniformly all over the skin.

Table 3. Characterization of the optimized Ginger extract niosomal gel and emulgel.

Properties	Ginger Niosomal Gel	Ginger Niosomal Emulgel
Visual inspection	Smooth and homogenous	Smooth and homogenous
pH	6.47± 0.31	6.62 ± 0.45
Spreadability (mm)	47.25± 2.4	36.54 ± 3.03 *
Viscosity (cP)	8686.67 ± 210.49	9510.33 ± 162.79 *

Values are stated as mean ± (SD). * $p < 0.05$ compared to Ginger niosomal gel.

2.5. In Vitro Release Investigations

The percentage of ginger extract released from ginger extract suspension and different generated preparations is illustrated in Figure 8. As shown in Figure 8, 96.20 ± 2.44% of the ginger extract in suspension was released within 4 h. On the other hand, a much slower release of ginger extract from various gel formulations was observed. The cumulative release percentage of ginger extract at 6 h from niosomal emulgel, drug emulgel, niosomal gel, and drug gel was 64.84 ± 2.35%, 70.52 ± 2.41%, 77.59 ± 3.02%, and 83.46 ± 2.86%, respectively. The significantly lower cumulative ginger extract release from different gel formulations, compared to free drug, might be accounted for by the existence of a gelling agent, which enhances the thickness or viscosity of the formulation, and thereby restrains drug release from gel formulations. In addition, the amount of Ginger extract released from the gel was substantially greater than that released from emulgel ($p = 0.004$), presumably owing to the higher aqueous content of gel formulations, which facilitates the migration of the drug into the release media. On the other hand, entrapment of the ginger extract within niosomal formulations was found to slow down drug release from niosomal emulgel formulation, compared to Ginger extract-loaded emulgel. The ability of cholesterol in the niosomal formulations to generate cement layer in the voids of the bilayers, resulting in more intact bilayers, could influence the drug diffusion, and thereby slow down drug release [27,28].

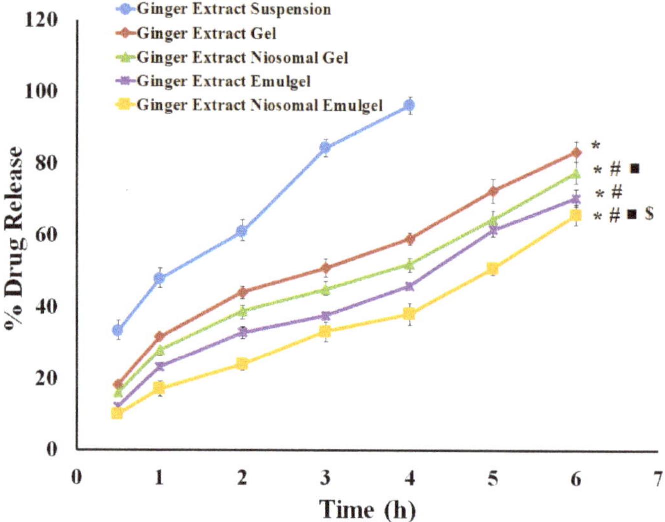

Figure 8. In vitro release study of ginger extract from different formulations compared to ginger extract suspension in phosphate buffer pH 7.4 at 37 °C. Results are expressed as the mean ± SD of three experiments. * $p < 0.05$ compared to ginger extract suspension; # $p < 0.05$ compared to ginger extract gel; ■ $p < 0.05$ compared to ginger extract emulgel and $ $p < 0.05$ compared to ginger extract niosomal gel.

2.6. Skin Permeation Study of Ginger Extract from Different Formulations

The in vitro investigation of ginger extract permeability across excised rabbit skin from the generated formulations was performed and compared to ginger extract suspension. As shown in Figure 9, the highest transdermal permeability was observed with ginger extract niosomal emulgel; the J_{ss} value was (81.84 µg/cm^2.h), while the lowest transdermal permeability was observed with ginger extract suspension (J_{ss} 30.95 µg/cm^2.h). In addition, the ex vivo permeation of ginger extract from ginger extract-loaded emulgel was higher than that from gel formulation. The J_{ss} values of ginger extract-loaded gel and ginger extract-loaded emulgel were 62.59 µg/cm^2.h and 51.24 µg/cm^2.h, respectively. The higher enhancement ratio of ginger extract from the emulgel formulation, compared to the gel formulation, might be ascribed to the existence of oil and S_{mix}, which could act as penetration enhancers [29,30]. Interestingly, the entrapment of ginger extract within niosomal vesicles was found to enhance drug permeability from niosomal gel and emulgel, with ERs of 2.30 and 2.64, respectively. This result could be ascribed to the ability of niosomes to modify the stratum corneum structure and make it looser and more permeable [31]. Furthermore, compared to niosomal gel formulation, incorporating niosomes within emulgel formulation greatly enhances the drug flow as a result of the dual action of surfactant in niosomes and nanoemulsion in emulgel, which further enhances the drug penetration when niosomal preparation is included in the emulgel formulation [32]. Therefore, the aforementioned argument could be used to explain why ginger extract niosomal emulgel is more permeable.

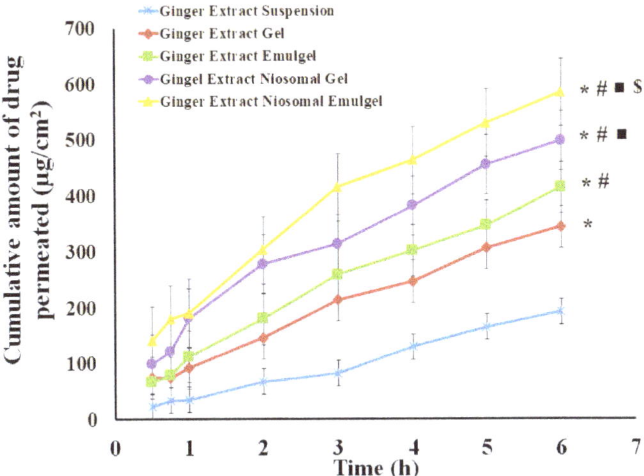

Figure 9. Permeation study of ginger extract from different formulations through excised rabbit skin compared to ginger extract suspension (control). Results are expressed as mean $\pm$ SD (n = 3). * $p < 0.05$ compared to ginger extract suspension; # $p < 0.05$ compared to ginger extract gel; ■ $p < 0.05$ compared to ginger extract emulgel and \$ $p < 0.05$ compared to ginger extract niosomal gel.

2.7. Anti-Inflammatory Testing; Carrageenan-Induced Rat Paw Edema Test

Figure 10 displays the changes of size of edema and the percent of rat hind paw edema inhibition induced by carrageenan injection compared to inflamed animals (non-treated control group). The edema size reached its peak within three hours post-injection of carrageenan in the inflamed groups. Similarly, the animals receiving plain niosomal gel formulation (placebo I) reached the greatest inflammation following three hours of starting the test and did not differ significantly if compared to control animals ($p < 0.05$). On the other hand, rats handled with either plain niosomal emulgel (placebo II), ginger extract orally, ginger extract niosomal gel, or ginger extract niosomal emulgel demonstrated a significant reduction in paw edema thickness (Figure 10b) when compared to control animals ($p < 0.05$), where they demonstrated $14.88 \pm 1.27\%$ (plain niosomal emulgel), $8.49 \pm 1.61\%$ (ginger extract orally), $26.96 \pm 2.62\%$ (ginger extract niosomal gel), and $60.27 \pm 4.19\%$ (ginger extract niosomal emulgel) edema inhibition after six hours. The difference between the groups treated with ginger extract niosomal gel and emulgel was found to be statistically significant ($p < 0.05$), demonstrating the importance of sesame oil and emulgel in enhancing anti-inflammatory effects. The most notable reduction in edema size was noticed in animals treated with ginger extract niosomal emulgel, which was statistically significant ($p < 0.05$) when compared to the groups treated with other administered formulations. Therefore, the ginger extract niosomal emulgel has the highest potential for anti-inflammatory activity. The current study provides significant insights for the enhanced effect and improved permeability of topical formulations using sesame oil and ginger extract in niosomal emulgel and suggesting a synergistic interaction between them. In a previous study, the potential of sesame oil to reduce inflammation was demonstrated [33].

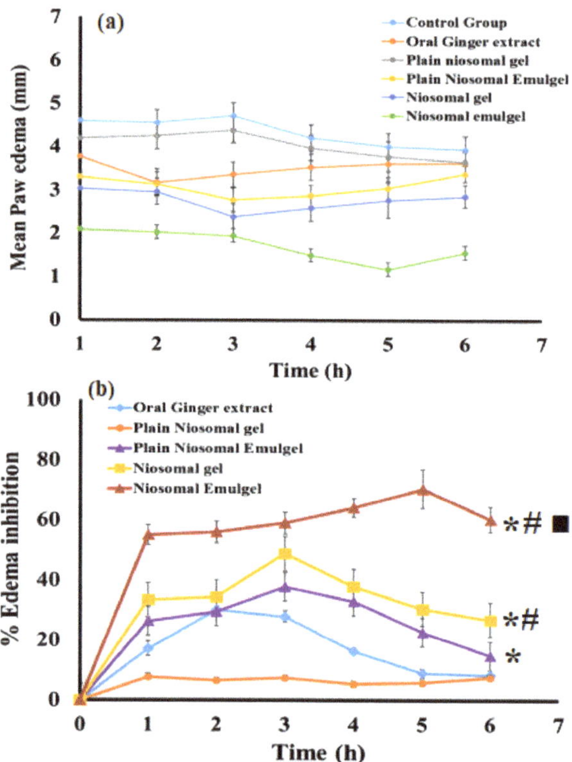

Figure 10. Effects of ginger extract prepared in various formulations on mean hind paw edema (**a**) and percentage of edema inhibition of carrageenan-induced paw edema rats (**b**). Results are expressed as mean with the bar showing SD ($n = 5$). * $p < 0.05$ compared to plain niosomal gel; # $p < 0.05$ compared to ginger extract orally and ■ $p < 0.05$ compared to ginger extract niosomal gel.

3. Conclusions

In the current investigation, ginger extract-loaded niosomal formulations were created, utilizing the thin film hydration approach. The ginger extract emulgels were produced by combining the formulation with sesame oil-based emulgel using the quality by design methodology. The formulation demonstrated satisfactory physical characteristics, enhanced skin permeability performance, and had considerable anti-inflammatory effects. The results demonstrated the remarkable impact of sesame oil on the in vivo behavior of niosomal formulation, suggesting a synergistic interaction between sesame oil and ginger extract. Finally, niosomal emulgel might be taken into consideration as a potential nanocarrier that effectively delivers the bioactive agents topically.

4. Materials and Methods

4.1. Materials

Tween 20, Tween 80, Span 20, Span 60, Span 80, polyethylene glycol (PEG) 400, propylene glycol, cholesterol, hydroxypropyl methylcellulose (HPMC), chloroform, ethanol, methanol, and sodium azide were procured from Sigma Chemical Co. (St. Louis, MO, USA). Solvents and chemicals of analytical grade were purchased from Sigma, USA. Sesame oil was provided by NOW® Essential Oils (NOW Foods, Bloomingdale, IL, USA).

4.2. Preparation of Ginger Extract

Ginger rhizome, obtained from a local market, was washed with clean water, sliced, and dried at 45 °C for 48 h in an oven and then a fine powder was made. Ginger extract was produced by using a maceration process. Briefly, 100 g of the dried powdered ginger was extracted by 500 mL of organic solvent (methanol), then stirred at 500 rpm by using a magnetic stirrer at room temperature for 24 h, then left to stand overnight at room temperature. Solution was filtered through muslin cloth and then re-filtered through Whatman No.1. The solvent was then evaporated at ambient temperature to produce ginger extract [34]. A full UV spectrum graph for the ginger extract is represented in Figure S1.

4.3. Preparation of Ginger Extract Loaded Niosomal Vesicles

The niosomal vesicle formulation was developed using a thin film hydration technique, as described previously [35]. Briefly, in a 50 mL round bottom flask, 100 mg ginger extract was mixed with non-ionic surfactant (Span 60) and cholesterol in (1:1) molar ratio, and then dissolved in 10 mL chloroform/ethanol mixture (2:1 v/v). Using a rotary evaporator (BÜCHI Labortechnik, Meierseggstrasse, Flawil, Switzerland), the organic solvents were then evaporated at 60 °C under reduced pressure. The dried thin lipid film developed on the wall of the round flask was hydrated with 10 mL phosphate buffer (pH 7.4), then the flask was agitated for an additional hour at 60 °C to obtain niosomal dispersion.

4.4. Characterization of the Developed Niosomal Vesicle Loaded with Ginger Extract
4.4.1. Entrapment Efficiency Determination

The centrifugation method was used to determine the entrapment efficiency of the developed ginger extract-loaded niosomes. In brief, the niosomal dispersion was centrifuged at 6000 rpm for 1 h at 4 °C using a cooling centrifuge [36]. The concentration of the un-entrapped drug in the supernatant was determined spectrophotometrically at λ_{max} of 273 nm [37]. Drug entrapment efficiency percentage was computed using the following equation [35]:

$$\text{Drug entrapment efficiency } (\%) = [(AT - AU)/AT] \times 100 \qquad (4)$$

where AT is the total amount of dug added and AU is the amount of the un-entrapped drug.

4.4.2. Vesicular Size and Size Distribution

Dynamic light scattering technique was used to assess the vesicular size and polydispersity (PDI) of ginger extract-loaded niosomal vesicles using Malvern Zetasizer (Worcestershire, UK) [38].

4.5. Stability Studies

The stability of ginger extract-loaded niosomes was investigated in accordance with the ICH guiding principles. The investigation was conducted at two distinct temperatures: 25 ± 1 °C and 4 ± 1 °C, with humidity of 60% [7]. The efficiency of drug entrapment and size and PDI of niosomal vesicles were measured after one and three months.

4.6. Development of Ginger Extract Niosomal Emulgel
4.6.1. Solubility Studies

The saturated solubility of ginger extract in different surfactants (Tween 20, Tween 80, Span 20, and Span 80) and co-surfactants (Propylene glycol and PEG 400) was studied in order to identify solvents with good solubilizing capacity. Extra amounts of ginger extract were mixed in capped 10 mL vials with 2 mL of each vehicle, using a vortex mixer (Fisher Scientific, Waltham, MA, USA), and agitated for 72 h in a thermostatically controlled shaking water bath (Julbo, SW23, Seelbach, Germany) at 25 ± 1 °C. After that, the mixtures were centrifuged at 25 °C at 5000 rpm for fifteen minutes to separate the undissolved ginger extract. Finally, the drug concentration (mg/mL) in the supernatant was detected spectrophotometrically at a maximum wavelength of 273 nm using methanol for dilution. For further investigations, components with the maximum solubility of ginger extract were employed.

4.6.2. Construction of Phase Diagrams

The concentration ranges of the components that potentially result in an extended nanoemulsion area were determined using pseudoternary phase plots. The construction of pseudoternary phase diagrams was employed using a water titration method. Surfactant/co-surfactant mixture (S_{mix}) composed of Tween 80 (surfactant) and PEG 400 (co-surfactants) were blended in various volume ratios (1:1, 1:2, 1:3, 2:1, and 3:1). Then, the S_{mix} mixture was blended with sesame oil in various volume ratios (1: 9 to 9:1), using a vortex mixer until the formation of homogenous mixture. Distilled water was added dropwise to the various mixtures until a turbidity was observed. Autodesk® Autocad® program was used to cosruct pseudoternary plots. S_{mix} ratio showed maximum nanoemulsion area was chosen for further studies.

4.6.3. Experimental Modelling Using Central Composite Design

Central composite (CC) design as a tool of response surface methodology (RSM) were implemented for the purpose of formulation optimization. The impact of independent factors on the observed dependent responses was evaluated using Design-Expert software version 12.0 (Stat-Ease, Minneapolis, MN, USA) [39]. As shown in Table 4, a design of three factors with two levels (2^3) was created utilizing three independent factors depicting concentration of polymer (A), concentration of oil (B) and S_{mix} concentration (C), with 2 levels comprising high (+1) and low (−1). Viscosity, cP (Y_1), in vitro drug release percent (Y_2), and percent of drug content (Y_3) were the dependent responses investigated. To acquire the optimum formula with the intended outputs, eighteen experiments were performed. The analysis of variance (ANOVA) test was used to evaluate the data in order to determine the model's significance and to demonstrate the data statistical analysis. The influence of the independent factors on the explored dependent responses was analyzed using ANOVA test, which was followed by the creation of model graphs in the form of 3D surface and 2D contour plots, as well as precise mathematical equations.

Table 4. Variables of central composite design for ginger extract emulgel formulations showing independent variables and their level of variation.

Independent Variable	Symbol	Level of Variation	
		−1	+1
Polymer concentration (%w/w)	A	4	6
Oil concentration (%v/v)	B	10	20
S_{mix} concentration (%v/v)	C	5	10
Independent Variable	**Symbol**	**Constrains**	
Viscosity (cP)	Y_1	In range	
In vitro drug release (%)	Y_2	Maximize	
Drug content (%)	Y_3	Maximize	

4.6.4. Preparation of Niosomal Sesame Oil Based Emulgel

Eighteen formulations were made with different concentrations of gel-forming polymer (hydroxypropyl methylcellulose (HPMC); 4–6%), sesame oil (10–20%), and surfactant/co-surfactant mixture (Tween 80/PEG 400; S_{mix} (2:1); 5–10% v/v). In brief, different amounts of gel-forming polymer (HPMC) were immersed in 10 mL distilled water and agitated until a uniform hydrogel was developed. On the other hand, nanoemulsion loaded with ginger extract was developed by dissolving the required amount of ginger extract in the required amount of sesame oil with stirring. Then the obtained mixture was mixed with the calculated amount of S_{mix} (Tween 80/PEG 400) for 5 min. Following that, by using a vortex mixer, the aqueous mixture was dripped over the oily mixture with continuous vortexing for 10 min until a nanoemulsion was achieved. The formulated nanoemulsion was mixed with the pre-formulated hydrogel and stirred with the help of a homogenizer mixer (Ika-eurostar 20 high speed digital, Germany) until a consistent sesame oil-based emulgel was generated [40]. The niosomal dispersion was blended with the developed emulgel for achieving the niosomal emulgel encapsulating ginger extract. A niosomal gel containing ginger extract was created to validate the influence of nanoemulsion and sesame oil in our study. A certain amount of HPMC was mixed with ten milliliters distilled water, then agitated thoroughly until a gel was formed, which was then blended with the niosomes to produce a niosomal gel loaded with ginger extract.

4.7. Characterization of the Investigated Ginger Loaded Niosomal Emulgel

4.7.1. Visual Inspection

The homogeneity of the generated formulations loaded with ginger extract was visually assessed [41].

4.7.2. pH Value Estimation

At room temperature, a calibrated digital pH meter (PCT-407 Portable pH Meter, Taipei City, Taiwan) was used to monitor the pH of the tested formulation [42]. The pH reading was taken three times and the average was recorded.

4.7.3. Spreadability Test

The goal of this study was to assess the tested formulation's spreadability by measuring the spreading diameters when they were introduced to the affected area. A specified weight of the tested formulations (1 g) was kept between two glass slides. A definite load of about 0.5 g was applied over the upper slide for one minute. The diameter of the spreading circle was recorded and calibrated in centimeters to determine the spreadability [7].

4.7.4. Viscosity Measurement

The viscosity of the tested formulations was measured at 25 °C using Brookfield viscometer (Brookfield viscometer, Model DV-II, Middleboro, MA, USA). The viscosity measurement in (cP) was carried out three times and the average reading was recorded [42].

4.7.5. Determination of Drug Content

Specified weight of about one gram of the produced test formulations (equal to 50 mg of ginger extract) was accurately diluted to 100 mL using phosphate buffer saline (pH 7.4): methanol mixture (8:2). The drug content was then quantified spectrophotometrically, evaluated at λ_{max} 273 nm [37]. The following formula was used to calculate the percentage of drug content:

$$\% \text{ Drug content} = \frac{\text{Actual amount of the drug in the formuation}}{\text{Theoretical amount of the drug in the foemulation}} \times 100 \quad (5)$$

4.8. In Vitro Drug Release Studies

The percentage of the drug released from different manufactured formulations (Ginger extract suspension, ginger extract emulgel, niosomal gel, and niosomal emulgel loaded with ginger extract) was assessed. The in vitro drug release was conducted, using a locally fabricated diffusion cell, as previously described [43] with minor modifications. Cellophane membranes with Molcular Weight cut-off 12,000–14,000 Da, were presoaked in the release medium overnight before use. Accurately weighed amounts (1 g) of each formulation, equivalent to 50 mg of ginger extract, were placed over the dialysis membrane (donor compartment) that covered the diffusion cell and fixed with a rubber band. The diffusion cells were then submerged in PBS with a pH of 7.4, kept at $37 \pm 1\,°C$, and rotated at 50 rpm. Two mL samples were taken at specified periods (0.25, 0.5, 1, 2, 4, and 6 h) and evaluated spectroscopically at λ_{max} 273 nm. To maintain sink conditions, the samples were replaced with fresh buffer of the same volume.

4.9. Ex Vivo Drug Permeation Study

The abdominal skins of white albino male rabbits with thicknesses of 1.1–1.2 mm were employed. Animal skin was carefully removed and processed for the experiment. The full thickness skin was mounted to one end of a glass tube (2.8 cm diameter and 6.15 cm^2 surface area) with the stratum corneum side facing upwards, while the dermis facing the receptor media. One gram of the investigated formulation containing 50 milligrams of ginger extract were then applied to the donor compartment. Next, the glass tubes were dipped in 500 mL PBS, pH 7.4, containing sodium azide (0.02%), kept at $37 \pm 1\,°C$, and rotated at 50 rpm. At specified time periods throughout 6 h, two milliliter samples were collected and examined at λ_{max} 273 nm using a spectrophotometer. Fresh buffer was added in the same volume in place of the samples. The following equations were used to estimate the steady state transdermal flux (J_{ss}) and enhancement ratio (ER):

$$J_{ss} = \frac{Amount\ of\ permeated\ drug}{area\ of\ permeation \times time} \tag{6}$$

$$ER = \frac{J_{ss\ (test)}}{J_{ss\ (control)}} \tag{7}$$

4.10. In Vivo Experimental Studies
4.10.1. Animals

Male healthy Wistar rats (220–250 g) were maintained under standardized conditions of temperature, relative humidity, and lighting environment. All experiments were approved by the Research Ethics Committee (REC) at University of Ha'il, Saudi Arabia (Approval no. H-2022-291 at 13/06/2022).

4.10.2. Anti-Inflammatory Activity

A carrageenan-induced rat hind paw edema test was adopted to scrutinize the anti-inflammatory potential of ginger extract incorporated into various formulations. In this assay, paw edema was induced in all rats by subcutaneous injection of 0.1 mL of carrageenan (1% w/v) into the sub-plantar surface of rat right paw. Rats were then randomly categorized into six groups (n = 5) as follows: Group I: non treated positive control rats; Group II: rats orally fed with ginger extract (100 mg/kg); Group III: rats treated topically with plain niosomal gel; Group IV: rats treated topically with plain niosomal emulgel; Group V: rats treated topically with ginger extract niosomal gel; Group VI: rats treated topically with ginger extract niosomal emulgel. The anti-inflammatory potential of different formulations was assessed in terms of paw edema thickness, estimated by subtracting the initial paw thickness at zero time before carrageenan injection from the paw thickness measured at each hour after carrageenan injection [44] using a digital caliper (Quick Mini Gage 0.5, Mitutoyo Cooperation, Kawasaki, Japan). In addition, percentage edema inhibition was calculated as an index of acute anti-inflammatory effect using the following equation: ∘

$$\% \text{ edema inhibition} = \frac{V_t - V_0}{V_t} \times 100 \tag{8}$$

where V_t is the thickness of paw edema at time (t), and V_0 is the thickness of paw edema at zero time.

4.11. Statistical Analysis

All values were expressed as mean ± S.D. The differences were compared using one way analysis of variable (ANOVA). *p*-values (<0.05) were considered statistically significant.

Supplementary Materials: The following supporting information can be downloaded at: https://www.mdpi.com/article/10.3390/gels8110737/s1, Figure S1: A full UV spectrum graph for the ginger extract.

Author Contributions: M.H.A.: Conceptualization, methodology, writing—review and editing and supervision; A.S.A.L., N.E.K. and W.M.A.K.: software, data curation, validation, writing—review and editing; H.A.E., E.-S.K., H.E.-S.E.-H. and S.E.-H.: methodology, software and writing—original draft preparation, formal analysis, investigation. All authors have read and agreed to the published version of the manuscript.

Funding: This research has been funded by Scientific Research Deanship at University of Ha'il, Saudi Arabia, through project number RG-22 022.

Institutional Review Board Statement: The study was approved by the Institutional Research Ethics Committee (IREC), University of Ha'il, Saudi Arabia (approval no. H-2022-291 at 13 June 2022).

Informed Consent Statement: Not applicable.

Data Availability Statement: Not applicable.

Acknowledgments: The authors thank Scientific Research Deanship at University of Ha'il, Saudi Arabia for funding this research through the project number RG-22 022.

Conflicts of Interest: The authors declare no conflict of interest.

References

1. Pastore, M.N.; Kalia, Y.N.; Horstmann, M.; Roberts, M.S. Transdermal patches: History, development and pharmacology. *Br. J. Pharmacol.* **2015**, *172*, 2179–2209. [CrossRef] [PubMed]
2. Isaac, M.; Holvey, C. Transdermal patches: The emerging mode of drug delivery system in psychiatry. *Ther. Adv. Psychopharmacol.* **2012**, *2*, 255–263. [CrossRef] [PubMed]
3. Paudel, K.S.; Milewski, M.; Swadley, C.L.; Brogden, N.K.; Ghosh, P.; Stinchcomb, A.L. Challenges and opportunities in dermal/transdermal delivery. *Ther. Deliv.* **2010**, *1*, 109–131. [CrossRef]
4. Palmer, B.C.; DeLouise, L.A. Nanoparticle-enabled transdermal drug delivery systems for enhanced dose control and tissue targeting. *Molecules* **2016**, *21*, 1719. [CrossRef] [PubMed]
5. Waqas, M.K.; Sadia, H.; Khan, M.I.; Omer, M.O.; Siddique, M.I.; Qamar, S.; Zaman, M.; Butt, M.H.; Mustafa, M.W.; Rasool, N. Development and characterization of niosomal gel of fusidic acid: In-vitro and ex-vivo approaches. *Des. Monomers Polym.* **2022**, *25*, 165–174. [CrossRef]
6. Abdallah, M.H.; Sabry, S.A.; Hasan, A.A. Enhancing Transdermal Delivery of Glimepiride Via Entrapment in Proniosomal Gel. *J. Young Pharm.* **2016**, *8*, 335. [CrossRef]
7. Elsewedy, H.S.; Younis, N.S.; Shehata, T.M.; Mohamed, M.E.; Soliman, W.E. Enhancement of Anti-Inflammatory Activity of Optimized Niosomal Colchicine Loaded into Jojoba Oil-Based Emulgel Using Response Surface Methodology. *Gels* **2021**, *8*, 16. [CrossRef]
8. Sah, S.K.; Badola, A.; Nayak, B.K. Emulgel: Magnifying the application of topical drug delivery. *Indian J. Pharm. Biol. Res.* **2017**, *5*, 25–33. [CrossRef]
9. Aithal, G.C.; Narayan, R.; Nayak, U.Y. Nanoemulgel: A Promising Phase in Drug Delivery. *Curr. Pharm. Des.* **2020**, *26*, 279–291. [CrossRef]
10. Nastiti, C.M.; Ponto, T.; Abd, E.; Grice, J.E.; Benson, H.A.; Roberts, M.S. Topical nano and microemulsions for skin delivery. *Pharmaceutics* **2017**, *9*, 37. [CrossRef]
11. Akram, A.; Rasul, A.; Waqas, M.K.; Irfan, M.; Khalid, S.H.; Aamir, M.N.; Murtaza, G.; Ur Rehman, K.; Iqbal, M.; Khan, B.A. Development, characterization and evaluation of in-vitro anti-inflammatory activity of ginger extract based micro emulsion. *Pak. J. Pharm. Sci.* **2019**, *32*, 1327–1332. [PubMed]
12. Habib, S.H.M.; Makpol, S.; Hamid, N.A.A.; Das, S.; Ngah, W.Z.W.; Yusof, Y.A.M. Ginger extract (*Zingiber officinale*) has anti-cancer and anti-inflammatory effects on ethionine-induced hepatoma rats. *Clinics* **2008**, *63*, 807–813. [CrossRef] [PubMed]
13. El-Beltagi, H.S.; Maraei, R.W.; El-Ansary, A.E.; Rezk, A.A.; Mansour, A.T.; Aly, A.A. Characterizing the Bioactive Ingredients in Sesame Oil Affected by Multiple Roasting Methods. *Foods* **2022**, *11*, 2261. [CrossRef] [PubMed]
14. Majdalawieh, A.; Farraj, J.F.; Carr, R.I. *Sesamum indicum* (sesame) enhances NK anti-cancer activity, modulates Th1/Th2 balance, and suppresses macrophage inflammatory response. *Asian Pac. J. Trop. Biomed.* **2020**, *10*, 316–324. [CrossRef]

15. Wichitsranoi, J.; Weerapreeyakul, N.; Boonsiri, P.; Settasatian, C.; Settasatian, N.; Komanasin, N.; Sirijaichingkul, S.; Teerajetgul, Y.; Rangkadilok, N.; Leelayuwat, N. Antihypertensive and antioxidant effects of dietary black sesame meal in pre-hypertensive humans. *Nutr. J.* **2011**, *10*, 82. [CrossRef]

16. Hsu, E.; Parthasarathy, S. Anti-inflammatory and antioxidant effects of sesame oil on atherosclerosis: A descriptive literature review. *Cureus* **2017**, *9*, e1438. [CrossRef]

17. Jadhav, R.; Yadav, G.; Jadhav, V.; Jain, A. Formulation and evaluation of black sesame seed oil sunscreen emulgel using natural gelling agent. *Int. Res. J. Pharm.* **2018**, *9*, 197–201. [CrossRef]

18. Nazari, M.; Mehrnia, M.A.; Jooyandeh, H.; Barzegar, H. Preparation and characterization of water in sesame oil microemulsion by spontaneous method. *J. Food Process Eng.* **2019**, *42*, e13032. [CrossRef]

19. Ibrahim, T.M.; Abdallah, M.H.; El-Megrab, N.A.; El-Nahas, H.M. Upgrading of dissolution and anti-hypertensive effect of Carvedilol via two combined approaches: Self-emulsification and liquisolid techniques. *Drug Dev. Ind. Pharm.* **2018**, *44*, 873–885. [CrossRef]

20. Rao, M.; Sukre, G.; Aghav, S.; Kumar, M. Optimization of metronidazole emulgel. *J. Pharm.* **2013**, *2013*, 501082. [CrossRef]

21. Abdallah, M.H.; Abdelnabi, D.M.; Elghamry, H.A. Response Surface Methodology for Optimization of Buspirone Hydrochloride-Loaded In Situ Gel for Pediatric Anxiety. *Gels* **2022**, *8*, 395. [CrossRef] [PubMed]

22. Bolla, P.K.; Clark, B.A.; Juluri, A.; Cheruvu, H.S.; Renukuntla, J. Evaluation of formulation parameters on permeation of ibuprofen from topical formulations using Strat-M®membrane. *Pharmaceutics* **2020**, *12*, 151. [CrossRef] [PubMed]

23. Jagdale, S.; Pawar, S. Gellified emulsion of ofloxacin for transdermal drug delivery system. *Adv. Pharm. Bull.* **2017**, *7*, 229. [CrossRef]

24. Shehata, T.M.; Nair, A.B.; Al-Dhubiab, B.E.; Shah, J.; Jacob, S.; Alhaider, I.A.; Attimarad, M.; Elsewedy, H.S.; Ibrahim, M.M. Vesicular emulgel based system for transdermal delivery of insulin: Factorial design and in vivo evaluation. *Appl. Sci.* **2020**, *10*, 5341. [CrossRef]

25. Daood, N.M.; Jassim, Z.E.; Gareeb, M.M.; Zeki, H. Studying the effect of different gelling agent on the preparation and characterization of metronidazole as topical emulgel. *Asian J. Pharm. Clin. Res.* **2019**, *12*, 571–577. [CrossRef]

26. Ibrahim, T.M.; Abdallah, M.H.; El-Megrab, N.A.; El-Nahas, H.M. Transdermal ethosomal gel nanocarriers; a promising strategy for enhancement of anti-hypertensive effect of carvedilol. *J. Liposome Res.* **2019**, *29*, 215–228. [CrossRef] [PubMed]

27. Agarwal, R.; Katare, O.; Vyas, S. Preparation and in vitro evaluation of liposomal/niosomal delivery systems for antipsoriatic drug dithranol. *Int. J. Pharm.* **2001**, *228*, 43–52. [CrossRef]

28. Kasliwal, N. Development, characterization, and evaluation of liposomes and niosomes of bacitracin zinc. *J. Dispers. Sci. Technol.* **2012**, *33*, 1267–1273. [CrossRef]

29. Magnusson, B.M.; Walters, K.A.; Roberts, M.S. Veterinary drug delivery: Potential for skin penetration enhancement. *Adv. Drug Deliv. Rev.* **2001**, *50*, 205–227. [CrossRef]

30. Shah, H.; Nair, A.B.; Shah, J.; Bharadia, P.; Al-Dhubiab, B.E. Proniosomal gel for transdermal delivery of lornoxicam: Optimization using factorial design and in vivo evaluation in rats. *DARU J. Pharm. Sci.* **2019**, *27*, 59–70. [CrossRef]

31. Muzzalupo, R.; Pérez, L.; Pinazo, A.; Tavano, L. Pharmaceutical versatility of cationic niosomes derived from amino acid-based surfactants: Skin penetration behavior and controlled drug release. *Int. J. Pharm.* **2017**, *529*, 245–252. [CrossRef] [PubMed]

32. Ibrahim, M.M.; Shehata, T.M. The enhancement of transdermal permeability of water soluble drug by niosome-emulgel combination. *J. Drug Deliv. Sci. Technol.* **2012**, *22*, 353–359. [CrossRef]

33. Monteiro, É.M.H.; Chibli, L.A.; Yamamoto, C.H.; Pereira, M.C.S.; Vilela, F.M.P.; Rodarte, M.P.; De Oliveira Pinto, M.A.; Da Penha Henriques do Amaral, M.; Silvério, M.S.; De Matos Araújo, A.L.S.; et al. Antinociceptive and anti-inflammatory activities of the sesame oil and sesamin. *Nutrients* **2014**, *6*, 1931–1944. [CrossRef] [PubMed]

34. Dessai, P.; Mhaskar, G.M. Formulation and Evaluation of Ginger officinale Emulgel. *Res. J. Pharm. Technol.* **2019**, *12*, 1559–1565. [CrossRef]

35. Abdelnabi, D.M.; Abdallah, M.H.; Elghamry, H.A. Buspirone hydrochloride loaded in situ nanovesicular gel as an anxiolytic nasal drug delivery system: In vitro and animal studies. *AAPS PharmSciTech* **2019**, *20*, 134. [CrossRef] [PubMed]

36. Abdallah, M.H.; Lila, A.S.A.; Unissa, R.; Elsewedy, H.S.; Elghamry, H.A.; Soliman, M.S. Brucine-Loaded Ethosomal Gel: Design, Optimization, and Anti-inflammatory Activity. *AAPS PharmSciTech* **2021**, *22*, 269. [CrossRef]

37. Tawde, S.; Gawde, K.; Mahind, D.; Mhaprolkar, C.; Sawant, K.; Surve, C. GINGEL: Development and Evaluation of Anti-Arthritic Gel Containing Ginger (*Zingiber officinale*). *Int. J. Innov. Sci. Res. Technol.* **2020**, *5*, 7. [CrossRef]

38. Abdallah, M.H.; Elsewedy, H.S.; AbuLila, A.S.; Almansour, K.; Unissa, R.; Elghamry, H.A.; Soliman, M.S. Quality by Design for Optimizing a Novel Liposomal Jojoba Oil-Based Emulgel to Ameliorate the Anti-Inflammatory Effect of Brucine. *Gels* **2021**, *7*, 219. [CrossRef]

39. Abdallah, M.H. Box-behnken design for development and optimization of acetazolamide microspheres. *India* **2014**, *5*, 1228–1239. [CrossRef]

40. Abdallah, M.H.; Lila, A.S.A.; Unissa, R.; Elsewedy, H.S.; Elghamry, H.A.; Soliman, M.S. Preparation, characterization and evaluation of anti-inflammatory and anti-nociceptive effects of brucine-loaded nanoemulgel. *Colloids Surf. B Biointerfaces* **2021**, *205*, 111868. [CrossRef]

41. Abdallah, M.H.; Lila, A.S.A.; Anwer, M.K.; Khafagy, E.-S.; Mohammad, M.; Soliman, M.S. Formulation, development and evaluation of ibuprofen loaded nano-transferosomal gel for the treatment of psoriasis. *J. Pharm. Res* **2019**, *31*, 1–8. [CrossRef]

42. Thorat, Y.S.; Kote, N.S.; Patil, V.V.; Hosmani, A.H. Formulation and evaluation of liposomal gel containing extract of piprine. *Int. J. Curr. Pharm. Res.* **2020**, *12*, 126–129. [CrossRef]
43. Abdallah, M.H.; Abu Lila, A.S.; Shawky, S.M.; Almansour, K.; Alshammari, F.; Khafagy, E.-S.; Makram, T.S. Experimental Design and Optimization of Nano-Transfersomal Gel to Enhance the Hypoglycemic Activity of Silymarin. *Polymers* **2022**, *14*, 508. [CrossRef] [PubMed]
44. Zammel, N.; Saeed, M.; Bouali, N.; Elkahoui, S.; Alam, J.M.; Rebai, T.; Kausar, M.A.; Adnan, M.; Siddiqui, A.J.; Badraoui, R. Antioxidant and anti-Inflammatory effects of *Zingiber officinale* roscoe and Allium subhirsutum: In silico, biochemical and histological Study. *Foods* **2021**, *10*, 1383. [CrossRef]

 *gels*

Article

Response Surface Methodology for Optimization of Buspirone Hydrochloride-Loaded In Situ Gel for Pediatric Anxiety

Marwa H. Abdallah [1,2,*], **Dina M. Abdelnabi** [2] and **Hanaa A. Elghamry** [2]

[1] Department of Pharmaceutics, College of Pharmacy, University of Ha'il, Hail 81442, Saudi Arabia
[2] Department of Pharmaceutics and Industrial Pharmacy, Faculty of Pharmacy, Zagazig University, Zagazig 44519, Egypt; dinaabdelnabi@zu.edu.eg (D.M.A.); haelghamry@zu.edu.eg (H.A.E.)
* Correspondence: mh.abdallah@uoh.edu.sa

Abstract: The purpose of the current investigation was to formulate, assess, and optimize oral in situ gels of buspirone hydrochloride (BH) with the specific end goal of expanding the time the medication spends in the stomach, thereby ensuring an extended medication discharge. This would allow the use of a once-a-day dose of liquid BH formulations, which is ideal for the treatment of pediatric anxiety. In situ gels loaded with BH were prepared using various concentrations of sodium alginate (Na alg.), calcium chloride ($CaCl_2$), and hydroxypropyl methylcellulose (HPMC K15M). The in situ gels exhibited the desired consistency, drug distribution, pH, ability to form gel, and prolonged drug release in vitro. The (3^3) full factorial design was utilized for the revealing of the ideal figures for the selected independent variables, Na alg. (X_1), HPMC (X_2), and $CaCl_2$ (X_3) based on measurements of the viscosity (Y_1) and percentage drug release after 6 h (Y_2). A pharmacokinetic study of the optimum formulation on rabbits was also performed. The formulation containing 2% of Na alg., 0.9% of HPMC-K15M, and 0.1125% of $CaCl_2$ was selected as the ideal formulation, which gave the theoretical values of 269.2 cP and 44.9% for viscosity and percentage of drug released after 6 h, respectively. The pharmacokinetic study showed that the selected oral Na alg. in situ gel formulation displayed a prolonged release effect compared to BH solution and the marketed tablet (Buspar®), which was confirmed by the low C_{max} and high T_{max} values. The optimum oral Na alg. in situ gel showed a 1.5-fold increment in bioavailability compared with the drug solution.

Keywords: in situ gelling; buspirone hydrochloride; sodium alginate; HPMC-K15M; factorial experimental design

Citation: Abdallah, M.H.; Abdelnabi, D.M.; Elghamry, H.A. Response Surface Methodology for Optimization of Buspirone Hydrochloride-Loaded In Situ Gel for Pediatric Anxiety. *Gels* **2022**, *8*, 395. https://doi.org/10.3390/gels8070395

Academic Editors: Maddalena Sguizzato, Rita Cortesi and Rachel Yoon Chang

Received: 4 May 2022
Accepted: 19 June 2022
Published: 22 June 2022

Publisher's Note: MDPI stays neutral with regard to jurisdictional claims in published maps and institutional affiliations.

1. Introduction

Buspirone hydrochloride (BH) is an anxiolytic drug and an example of drugs that are easily absorbed from the GIT. It has a short half-life (2 to 4 h) due to first-pass metabolism [1] bringing about a quick exit from the blood flow; therefore, various dosages are required. To avoid this problem, oral extended-release formulations are typically utilized, as these release the drug slowly into the gastrointestinal tract (GIT) [2]. A comparison between a controlled-release and an immediate-release formulation of BH showed an almost 3.3-fold higher plasma concentration at a steady state following the extended-release dose and a relative bioavailability of 170–190% compared to a similar dose of an immediate-release formulation [3]. This was explained by the fact that BH is mainly metabolized by first-pass metabolism in the gut wall. Since BH is released from the immediate-release formulation at a much faster rate than from the extended-release formulation, more BH is metabolized [4]. This explains the need to develop an extended-release oral dosage form of BH.

Oral liquid dosage forms are ordinarily viewed as the most favored forms of drug administration [5]. The optimization of oral drug delivery systems for young patients is a major challenge. The majority of pediatric patients aged 6 to 11 years are unable to swallow solid oral dosage forms. As a result, finding an easily swallowable dosage form for children is critical [6]. Oral in situ gel is an innovative mucoadhesive drug delivery system that takes the form of low-viscosity liquid upon formulation but transforms into gel under certain conditions in the body (pH, temperature, etc) [7]. As a result, it not only extends the contact period between the medication and the absorptive sites in the stomach, but also allows the drug to be released slowly and continuously, making it particularly effective for chronically used treatments [8]. Drug delivery systems that are formed in situ are simple to manufacture and easily swallowed, especially by children [9].

Alginate polymers contain β-D-mannuronic acid and α-L-glucuronic acid residues linked by 1,4-glycosidic linkages. The interaction between glucuronic acid in alginate chains and Ca^{2+} ions causes the gelation of alginate solutions [7].

Experimental design is a form of design in which some variables can be evaluated at several levels in a definite number of investigations. Experimental designs are divided into two types: factorial design and response surface design. Factorial design is a form of screening design, which classified into full factorial design and fractional factorial design; response surface design includes central composite design and Box Behnken design [10]. Box Behnken design (BBD) is an operative software related to response surface methodology (RSM), which is based on designing experiments and studying models via mathematical and statistical equations in addition to specific graphical forms [11].

This study attempts to discuss the formulation and optimization of oral in situ gels for the sustained delivery of BH in order to achieve a reduced daily dose frequency. Gastro-retentive in situ gelling liquids were formulated using different concentrations of Na alg. and HPMC. Calcium chloride is the most commonly used chemical agent to bind alginate molecules together in the preparation of in situ alginate hydrogels from alginate solutions in water [12]. Sodium citrate is added to form a complex with free Ca^{2+} in the formulation to maintain its fluidity until it reaches the stomach, where Ca^{2+} starts to leach from the formulation in response to the acidic environment, causing Na alg. to shape into gel [13].

In the present study, a sustained oral delivery system of sodium alginate in situ gel for buspirone HCl was developed. Its viscosity, drug distribution, pH, ability to form gel, in vitro and in vivo animal study were explored.

2. Results and Discussion

2.1. DSC Studies

The DSC charts of the pure BH, pure Na alg., and pure HPMC K15M, as well as their physical mixture, are shown in Figure 1. The DSC thermogram of the pure drug showed two endothermic peaks at 189.56 °C and 204.45 °C, and one exothermic peak at 192.25 °C. The endothermic peak at 189.56 °C was due to the melting of the pure drug sample. The exothermic peak at 192.25 °C was due to the conversion of the polymorphic form-1 of the drug to polymorphic form2. The second endothermic peak, at 204.45 °C, was due to the melting of the recrystallized polymorph-2 of the drug. These results were in accordance with the data obtained in the literature [14,15]. The thermogram of the Na alg. showed a broad endothermic peak at 87.32 °C, which may have appeared due to the loss of the water content and moisture of the polysaccharide, and an exothermic peak at 250.42 °C, indicating the thermal decomposition of the Na alg. [16]. A broad endothermic peak was observed as a result of the dehydration process over a temperature range of 60–100 °C for HPMC K15M [17]. The DSC thermograph of the physical mixture consisting of BH, Na alginate, and HPMC K15M showed a broad endothermic peak at 75.61 °C and broad exothermic peak at 261.07 °C. Furthermore, the physical mixture did not retain the drug endothermic peaks, suggesting that the drug lost its crystalline properties and converted to an amorphous state. A DSC thermogram of the physical mixture showed no appearance of new peaks, indicating the compatibility between the drug and the polymers [17].

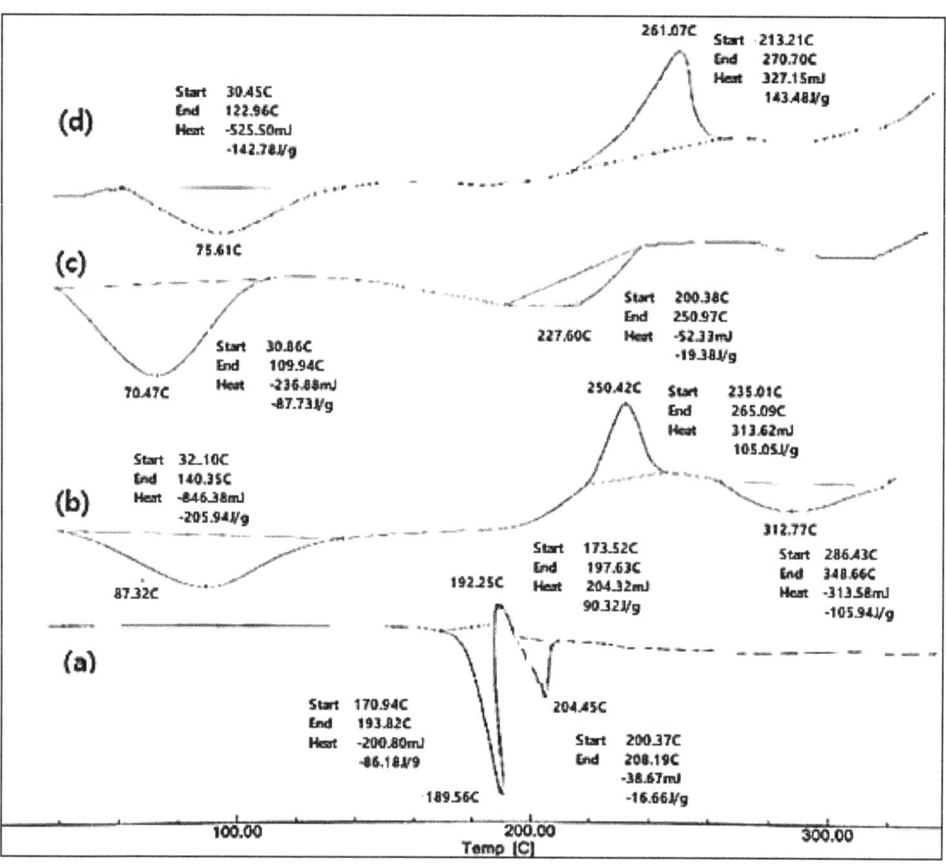

Figure 1. DSC thermograms of (**a**) pure drug (BH), (**b**) Na alginate, (**c**) HPMC K15M, and (**d**) physical mixture of the drug with Na alginate and HPMC K15M.

2.2. Physicochemical Evaluation of In Situ Gelling Solutions

At room temperature, all formulations were liquid and did not show any signs of gelation. It was found that the drug content percentage of the prepared formulations was between 85 and 99.52%, as shown in Table 1. These results indicate the homogenous drug distribution throughout the gelling solution. All the formulations had pH in the range of 6.9–7.7, which was found to be suitable for oral administration; therefore, there was no need to adjust the pH [18]. The in vitro gelling capacity of the in situ gelling formulations is demonstrated in Table 1. The viscosity of the formulations was taken into consideration during the selection of the in situ gelling system [19]. All the formulations remained liquid and suitable for oral administration. The formulations that contained 1% w/v of Na alg. with 0% and 0.3% HPMC and 1.5% w/v of Na alginate with 0% HPMC showed immediate gelation upon contact with the 0.1 N HCl and remained for 12 h (++) as they began to dissolve and erode, probably because of the weak cross-linking that resulted from the low polymer concentration [18]. All the other formulations showed immediate gelation and remained intact for more than 24 h (+++). As shown in Table 1, the formulations showed an increase in viscosity with increasing concentrations of both Na alginate and HPMC [13]. This can be attributed to the greater likelihood of chain interactions in the presence of high concentrations of the polymer [19,20].

Table 1. Composition of the in situ gel formulations, their viscosity, pH, drug content, and gelling capacity.

Na Alginate Conc. % *w/v*	HPMC K-15M Conc. % (*w/v*)	Viscosity (cP)	pH	Drug Content % (*w/v*)	Gelling Capacity
	0.0	30.00 ± 0.50	6.90	96.00 ± 0.20	++
	0.3	38.33 ± 1.53	7.25	98.96 ± 0.37	++
1	0.6	50.33 ± 3.06	7.30	97.20 ± 0.55	+++
	0.9	52.33 ± 0.58	7.41	95.89 ± 0.63	+++
	0.0	49.00 ± 3.60	7.32	99.00 ± 1.15	++
	0.3	103.6 ± 3.00	7.40	99.27 ± 0.24	+++
1.5	0.6	125.0 ± 2.55	7.46	98.85 ± 0.62	+++
	0.9	139.6 ± 1.40	7.50	89.60 ± 0.46	+++
	0.0	100.5 ± 0.44	7.44	98.00 ± 0.88	+++
	0.3	169.0 ± 1.00	7.53	96.40 ± 0.38	+++
2	0.6	199.7 ± 1.53	7.55	99.52 ± 0.42	+++
	0.9	227.0 ± 1.00	7.62	92.40 ± 0.49	+++
	0.0	177.0 ± 2.00	7.70	98.00 ± 1.36	+++
	0.3	308.0 ± 2.32	7.73	95.20 ± 0.57	+++
2.5	0.6	377.0 ± 2.88	7.75	92.70 ± 0.90	+++
	0.9	401.4 ± 3.00	7.77	90.50 ± 0.53	+++
	0.0	300.0 ± 4.00	7.70	97.00 ± 0.36	+++
	0.3	447.0 ± 6.08	7.76	85.00 ± 0.24	+++
3	0.6	555.0 ± 5.00	7.77	87.00 ± 0.78	+++
	0.9	575.7 ± 5.13	7.78	88.42 ± 0.73	+++

All formulations contained 0.25% sodium citrate and 0.075% calcium chloride. (++) formulations showed immediate gelation and remained for 12 h; (+++) formulations showed immediate gelation and remained intact for more than 24 h. Each result is the mean of three determinations ± standard deviation (SD).

2.3. In Vitro Release of BH from In Situ Forming Gels

According to the data of the in vitro release of BH from the different Na alg. in situ gels, the in vitro release rate of BH was slower than that detected for the free BH solution. Therefore, it was observed that the in situ gelling preparations had a high efficiency in decreasing the drug release rate compared with the free BH solution, which released about 95% within six hours, as shown in Figure 2.

2.3.1. Effect of Sodium Alginate Concentration on the In Vitro Release of BH from In Situ Gelling Formulations

The effect of the Na alg. concentration on the in vitro release of the drug from the in situ gels was illustrated in Figure 2a. A significant decrease ($p < 0.05$) in the drug release was observed with the increase in Na alg. concentration; this was related to the increased aggregation of the polymer molecules and the increase in the diffusional path length, which the drug molecules ought to traverse [18]. Moreover, as the viscosity increased with increasing concentrations of the polymer, the solvent's penetration into the core of the matrix was decreased, and the release of the drug was hindered [18].

2.3.2. Effect of HPMC Concentration on the In Vitro Release of BH from In Situ Gelling Formulations

From Figure 2b–d, it can be observed that the HPMC had a release-retarding effect on all the chosen sodium alginate concentrations (1, 2, and 3% *w/v*). This could be explained by the fact that with an increase in HPMC concentration, the number of polymer particles is increased, thereby increasing the viscosity and retarding the release [21].

Moreover, HPMC forms a gel layer that prevents the diffusion of the dissolved calcium chloride out of the matrix quickly. Therefore, there is enough time for calcium ions to attach to the swelled alginate chain, forming a viscous gel within the matrix [22].

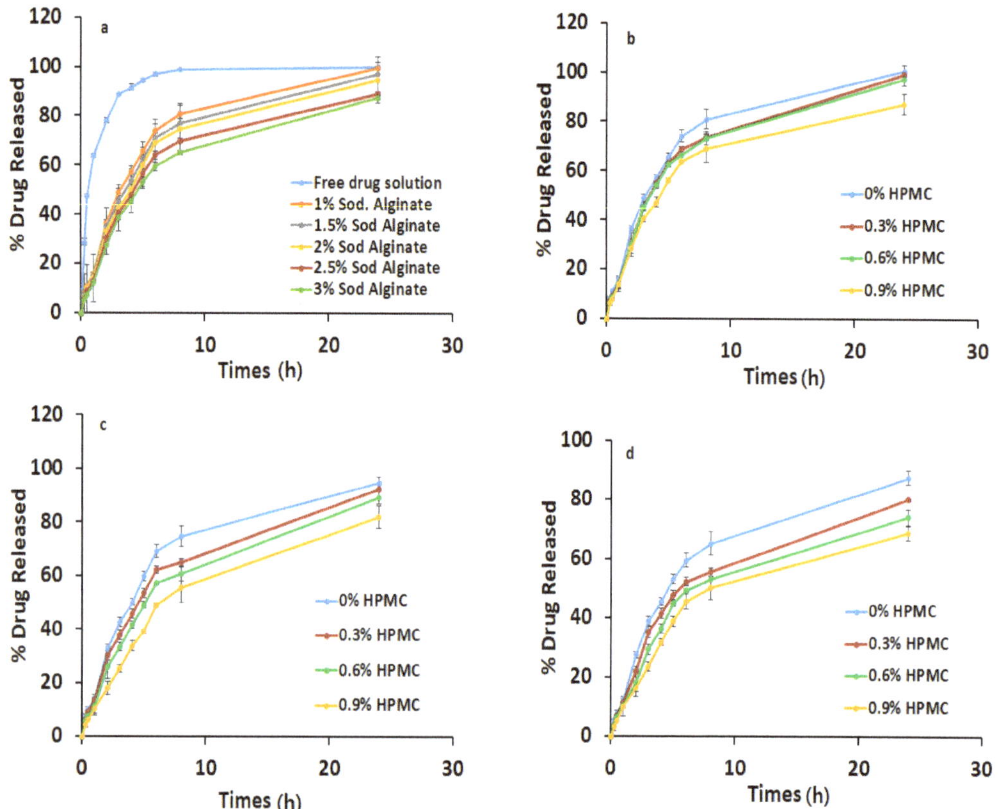

Figure 2. (**a**) Effect of Na alginate concentration on the in vitro release of BH from in situ gelling formulations. Effect of HPMC concentration on the in vitro release of BH from in situ gelling formulations containing (**b**) 1% Na alginate, (**c**) 2% Na alginate, and (**d**) 3% Na alginate.

2.4. Data Analysis

2.4.1. Full Factorial Experimental Design

A (3^3) full factorial experimental design with three independent variables at three different levels was used to investigate the effect of three factors, sodium alginate concentration (X_1), HPMC concentration (X_2), and CaCl$_2$ concentration (X_3), on the viscosity (Y_1) and percentage drug release from the in situ forming gel after 6 h (Y_2). The transformed values of all the formulations, along with their results, are shown in Table 2. The viscosity (Y_1) (dependent variable) values ranged from a minimum of 38.33 $\pm$ 1.53 cP to a maximum of 1660.0 $\pm$ 10.00), and the polynomial equation (full model) that described the response was:

$$Y_1(\text{viscosity}) = 192.67 + 314.06X_1 + 40.204X_2 + 54.926X_3 + 116.03X_1X_2 + 152.72X_1X_3 + 75.22X_2X_3 + 187X_1^2 + 36.33X_2^2 + 62.33X_3^2 \qquad R^2 = 0.9933 \text{ (good fit)} \tag{1}$$

The positive coefficients of X_1, X_2, and X_3 indicate that the viscosity increased with increases in the X_1, X_2, and X_3 concentrations. The statistical analysis of the full model shows that the independent variables had a significant effect on the responses ($p < 0.05$). The standardized effect of the independent variables and the effect of their interaction on the viscosity were easily described by preparing a Pareto chart (Figure 3a). The theoretical (predicted) values and observed values were in reasonably good agreement, as shown in Table 2.

Table 2. Observed responses in (3^3) factorial experimental design for BH in situ forming gel formulations.

Formulation No.	Independent Variables			Dependent Variables			
	X_1	X_2	X_3	Observed Value of Y_1	Predicted Value of Y_1	Observed Value of Y_2	Predicted Value of Y_2
1	3	0.9	0.15	1660 ± 135	1647.7	35.25 ± 0.68	35.69
2	3	0.9	0.11	1020 ± 98	991.73	41.41 ± 1.07	41.15
3	3	0.9	0.075	575 ± 45	613.00	45.48 ± 0.74	45.49
4	3	0.6	0.15	1130 ± 103	1113.1	40.61 ± 1.87	40.38
5	3	0.6	0.11	610 ± 40	693.72	45.41 ± 0.97	45.08
6	3	0.6	0.075	555 ± 37	494.63	49.19 ± 0.92	49.39
7	3	0.3	0.15	787 ± 30	814.33	45.88 ± 1.89	45.71
8	3	0.3	0.11	609 ± 29	559.21	48.20 ± 0.80	48.73
9	3	0.3	0.075	447 ± 16	467.54	52.26 ± 1.67	52.09
10	2	0.9	0.15	509 ± 21	526.24	39.38 ± 1.17	38.79
11	2	0.9	0.11	246 ± 12	269.20	45.00 ± 0.03	44.86
12	2	0.9	0.075	227 ± 12	193.72	48.98 ± 1.65	49.35
13	2	0.6	0.15	299 ± 12	309.93	45.40 ± 0.83	45.92
14	2	0.6	0.11	217 ± 10	192.67	50.95 ± 1.92	51.70
15	2	0.6	0.075	199 ± 13	200.07	57.17 ± 1.05	56.64
16	2	0.3	0.15	261 ± 11	238.50	53.35 ± 1.11	53.36
17	2	0.3	0.11	197 ± 13	188.79	58.42 ± 0.75	57.95
18	2	0.3	0.075	169 ± 13	206.87	62.33 ± 0.41	62.43
19	1	0.9	0.15	90 ± 1.00	102.07	56.24 ± 1.10	56.38
20	1	0.9	0.11	69 ± 1.00	40.73	60.38 ± 0.94	60.82
21	1	0.9	0.075	52 ± 0.58	65.28	63.63 ± 0.59	63.24
22	1	0.6	0.15	112 ± 11	83.93	59.12 ± 0.21	58.87
23	1	0.6	0.11	57 ± 1.00	65.61	63.99 ± 1.22	63.51
24	1	0.6	0.075	50 ± 3.00	76.29	66.48 ± 0.64	66.85
25	1	0.3	0.15	106 ± 11	119.83	61.21 ± 1.09	61.36
26	1	0.3	0.11	47 ± 0.58	72.32	65.31 ± 0.46	65.29
27	1	0.3	0.075	38 ± 2.00	38.07	68.59 ± 1.36	68.64

X_1: Na alg. concentration; X_2: HPMC concentration; X_3: $CaCl_2$ concentration; Y_1: viscosity cP; Y_2: percentage of drug released after 6 h.

The percentage of the drug released after 6 h was found to be in the range of 35.24 to 68.59%. A polynomial equation was also developed for the percentage of the drug released after 6 h:

$$Y_2(\% \text{ drug released after 6 h}) = 51.7 - 9.215X_1 - 6.545X_2 - 5.36X_3 - 0.78X_1X_2 - 0.26X_1X_3 - 0.37X_2X_3 + 2.595X_1{}^2 - 0.297X_2{}^2 - 0.423X_3{}^2 \qquad R^2 = 0.9886 \text{ (good fit)} \qquad (2)$$

The negative coefficients for X_1, X_2, and X_3 and the interactions between the two variables, X_1X_2, X_1X_3, X_2X_3, $X_2{}^2$, and $X_3{}^2$ indicated an unfavorable effect on the percentage of the drug released after 6 h, while the positive coefficient for $X_1{}^2$ indicates a favorable effect on the percentage of the drug released after 6 h.

The statistical analysis of the full model shows that among the independent variables selected and their interactions, X_1, X_2, X_3, X_1X_2, X_2X_3, $X_1{}^2$ were found to be significant ($p < 0.05$), indicating their major contribution to the percentage of the drug released after 6 h. The significance level of the coefficients b_5, b_8, and b_9 was found to be more than 0.05 ($p > 0.05$); hence, they were omitted from the full model to generate the reduced model second-order polynomial Equation (3):

$$Y_2 = 51.7 - 9.215X_1 - 6.545X_2 - 5.36X_3 - 0.78X_1X_2 - 0.37X_2X_3 + 2.595X_1{}^2 \qquad (3)$$

The main effects of the independent variables and their interaction on the percentage of the drug released after 6 h were illustrated in a Pareto chart (Figure 3b). The theoretical (predicted) values and observed values were also in reasonably good agreement, as shown in Table 2.

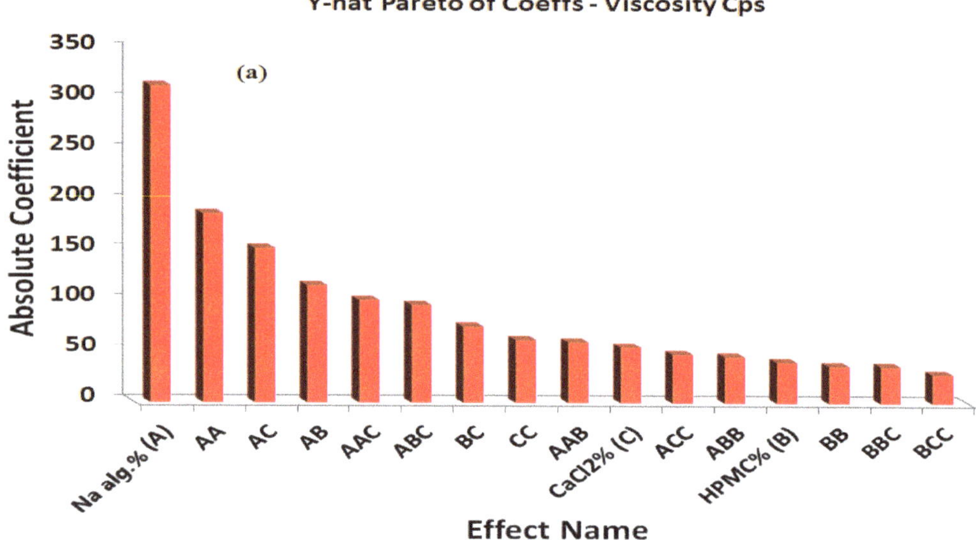

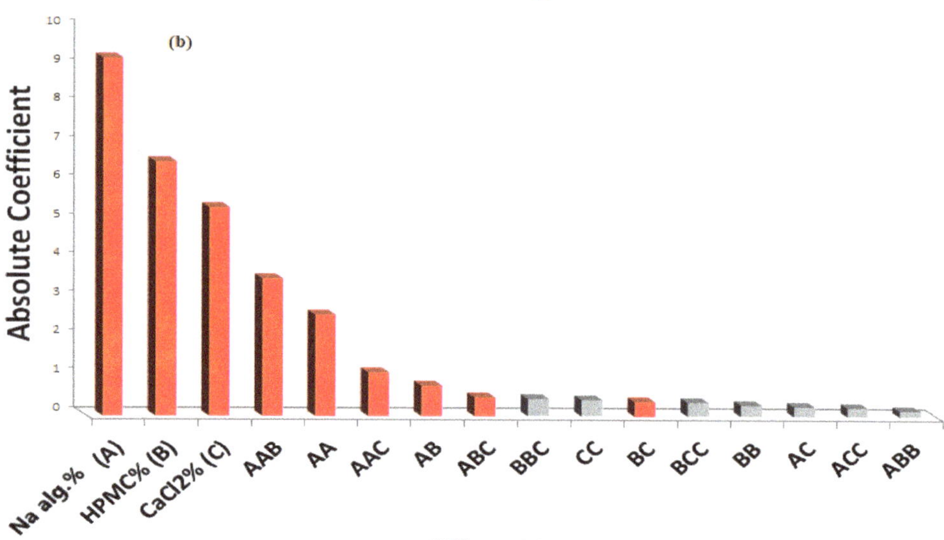

Figure 3. Y-hat Pareto chart showing (**a**) the standardized effect of independent variables and their interaction on the viscosity of the in situ forming gels, (**b**) the standardized effect of independent variables and their interaction on the percentage of drug released after 6 h from the in situ forming gels; (red color indicates significant effect; grey color indicates insignificant effect).

From the analysis of variance (ANOVA) shown in Table S1, we can conclude that the model is highly significant. The F values of 0.000 for Y_1 and 0.000 for Y_2 indicated a significant effect of the independent factors on the responses Y_1 and Y_2. This implies that the main effect of the sodium alginate concentration percentage, the HPMC concentration percentage, and the $CaCl_2$ concentration percentage is significant. Moreover, the non-significant lack-of-fit values for the two responses ($p > 0.05$), 30.6491 and 0.8361, and the consistent p-values of 0.1375 and 0.5964 for Y_1 and Y_3, respectively, indicated a good relation between the experimental and predicted values.

The surface plots were constructed for further clarification of the relationship between the dependent and independent variables. The effects of the interaction of each pair of factors on the viscosity at a fixed level of the third one (medium level) are shown in Figure 4. It was determined from the surface plots that a lower viscosity could be obtained with an X_1 range from 1.0 to 2.4%, with all the X_2 and X_3 ranges at medium X_3 and X_2 levels, respectively, with a viscosity ranging from 10 to 310 cP. It is evident that increasing the level of X_1 was the main cause of the increasing the viscosity. It was obvious that the sodium alginate concentration contributed more than the HPMC and CaCl$_2$ concentrations to controlling the viscosity of the formulation. The percentage contribution of the sodium alginate concentration was found to be 68.15% versus 4.57% for the HPMC concentration and 10.15% for CaCl$_2$ concentration.

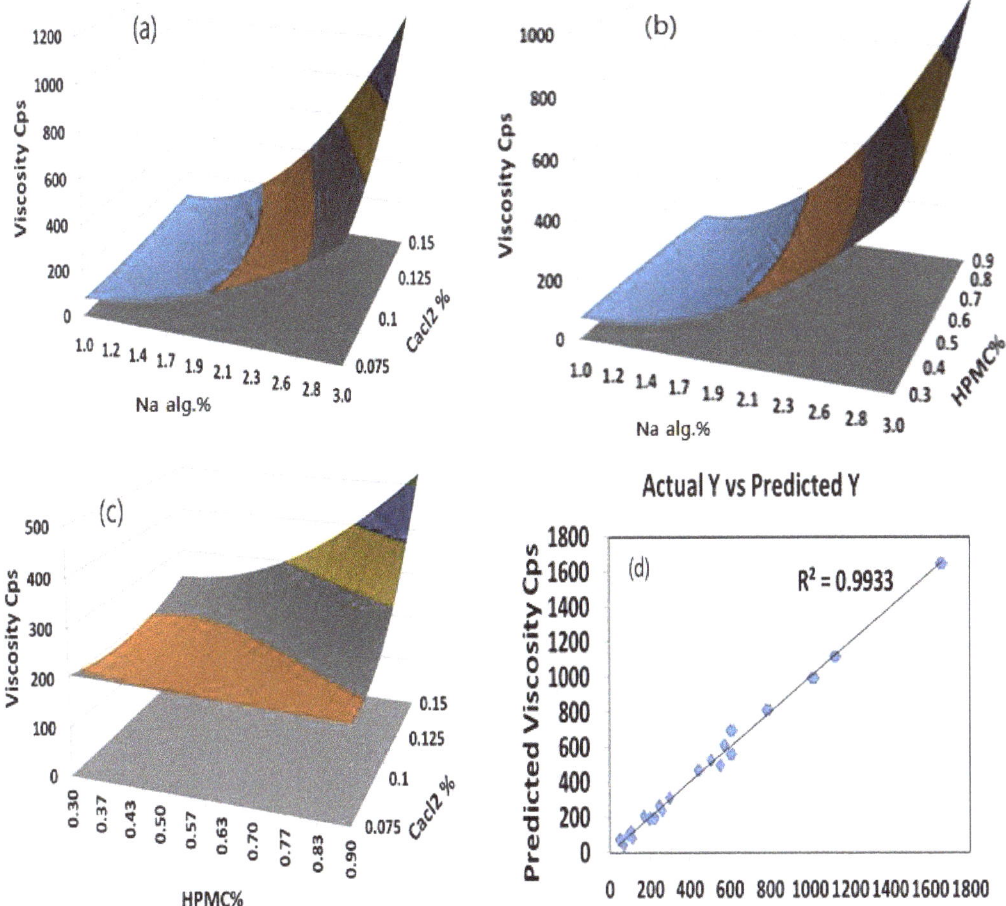

Figure 4. (**a**) Y-hat surface plot showing the effect of X_1 and X_3 on Y_1 at constant $X_2 = 0$; (**b**) Y-hat surface plot showing the effect of X_1 and X_2 on Y_1 at constant $X_3 = 0$; (**c**) Y-hat surface plot showing the effect of X_2 and X_3 on Y_1 at constant $X_1 = 0$; (**d**) linear correlation plot showing predicted against actual values to demonstrate the influence of independent variables X_1, X_2, and X_3 on viscosity cP.

As far as the percentage of the drug released after 6 h is concerned, the effects of the interaction of each pair of factors on the percentage of the drug released after 6 h at a fixed level of the third one (medium level) are shown in Figure 5. It was determined from the surface plots that a lower value of Y_2 could be obtained with an X_1 level ranging from 1.8 to 3% in combination with an X_2 level ranging from 0.36 to 0.9%, at a medium level of X_3 and in combination with an X_3 level in the range of 0.09 to 0.15%, and at a medium level of X_2 with a percentage of drug released after 6 h ranging from 40% to 48%. It was obvious that the sodium alginate concentration contributed more than the HPMC and $CaCl_2$ concentrations to controlling the percentage of the drug released after 6 h from the formulation. The percentage contribution of the sodium alginate concentration was found to be 65.43% versus 15.61% for the HPMC concentration and 14.82% for the $CaCl_2$ concentration.

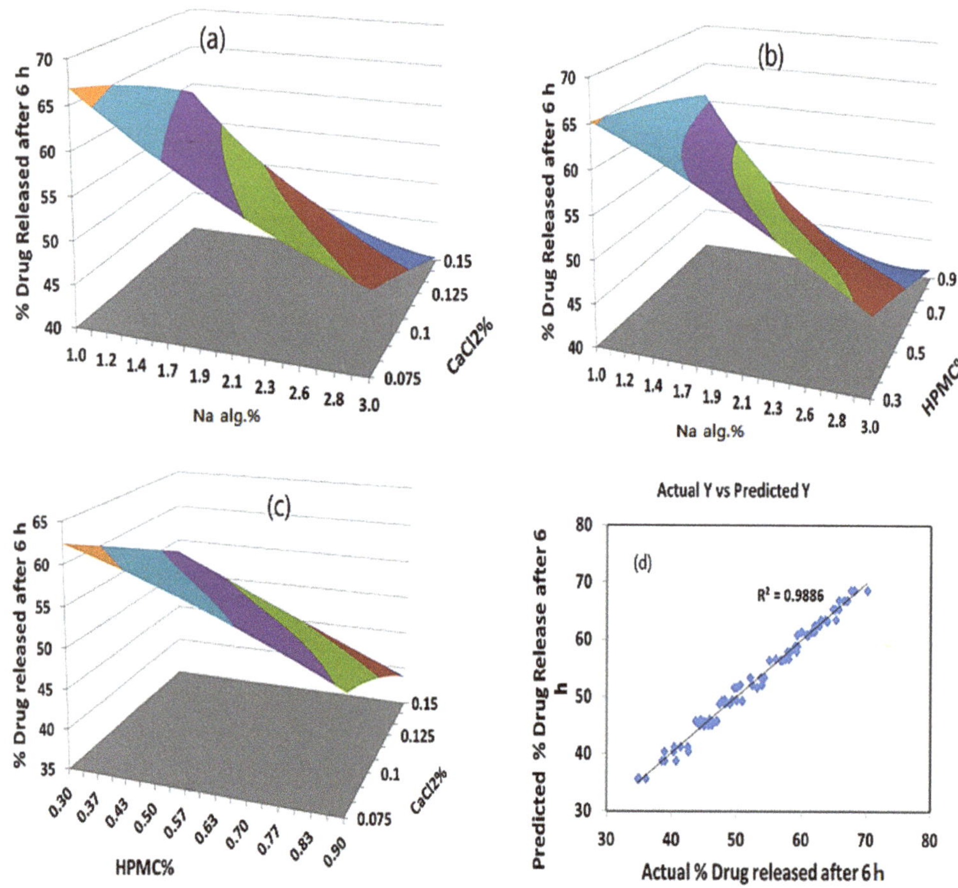

Figure 5. (**a**) Y-hat surface plot showing the effect of X_1 and X_3 on Y_2 at medium level of X_2; (**b**) Y-hat surface plot showing the effect of X_1 and X_2 on Y_2 at medium level of X_3; (**c**) Y-hat surface plot showing the effect of X_2 and X_3 on Y_2 at medium level of X_1; (**d**) linear correlation plot showing predicted against actual values to demonstrate the influence of independent variables X_1, X_2, and X_3 on percentage of rug released after 6 h.

2.4.2. The Optimum Formulation

The optimum formulation is the formulation that gives the minimum viscosity and a controlled drug release. It is evident from the polynomial equations and surface plots in Figures 4 and 5 that increasing the sodium alginate concentration increased the viscosity and decreased the percentage of the drug released after 6 h. Hence, the medium level was selected as the optimum for the sodium alginate concentration percentage (X_1). Using a computer optimization process, a 2% sodium alginate concentration (X_1), a 0.9% HPMC concentration (X_2), and a 0.1125 $CaCl_2$ concentration (X_3) were selected as the optimum formulation, which produced theoretical values of 269.2 cP and 44.9% for the viscosity and percentage of the drug released after 6 h, respectively.

2.5. Bioavailability of Orally Administered BH

The mean plasma concentrations of BH per duration of the three preparations, the BH solution, the BH marketed tablet, and the optimized BH in situ gel formulation, are illustrated in Figure 6. From the obtained results, it was evident that there was a difference between the mean plasma concentrations of the in situ gel formulation at all the time intervals compared to the plain drug and the marketed tablet. Furthermore, there was a marked difference in the T_{max} between the plain drug and the tested formulation.

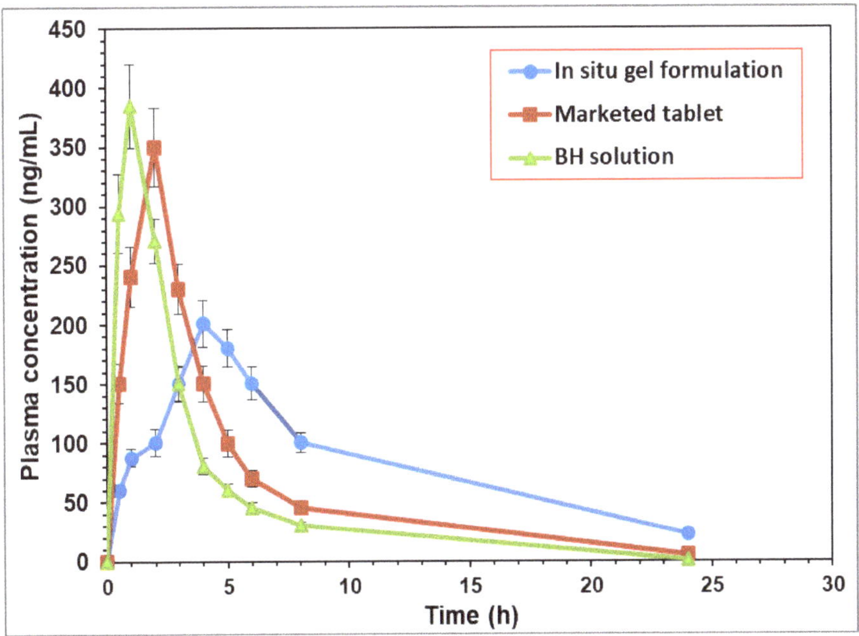

Figure 6. Mean plasma concentrations of BH after oral administration of different formulations (equivalent to 10 mg/kg) to rabbits.

The mean pharmacokinetic parameters of the BH from the different formulations are summarized in Table S2 and represented by the value of C_{max} (ng/mL), T_{max} (h), K_{el} (h^{-1}), $t_{1/2}$ (h), AUC_{0-24} (ng·h·mL^{-1}), $AUC_{0-\infty}$ (ng·h·mL^{-1}), $AUMC_{0-\infty}$ (ng·h^2·mL^{-1}), and MRT (h). From the observed data, it was noticed that the absorption of the BH from the BH solution was fast and reached its peak plasma concentration in 0.75 ± 0.04 h, whereas the mean T_{max} for the marketed tablet and the tested formulation were 2.00 ± 0.27 h and 4.00 ± 0.30 h, respectively. The mean peak plasma concentrations (C_{max}) were 350.36 ± 33.22 ng/mL for the marketed product and 200.95 ± 19.43 ng/mL for the in situ gel formulation compared to 384.82 ± 35.48 ng/mL for the BH solution. The obtained results showed a decrease in the mean C_{max} and an increase in the mean T_{max} of the drug-loaded in situ gel formulation compared to the plain drug solution, which explained the controlled release of the BH in situ gel formulation. The mean AUC_{0-24} was found to be 2041.19 ± 200.16 ng·h·mL^{-1} for the in situ gel formulation compared to 1350.87 ± 136.05 ng·h·mL^{-1} for the BH solution. The mean residence time for the BH when released from the in situ gel was significantly longer than that following the oral administration of the drug in the solution [23]. As seen in Table S2, the in situ gel formulation increased the MRT of the BH by 3.33-fold compared to free drug solution and resulted in an improved bioavailability.

From the obtained data, it was concluded that the relative bioavailability of the BH-loaded in situ gel was higher than the drug solution by more than 1.5-fold. The enhanced relative bioavailability of the BH may have been due to the decreased release of the drug from the in situ gel, which led to a more sustained release and absorption of the drug compared to the free-solution form [15,24].

3. Conclusions

A novel oral in situ gel system for the sustained delivery of BH was developed utilizing polymers that exhibit solution-to-gel phase transition due to changes in pH. In situ gel formation depends on the presence of sodium citrate, which complexes with free Ca ions to maintain the in situ gel fluidity until it reaches the stomach. The in situ gel formulation viscosity showed a marked increase with increases in the Na alginate concentration, with a significant decrease in the rate and extent of drug release, as proven by the (3^3) full factorial design. The derived polynomial equations, contour, and surface plots helped to predict the values of the selected independent variables for the preparation of optimum in situ gelling formulations with the desired properties. The formulation containing 2% Na alginate concentration (X_1), 0.9% HPMC concentration (X_2), and 0.11 CaCl$_2$ concentration (X_3) was selected as the optimum formulation, which showed a 1.5-fold increase in bioavailability in comparison to the drug solution.

4. Materials and Methods

4.1. Materials

Buspirone HCl (BH) and sodium alginate (Na alg.) were gift samples kindly supplied by Sigma Pharmaceuticals, Quesna, Egypt. Hydroxypropyl Methylcellulose (HPMC K15M) was supplied as a gift sample by the Egyptian International Pharmaceutical Industries Co., (EPICO), El-Asher of Ramadan city, Egypt. Sodium citrate, calcium chloride, and hydrochloric acid were purchased from El-Nasr Pharmaceutical Chemical Co., Cairo, Egypt.

4.2. Preparation of In Situ Gelling Solution

Different HPMC amounts to produce final concentrations of 0.3, 0.6, and 0.9% *w/v* were dissolved in around 50% of the total amount of distilled water containing calcium chloride (0.075, 0.1, 0.15% *w/v*), sodium citrate (0.25% *w/v*), and BH (1 mg/mL), so that there was a proper and homogenous dispersion of BH in the solution. Sodium alginate at different concentrations (1.0, 1.5, 2.0, 2.5, and 3.0% *w/v*) was added to the other half of distilled water and then heated to 60 °C while stirring. After cooling, these two solutions were thoroughly mixed using magnetic stirrer (AREC Digital Ceramic Hot Plate Stirrer, Usmate Velate (MB)-Italy) [25,26].

4.3. Differential Scanning Calorimetry Studies (DSC)

The DSC thermograms were recorded using a Differential scanning calorimeter (Model DSC-50, Shimadzu Corporation, Kyoto, Japan). About 2 mg of samples were sealed in aluminum pans and heated over a temperature range of 0–300 °C at a constant rate of 10 °C/min under a nitrogen purge (30 mL/min) [27,28]. DSC thermograms of pure BH, Na alg., and HPMC were taken to identify their characteristic endothermic peaks in order to determine any possible interactions in the physical mixtures of the drug and polymer.

4.4. Determination of Drug Content

In total, 1 mL of the in situ gelling solution (equivalent to 1 mg of BH) was diluted in 100 mL of distilled water to yield a solution containing a theoretical strength of 10 μg/mL. The UV absorbance of the sample was determined at a wavelength of 239 nm using a blank containing the same components of the gelling solution, exempt the drug [29]. The test was repeated 3 times, and percentage drug content was detected according to the following equation [30,31]:

$$\% \text{ Drug content} = \frac{\text{Actual amount of the drug in the formulation}}{\text{Theoretical amount of the drug in the formulation}} \times 100 \qquad (4)$$

4.5. Measurement of pH

The pH measurement of sodium-alginate-based in situ gelling solutions were performed using a calibrated digital pH meter (Cole-parmer instrument Co., Vernon Hills, IL, USA) at room temperature [32,33].

4.6. Gelling Capacity

In vitro gelling capacity was assessed visually by transferring 5 mL of each formulation into 25 mL of the gelation solution (0.1 N HCL, pH 1.2) in a beaker and observing the gelation time and how long the formed gel remained intact. Formulations were graded as follows [18,34]:

(+) Gels after a few minutes, dispersed rapidly.

(++) Gelation is immediate and remains for 12 h.

(+++) Gelation is immediate and remains for more than 12 h.

4.7. Measurement of Viscosity

The viscosity of the prepared in situ gelling solutions was determined by viscostar-R rotational viscometer (Fungilab S.A., Barcelona, Spain) using 100-milliliter sample. The study was carried out at 25 °C and 100 rpm using suitable spindle number R2 [26].

4.8. In Vitro Drug Release Study from In Situ Gels

The 10-milliliter in situ gelling solution containing 10 mg BH was poured into a glass tube with a cellophane membrane (mol.wt cutoff = 10.000 D) on one side and suspended in a beaker containing 100 mL of 0.1 N HCL, pH 1.2 to maintain sink condition. The beaker was placed in a mechanical shaker water bath (Julabo Shaking water bath SW–20C, Berlin, Germany) and agitated at 100 rpm while maintaining a temperature of 37 ± 1 °C [11]. Two-milliliter samples were taken at different time intervals for 24 h and the amount of drug was spectrophotometrically determined at 239 nm using buffer pH 1.2 as a blank. Each sample was replaced with an equal volume of fresh buffer solution, pH 1.2, at 37 ± 0.5 °C [13]. First, the effect of different concentrations of Na alg. on the drug release was investigated; next, a preliminary investigation of the influence of changing HPMC concentration on three concentrations of Na alg. (1%, 2% and 3%) was also performed.

4.9. Data Analysis

4.9.1. Factorial Experimental Design

The approach of changing one variable at a time is traditionally used for the development of pharmaceutical formulations, but there are drawbacks to this approach, such as its high consumption of time and raw materials. Furthermore, it may be difficult to predict the combined effects of various independent variables on the end product of the formulation process. This is why it is essential to comprehend the intricacy of pharmaceutical formulations by using a factorial-design statistical program, which is an effective method of demonstrating the relative significance of various variables and their associations [35].

Quantum XL® Version 5.50 software (SigmaZone, Orlando, FL, USA, accessed on 4 June 2021). was used to conduct the study. The present study used a three-level three-factorial (3^3) design for experimentation with 3 factors, 3 levels, and 27 runs to study the effect of independent variables, concentration of sodium alginate (X_1), concentration of $CaCl_2$ (X_2), and concentration of HPMC (X_3), on the dependent variables' viscosity (Y_1) and percentage drug release after 6 h (Y_2).

The independent variables and their levels were listed in Table 3. The concentrations of X_1, X_2, and X_3 were selected based on the results of preliminary experiments. The following quadratic mathematical model equation is used to clarify the effects of independent variables on the responses:

$$Y = b_0 + b_1X_1 + b_2X_2 + b_3X_3 + b_4X_1X_2 + b_5X_1X_3 + b_6X_2X_3 + b_7X_1{}^2 + b_8X_2{}^2 + b_9X_3{}^2 \quad (5)$$

where Y is the dependent variable, b_0 is the intercept, b_1 to b_9 are the regression coefficients measured from the observed experimental values of Y during the experimental runs, and X_1, X_2, and X_3 represent the average results of changing one variable at a time from its lowest level to its highest level. X_1X_2, X_1X_3, and X_2X_3 show how the dependent variable changes when two variables are changed. The terms (b_4, b_5, b_6) and (b_7, b_8, b_9) represent the interaction and quadratic terms, respectively [36,37]. Optimization was performed to determine the levels of the independent variables (X_1, X_2, and X_3) that would produce a minimum value of viscosity and extended drug release.

Table 3. Variables and their levels in the 3^3 factorial design.

Independent Variables	Levels		
	Low (−1)	Medium (0)	High (1)
X_1 = Sodium alginate concentration (%).	1	2	3
X_2 = Cacl$_2$ concentration (%).	0.075	0.1125	0.15
X_3 = HPMC concentration (%).	0.3	0.6	0.9
Dependent variables	Constraints		
Y_1 = Viscosity.	Minimize		
Y_2 = % Drug released after 6 h.	Prolong		

4.9.2. Statistics

All results were expressed as mean $\pm$ SD. Three runs were used to calculate the mean value. ANOVA test was used for comparison of sample means and for determination of statistical significance. Statistically, all the results were considered significant if $p < 0.05$.

4.10. Bioavailability Study of BH after Oral Administration to Experimental Animals

White male albino rabbits (weighing 2–2.5 kg) were selected for the bioavailability studies. All animal approaches were in accordance with the accepted protocol for experimental animals organized by the Research Ethics Committee (REC) of Ha'il University (20455/5/42). All animals were fasted for 12 h before the experiments with free water access. This examination was designed as a single oral dose. All animals received 10 mg BH/kg of body weight [38]. Animals were divided into three groups of three rabbits each, as follows:

Group 1 received BH solution in distilled water.

Group 2 received BH marketed product (Buspar $^{\circledR}$ tablet).

Group 3 received the optimized oral BH in situ gel.

About 2.5 mL of blood samples were withdrawn from the sinus orbital into heparinized tubes at different time durations: 0.5, 1, 2, 3, 4, 5, 6, 8, and 24 h. The blood samples were centrifuged immediately using centrifuge (Hermle Labortechnik GmbH-vZ 300 K, Wehingen, Germany) at 4000 rpm for 10 min at 4 °C to obtain the plasma samples, which were then stored at -20 ± 0.5 °C until HPLC analysis.

For analysis, BH content was assessed using HPLC with a variable -wavelength PDA detector, according to the technique described by Bshara et al. [39], with slight modifications. Aliquot of 500 µL of each thawed plasma sample was mixed with one mL of acetonitrile. The mixture was vortex-mixed for 30 s and then centrifuged at 6000 rpm for 15 min. The HPLC system (ThermoScientific Surveyor Plus HPLC system, ThermoScientific company, Waltham, MA, USA) consisting of Hypersil gold C18 column (particle size 5 µm, 150×4.6 mm) was conditioned at 30 °C and eluted with a mobile phase consisting of acetonitrile and a potassium phosphate buffer (10 mM) 30:70 (v/v) adjusted to pH 4.6 with orthophosphoric acid at a flow rate of 0.7 mL/min and an injection volume of 25 µL. The effluent was monitored at 235 nm.

Pharmacokinetic parameters (C_{max}, T_{max}, AUC_{0-24}, $AUC_{0-\infty}$, $AUMC_{0-\infty}$, $t_{1/2}$, and mean residence time (MRT)) in plasma were calculated using the noncompartmental model in the WinNonlin Standard Edition Version1.1 program (Pharsight, Mountain View, CA, USA) [40]. Bioavailability of BH nanovesicular in situ gel formulation relative to the BH solution was calculated using the following equation:

$$\text{Relative bioavailability} = \frac{\text{AUC } 0-24 \text{ (tested formulation)}}{\text{AUC } 0-24 \text{ (control)}} \times 100 \qquad (6)$$

Supplementary Materials: The following supporting information can be downloaded at: https://www.mdpi.com/article/10.3390/gels8070395/s1. Table S1: Results of ANOVA test for viscosity and percentage of drug released after 6 h of in situ forming gels. Table S2: Pharmacokinetics parameters after oral administration of BH in various formulations.

Author Contributions: M.H.A.: Conceptualization, methodology, writing—review and editing, and supervision; D.M.A.: methodology, software, data curation, validation, and writing—original draft; H.A.E.: writing—review and editing and supervision All authors have read and agreed to the published version of the manuscript.

Funding: This research received no external funding.

Institutional Review Board Statement: The study was conducted according to the guidelines of the Declaration of Helsinki, and approved by the Institutional Research Ethics Committee (IAEC), University of Ha'il, Saudi Arabia (approval no. 20455/5/42 at 27 November 2020).

Informed Consent Statement: Not applicable.

Data Availability Statement: Not applicable.

Conflicts of Interest: The authors declare no conflict of interest.

References

1. Sweetman, S.C. *Martindale: The Complete Drug Reference*; Pharmaceutical Press: London, UK, 2005.
2. Rathod, H.; Patel, V.; Modasiya, M. In Situgel: Development, Evaluation and Optimization Using 3² Factorial Design. *J. Pharm. Sci. Res.* **2011**, *3*, 1156.
3. Takka, S.; Sakr, A.; Goldberg, A. Development and validation of an in vitro–in vivo correlation for buspirone hydrochloride extended release tablets. *J. Control. Release* **2003**, *88*, 147–157. [CrossRef]
4. Andrade, C. Sustained-release, extended-release, and other time-release formulations in neuropsychiatry. *J. Clin. Psychiatry* **2015**, *76*, 20558. [CrossRef] [PubMed]
5. Remya, P.; Damodharan, N.; Venkata, M. Oral Sustained Delivery of Ranitidine From In-Situ Gelling Sodium-Alginate Formulation. *J. Chem. Pharm. Res.* **2011**, *3*, 814–821.
6. Batchelor, H.K.; Marriott, J.F. Formulations for children: Problems and solutions. *Br. J. Clin. Pharmacol.* **2015**, *79*, 405–418. [CrossRef] [PubMed]
7. Kaur, P.; Garg, T.; Rath, G.; Goyal, A.K. In situ nasal gel drug delivery: A novel approach for brain targeting through the mucosal membrane. *Artif. Cells Nanomed. Biotechnol.* **2016**, *44*, 1167–1176. [CrossRef] [PubMed]
8. Vigani, B.; Rossi, S.; Sandri, G.; Bonferoni, M.C.; Caramella, C.M.; Ferrari, F. Recent advances in the development of in situ gelling drug delivery systems for non-parenteral administration routes. *Pharmaceutics* **2020**, *12*, 859. [CrossRef]
9. Modasiya, M.K.; Prajapati, B.G.; Patel, V.M.; Patel, J.K. Sodium alginate based insitu gelling system of famotidine. *J. Sci. Technol.* **2010**, *5*, 27–42.

10. Supare, V.; Wadher, K.; Umekar, M. Experimental Design: Approaches and Applications in Development of Pharmaceutical Drug Delivery System. *J. Drug Deliv. Ther.* **2021**, *11*, 154–161. [CrossRef]
11. Abdallah, M.H.; Abu Lila, A.S.; Shawky, S.M.; Almansour, K.; Alshammari, F.; Khafagy, E.-S.; Makram, T.S. Experimental Design and Optimization of Nano-Transfersomal Gel to Enhance the Hypoglycemic Activity of Silymarin. *Polymers* **2022**, *14*, 508. [CrossRef]
12. Kim, D.Y.; Kwon, D.Y.; Kwon, J.S.; Kim, J.H.; Min, B.H.; Kim, M.S. Stimuli-responsive injectable in situ-forming hydrogels for regenerative medicines. *Polym. Rev.* **2015**, *55*, 407–452. [CrossRef]
13. Swathi, G.; Lakshmi, P. Design and optimization of hydrodynamically balanced oral in situ gel of glipizide. *J. Appl. Pharm. Sci.* **2015**, *5*, 31–38. [CrossRef]
14. Jaipal, A.; Pandey, M.; Charde, S.; Raut, P.; Prasanth, K.; Prasad, R. Effect of HPMC and mannitol on drug release and bioadhesion behavior of buccal discs of buspirone hydrochloride: In-vitro and in-vivo pharmacokinetic studies. *Saudi Pharm. J.* **2015**, *23*, 315–326. [CrossRef] [PubMed]
15. Abdelnabi, D.M.; Abdallah, M.H.; Elghamry, H.A. Buspirone hydrochloride loaded in situ nanovesicular gel as an anxiolytic nasal drug delivery system: In vitro and animal studies. *AAPS PharmSciTech* **2019**, *20*, 134. Available online: https://link.springer.com/article/10.1208/s12249-018-1211-0 (accessed on 4 March 2019). [CrossRef] [PubMed]
16. Rao, K.M.; Rao, K.K.; Sudhakar, P.; Rao, K.C.; Subha, M. Synthesis and Characterization of biodegradable Poly (*Vinyl caprolactam*) grafted on to sodium alginate and its microgels for controlled release studies of an anticancer drug. *J. Appl. Pharm. Sci.* **2013**, *3*, 61–69.
17. Chandak, A.R.; Verma, P.R.P. Design and development of hydroxypropyl methycellulose (HPMC) based polymeric films of methotrexate: Physicochemical and pharmacokinetic evaluations. *Yakugaku Zasshi* **2008**, *128*, 1057–1066. [CrossRef]
18. Alhamdany, A.T.N.; Maraie, N.K.; Msheimsh, B.A.R. Development and in vitro/in vivo evaluation of floating in situ gelling oral liquid extended release formulation of furosemide. *Pharm. Biosci. J.* **2014**, *2*, 1–11. [CrossRef]
19. El Maghraby, G.M.; Elsisi, A.E.; Elmeshad, G.A. Development of liquid oral sustained release formulations of nateglinide: In Vitro and In Vivo evaluation. *J. Drug Deliv. Sci. Technol.* **2015**, *29*, 70–77. [CrossRef]
20. Pandya, K.; Aggarwal, P.; Dashora, A.; Sahu, D.; Garg, R.; Pareta, L.K.; Menaria, M.; Joshi, B. Formulation and evaluation of oral floatable in-situ gel of ranitedine hydrochloride. *J. Drug Deliv. Ther.* **2013**, *3*, 90–97. [CrossRef]
21. Madan, J.R.; Adokar, B.R.; Dua, K. Development and evaluation of in situ gel of pregabalin. *Int. J. Pharm. Investig.* **2015**, *5*, 226. [CrossRef]
22. Nokhodchi, A.; Tailor, A. In situ cross-linking of sodium alginate with calcium and aluminum ions to sustain the release of theophylline from polymeric matrices. *Il Farm.* **2004**, *59*, 999–1004. [CrossRef] [PubMed]
23. Xu, H.; Shi, M. A novel in situ gel formulation of ranitidine for oral sustained delivery. *Biomol. Ther.* **2014**, *22*, 161. [CrossRef] [PubMed]
24. Pandey, M.; Choudhury, H.; Binti Abd Aziz, A.; Bhattamisra, S.K.; Gorain, B.; Su, J.S.T.; Tan, C.L.; Chin, W.Y.; Yip, K.Y. Potential of stimuli-responsive in situ gel system for sustained ocular drug delivery: Recent progress and contemporary research. *Polymers* **2021**, *13*, 1340. [CrossRef] [PubMed]
25. Sharma, A.; Sharma, J.; Kaur, R.; Saini, V. Development and characterization of in situ oral gel of spiramycin. *BioMed Res. Int.* **2014**, *2014*, 876182. [CrossRef] [PubMed]
26. Abdallah, M.H.; Sabry, S.A.; Hasan, A.A. Enhancing Transdermal Delivery of Glimepiride Via Entrapment in Proniosomal Gel. *J. Young Pharm.* **2016**, *8*, 335. [CrossRef]
27. Husseiny, R.A.; Lila, A.S.A.; Abdallah, M.H.; Hamed, E.E.; El-ghamry, H.A. Design, in vitro/in vivo evaluation of meclizine HCl-loaded floating microspheres targeting pregnancy-related nausea and vomiting. *J. Drug Deliv. Sci. Technol.* **2018**, *47*, 395–403. [CrossRef]
28. Ibrahim, T.M.; Abdallah, M.H.; El-Megrab, N.A.; El-Nahas, H.M. Upgrading of dissolution and anti-hypertensive effect of Carvedilol via two combined approaches: Self-emulsification and liquisolid techniques. *Drug Dev. Ind. Pharm.* **2018**, *44*, 873–885. [CrossRef]
29. Wamorkar, V.; Varma, M.M.; Manjunath, S. Formulation and evaluation of stomach specific in-situ gel of metoclopramide using natural, bio-degradable polymers. *Int. J. Res. Pharm. Biomed. Sci.* **2011**, *2*, 193–201.
30. Abdallah, M.H.; Lila, A.S.A.; Anwer, M.K.; Khafagy, E.-S.; Mohammad, M.; Soliman, M.S. Formulation, development and evaluation of ibuprofen loaded nano-transferosomal gel for the treatment of psoriasis. *J. Pharm. Res* **2019**, *31*, 1–8. [CrossRef]
31. Abdallah, M.H.; Lila, A.S.A.; Unissa, R.; Elsewedy, H.S.; Elghamry, H.A.; Soliman, M.S. Brucine-Loaded Ethosomal Gel: Design, Optimization, and Anti-inflammatory Activity. *AAPS PharmSciTech* **2021**, *22*, 269. [CrossRef]
32. Ayoub, A.M.; Ibrahim, M.M.; Abdallah, M.H.; Mahdy, M.A. Sulpiride microemulsions as antipsychotic nasal drug delivery systems: In-vitro and pharmacodynamic study. *J. Drug Deliv. Sci. Technol.* **2016**, *36*, 10–22. [CrossRef]
33. Abdallah, M.H.; Elsewedy, H.S.; AbuLila, A.S.; Almansour, K.; Unissa, R.; Elghamry, H.A.; Soliman, M.S. Quality by Design for Optimizing a Novel Liposomal Jojoba Oil-Based Emulgel to Ameliorate the Anti-Inflammatory Effect of Brucine. *Gels* **2021**, *7*, 219. [CrossRef] [PubMed]
34. Karemore, M.N.; Avari, J.G. In-situ gel of nifedipine for preeclampsia: Optimization, in-vitro and in-vivo evaluation. *J. Drug Deliv. Sci. Technol.* **2019**, *50*, 78–89. [CrossRef]

35. Bhattacharjee, A.; Das, P.J.; Dey, S.; Nayak, A.K.; Roy, P.K.; Chakrabarti, S.; Marbaniang, D.; Das, S.K.; Ray, S.; Chattopadhyay, P. Development and optimization of besifloxacin hydrochloride loaded liposomal gel prepared by thin film hydration method using 32 full factorial design. *Colloids Surf. A Physicochem. Eng. Asp.* **2020**, *585*, 124071. [CrossRef]

36. Abdallah, M.H. Box-behnken design for development and optimization of acetazolamide microspheres. *India* **2014**, *5*, 1228–1239. [CrossRef]

37. Abdallah, M.H.; Lila, A.S.A.; Unissa, R.; Elsewedy, H.S.; Elghamry, H.A.; Soliman, M.S. Preparation, characterization and evaluation of anti-inflammatory and anti-nociceptive effects of brucine-loaded nanoemulgel. *Colloids Surf. B Biointerfaces* **2021**, *205*, 111868. [CrossRef] [PubMed]

38. Kumar, Y.S.; Adukondalu, D.; Latha, A.B.; Vishnu, Y.V.; Ramesh, G.; Kumar, R.S.; Rao, Y.M.; Sarangapani, M. Effect of pomegranate pretreatment on the oral bioavailability of buspirone in male albino rabbits. *DARU J. Fac. Pharm. Tehran Univ. Med. Sci.* **2011**, *19*, 266.

39. Bshara, H.; Osman, R.; Mansour, S.; El-Shamy, A.E.-H.A. Chitosan and cyclodextrin in intranasal microemulsion for improved brain buspirone hydrochloride pharmacokinetics in rats. *Carbohydr. Polym.* **2014**, *99*, 297–305. [CrossRef]

40. Ibrahim, M.M.; Ayoub, A.M.; Mahdy, M.A.E.; Abdallah, M.H. Solid Lipid Nanoparticles of Sulpiride: Improvement of Pharmacokinetic Properties. *Int. J. Pharm. Investig.* **2019**, *9*, 122–127. [CrossRef]

Article

Formulation and Evaluation of Nano Lipid Carrier-Based Ocular Gel System: Optimization to Antibacterial Activity

Sadaf Jamal Gilani [1], May Nasser bin Jumah [2,3,4], Ameeduzzafar Zafar [5,*], Syed Sarim Imam [6,*], Mohd Yasir [7], Mohammad Khalid [8], Sultan Alshehri [6], Mohammed M. Ghuneim [9] and Fatima M. Albohairy [10]

[1] Department of Basic Health Sciences, Preparatory Year, Princess Nourah bint Abdulrahman University, Riyadh 11671, Saudi Arabia; sjglani@pnu.edu.sa
[2] Biology Department, College of Science, Princess Nourah bint Abdulrahman University, Riyadh 11671, Saudi Arabia; mnbinjumah@pnu.edu.sa
[3] Environment and Biomaterial Unit, Health Sciences Research Center, Princess Nourah bint Abdulrahman University, Riyadh 11671, Saudi Arabia
[4] Saudi Society for Applied Science, Princess Nourah bint Abdulrahman University, Riyadh 11671, Saudi Arabia
[5] Department of Pharmaceutics, College of Pharmacy, Jouf University, Sakaka 72341, Saudi Arabia
[6] Department of Pharmaceutics, College of Pharmacy, King Saud University, Riyadh 11451, Saudi Arabia; salshehri1@ksu.edu.sa
[7] Department of Pharmacy, College of Health Science, Arsi University, Asella 396, Ethiopia; mohdyasir31@gmail.com
[8] Department of Pharmacognosy, College of Pharmacy, Prince Sattam Bin Abdulaziz University, Al-Kharj 11942, Saudi Arabia; drkhalid8811@gmail.com
[9] Department of Pharmacy Practice, College of Pharmacy, AlMaarefa University, Ad Diriyah 13713, Saudi Arabia; mghoneim@mcst.edu.sa
[10] Electron Microscope Research Unit, Health Sciences Research Center, Princes Nourah bint Abdulrahman University, Riyadh 11671, Saudi Arabia; fmalbohairy@pnu.edu.sa
* Correspondence: azafar@ju.edu.sa (A.Z.); simam@ksu.edu.sa (S.S.I.)

Citation: Gilani, S.J.; Jumah, M.N.b.; Zafar, A.; Imam, S.S.; Yasir, M.; Khalid, M.; Alshehri, S.; Ghuneim, M.M.; Albohairy, F.M. Formulation and Evaluation of Nano Lipid Carrier-Based Ocular Gel System: Optimization to Antibacterial Activity. *Gels* **2022**, *8*, 255. https://doi.org/10.3390/gels8050255

Academic Editors: Maddalena Sguizzato, Rita Cortesi and Rachel Yoon Chang

Received: 1 March 2022
Accepted: 12 April 2022
Published: 21 April 2022

Publisher's Note: MDPI stays neutral with regard to jurisdictional claims in published maps and institutional affiliations.

Abstract: The present research work was designed to prepare Azithromycin (AM)-loaded nano lipid carriers (NLs) for ocular delivery. NLs were prepared by the emulsification–homogenization method and further optimized by the Box Behnken design. AM-NLs were optimized using the independent constraints of homogenization speed (A), surfactant concentration (B), and lipid concentration (C) to obtain optimal NLs (AM-NLop). The selected AM-NLop was further converted into a sol-gel system using a mucoadhesive polymer blend of sodium alginate and hydroxyl propyl methyl cellulose (AM-NLopIG). The sol-gel system was further characterized for drug release, permeation, hydration, irritation, histopathology, and antibacterial activity. The prepared NLs showed nano-metric size particles (154.7 ± 7.3 to 352.2 ± 15.8 nm) with high encapsulation efficiency (48.8 ± 1.1 to 80.9 ± 2.9%). AM-NLopIG showed a more prolonged drug release (98.6 ± 4.6% in 24 h) than the eye drop (99.4 ± 5.3% in 3 h). The ex vivo permeation result depicted AM-NLopIG, AM-IG, and eye drop. AM-NLopIG exhibited significant higher AM permeation (60.7 ± 4.1%) than AM-IG (33.46 ± 3.04%) and eye drop (23.3 ± 3.7%). The corneal hydration was found to be 76.45%, which is within the standard limit. The histopathology and HET-CAM results revealed that the prepared formulation is safe for ocular use. The antibacterial study revealed enhanced activity from the AM-NLopIG.

Keywords: NLs; in situ gel; ocular delivery; azithromycin; antibacterial; HET CAM

1. Introduction

The treatment of ocular diseases is challenging due to the typical anatomy and physiology of the eye [1]. To attain the required amount of drug at the site of action at an appropriate time is difficult due to the dilution of formulations by tear fluid, lachrymal fluid, eyelid blinking, tear fluid turnover, and nasolacrimal drainage. Most ocular diseases are treated by eye drops, but only 5% of an administered dose is available for therapeutic action. The remaining dose was excreted out by various protection mechanisms. Therefore, the bioavailability of eye drops is low and they require multiple and frequent dosing, which may produce side effects [2,3].

The administration of topical non-invasive ocular formulations is the more preferable way to treat the anterior segment of the eye. To enhance the ocular bioavailability of the drugs, various novel drug carriers were investigated by the researchers. The delivery systems include thymoquinone liposomes [4], pilocarpine niosomes [5], acetonide polymeric NPs [6], clarithromycin SLNs [7], methazolamide NLCs [8], besifloxacin nanoemulsion [9], and tacrolimus in situ gel system [10].

Among these nano-delivery systems, NLCs have been widely used as potential carriers belonging to lipid base nano-system [11]. NLCs are nano-metric size particles and give high entrapment efficiency due to lesser crystallinity of the lipid at low surfactant concentrations [12]. It also increases corneal permeability and bioavailability as well as reduces the local and systemic side effects. Various research studies have investigated NLCs for the treatment of ocular diseases [13,14]. Lakhani et al. formulated amphotericin B-loaded NLCs for ocular delivery and exhibited higher entrapment efficiency and higher antifungal activity than the marketed formulation [15]. In another study, Kiss et al. prepared dexamethasone NLCs for ocular inflammation. The prepared formulations exhibited a significant drug concentration in the stroma layer, confirmed by the porcine cornea study [16]. The application of NLCs was further explored by Seyfoddin and his research team and they developed acyclovir NLCs that exhibited higher entrapment efficiency than solid lipid nanoparticles as well as depicted high and faster permeation across the corneal membrane [17]. Besifloxacin-loaded NLC formulations were evaluated for different parameters [18]. The prepared formulation showed nano-metric size, high entrapment efficiency, and significant enhancement in corneal permeation.

Due to the low viscosity of the NLC formulations, their efficacy can be enhanced by transforming them into a sol-gel system. It is prepared in solution form and changes to gel phase in a cul-de-sac by temperature, ion, and pH [19]. Due to conversion into the gel phase, the increase in corneal residence time takes place. It leads to decreased dosing frequency and increases therapeutic efficacy and may reduce the systemic side effects by minimizing the nasolacrimal outflow. Different research designs were reported for the NLC-based sol-gel system by the different administration routes. Ciprofloxacin-loaded NLC in situ gel was prepared and evaluated for the different parameters. The prepared formulations depicted 3.5-fold and 1.9-fold enhancements in the flux and permeation compared to the marketed ciprofloxacin formulation [20]. In another study, moxifloxacin NLC-laden in situ gel was developed and evaluated for endophthalmitis infection. The formulation exhibited a twofold higher permeation than pure drug solution [21]. Natamycin-loaded NLC in situ gel was prepared and optimized by the factorial design method. The formulations were prepared using guar gum, boric acid, and Carbopol® 940 as a gelling agent. The prepared formulations depicted lower flux value in comparison to Natacyn® suspension [22].

Alginates are natural, nontoxic, mucoadhesive, viscosity enhancing polymers derived from different brown seaweeds [23]. They are used for ocular in situ gel systems due to their mucoadhesive, biocompatible, and non-toxic nature [24,25]. They have the properties of effective gelling capacity and viscosity and are commonly used in in situ gelling systems [26]. The prepared ocular solution has a low viscosity at room temperature for easy administration and interacts with the calcium ions in the tear fluid to form a gel matrix and lead to controlled drug release [27]. The alginates interact with divalent cations and bind to the guluronate blocks of the sodium alginate to facilitate gelation. The cross-linking of the cation–guluronate blocks then yields a gel-like structure, with a slow and steady drug release at the application site [28]. Draize ocular irritation studies on rabbits also reported sodium alginates to be nonirritating to the eye [23,29]. Hydroxypropyl methylcellulose (HPMC) is added into the formulation as a viscosity enhancer. It is a non-ionic, nontoxic, viscoelastic polymer with good swelling capacity [27,30].

Azithromycin (AM) is a semisynthetic macrolide antibiotic. It is a broad-spectrum antibiotic and is highly stable in an acidic environment compared to other macrolides. It acts by inhibiting the 50s ribosomal subunit of bacteria and inhibits protein synthesis. It is commercially available on the market as eye drops.

The present research work was designed with two steps. In the first step, AM-NLs were prepared by the emulsification homogenization method and further optimized by the Box Behnken design (BBD) using different independent variables. BBD-based optimized AM-NLop was further converted into an in situ gel (sol-gel) system by using a thermosensitive gelling agent. The in situ gel was further characterized for viscosity, gelling capacity, in vitro release, ex vivo corneal permeation, ocular tolerance, and antimicrobial evaluation.

2. Result and Discussion

2.1. Screening of Solid and Liquid Lipids

Screening of the solid lipids, liquid lipids, and surfactants used for the development of NLs was performed based on the maximum solubility of AM, and the results are expressed in Figure 1. The order of solubility of AM in various solid lipids is Glyceryl behenate > Stearic acid > Precirol ATO-5 > Tripalmitin > Glycerol monostearate > Myristic acid > Glyceryl monooleate. The highest solubility of AM was found in GB (92.8 ± 6.2 mg/g), and this was selected for further use as a solid lipid. In the case of liquid lipids, the order of solubility of AM was found as follows: Miglyol > Labrasol > Sunflower oil > Sesame oil > Coconut oil > Isopropyl myristate. The maximum solubility of AM was found to be in Miglyol (76.3 ± 4.8 mg/mL), and this was selected as the optimal liquid lipid. The order of solubility of AM in various surfactants is Kolliphor EL > Cremophor RH60 > Span 20 > Span 60. The highest solubility of AM was found to be in Kolliphor EL (83.1 ± 4.5 mg/mL). Based on the maximum solubility of AM, Glyceryl behenate, Miglyol, and Kolliphor EL were used for the formulation of NLs.

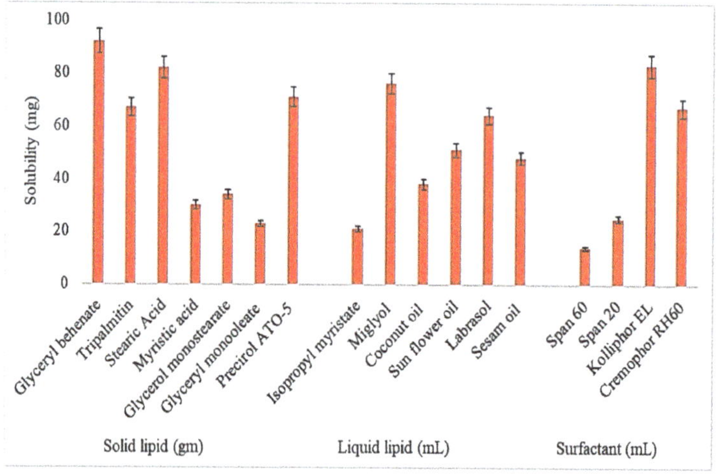

Figure 1. Solubility profile of Azithromycin in different lipids (solid and liquid) and surfactants. Study performed in triplicate and results shown as mean ± SD.

2.2. Screening of Miscibility Ratio of Solid and Liquid Lipid

The miscibility ratio (solid and liquid lipid) was determined by mixing the melted solid and liquid lipid in different ratios, i.e., 9:1, 8:2, 7:3, 6:4, 5:5, and 2:8. Among the tested ratios, the combination 6:4 ratio was found to be the best blend because it did not show any phase separation, oil droplets, or crystallization. So, this 6:4 ratio of solid and liquid lipids was selected for the development of NLs.

2.3. Optimization

The Box Behnken design (BBD) depicted a total of seventeen compositions (five common) using the independent variables of homogenization speed (A), surfactant concentration (B), and lipid concentration (C) at three different levels (low, medium, high), as shown in Table 1. The findings of the prepared AM-NLs were added to the software to obtain the optimized composition of independent constraints. Their effects were observed on the different models to obtain an optimized combination with desirable particle size and encapsulation efficiency. The impact of homogenization speed (A), surfactant concentration (B), and lipid concentration (C) was chosen based on a non-significant lack of fit and low predicted residual error sum of squares value (PRESS value) (Table 2). The polynomial equations were generated by ANOVA to explore the effect of homogenization speed (A), surfactant concentration (B), and lipid concentration (C) on responses (particle size and encapsulation efficiency). The positive and negative value of the polynomial equation indicates the synergistic and antagonistic effect on the dependent factors (responses). Among the different models of the responses, namely particle size (Y_1) and entrapment efficiency (Y_2), the quadratic model was found to be the best fitting model [16].

Table 1. Composition of Azithromycin nano lipid carrier with experimental results.

Code	Independent Variables			Dependent Variables	
	Homogenization Speed (rpm)	Surfactant (%, *w/v*)	Lipid (%, *w/v*)	Particle Size (nm)	Entrapment Efficiency (%)
	A	B	C	Y_1	Y_2
AM-NL1	12,000	1.00	3.00	352.2 ± 15.8	56.1 ± 2.5
AM-NL2	20,000	1.00	3.00	228.5 ± 09.2	54.3 ± 2.1
AM-NL3	12,000	3.00	3.00	282.8 ± 13.5	67.9 ± 3.3
AM-NL4	20,000	3.00	3.00	210.3 ± 08.7	63.3 ± 1.9
AM-NL5	12,000	2.00	1.50	260.9 ± 14.2	57.9 ± 1.4
AM-NL6	20,000	2.00	1.50	165.6 ± 6.8	55.7 ± 1.8
AM-NL7	12,000	2.00	4.50	330.9 ± 15.9	80.6 ± 3.6
AM-NL8	20,000	2.00	4.50	220.0 ± 11.2	76.8 ± 2.9
AM-NL9	16,000	1.00	1.50	205.8 ± 9.6	48.8 ± 1.1
AM-NL10	16,000	3.00	1.50	154.7 ± 07.3	56.9 ± 2.1
AM-NL11	16,000	1.00	4.50	257.9 ± 13.8	62.3 ± 2.4
AM-NL12	16,000	3.00	4.50	222.2 ± 11.7	80.9 ± 2.9
* AM-NL13	16,000	2.00	3.00	176.1 ± 12.4	75.4 ± 2.6
* AM-NL14	16,000	2.00	3.00	176.1 ± 12.4	73.4 ± 2.4
* AM-NL15	16,000	2.00	3.00	180.5 ± 11.9	75.4 ± 2.6
* AM-NL16	16,000	2.00	3.00	176.1 ± 12.4	73.4 ± 2.4
* AM-NL17	16,000	2.00	3.00	180.5 ± 11.9	75.4 ± 2.6

* Center point.

2.4. Influence of Independent Variables on Particle Size (Y_1)

The mean particle size of the prepared AM-NLs was between 154.7 ± 7.3 (AM-NL10) and 352.2 ± 15.8 nm (AM-NL1), as shown in Table 1. The minimum particle size (Y_1) was found from the formulation (AM-NL10) prepared with homogenization speed of 16,000 rpm, surfactant concentration of 3% *w/v*, and lipid concentration of 1.5% *w/v*. The maximum particle size (Y_1) was found from the formulation (AM-NL1) prepared with composition homogenization speed of 12,000 rpm, surfactant concentration of 1% *w/v*, and lipid concentration of 3% *w/v*. From the results, there is a significant difference in the particle size was observed by changing the composition of independent variables. The effects of independent variables (homogenization speed, surfactant concentration, and lipid concentration) were evaluated on the 3D response surface plot (Figure 2). As shown in Figure 2, homogenization speed (A) showed a dual effect on particle size. The increase in the homogenization speed from a low level to a high level caused the particle size to decrease. After reaching an intermediate level, the further increase in the homogenization speeds results in increased particle size. The increase in size may take place due to the aggregation of particles at a higher homogenization speed. The small globules of emulsion aggregate in the inter-particular space of big globules, leading to the re-coalescence or Ostwald ripening. Besides this, surfactant (B) was responsible for the decrease in interfacial tension, and as per the Laplaces pressure theory, as the interfacial tension decrease, more droplets are disrupted into fine particles. Hence, the surfactant concentration exhibited a negative effect on particle size. Similar results are reported in the literature [21]. At a fixed homogenization speed (A) and surfactant concentration (B), the particle size (Y_1) increases with increased lipid concentration. This might be due to an enhancement in the viscosity of lipid dispersion [31]. Moreover, there was a high chance of incomplete emulsification at a high lipid concentration and low surfactant concentration (due to lack of surfactant), which was responsible for the aggregation of lipid nanoparticles. The controlled optimal particle size is good for the stability of nano-dispersion as it promotes the Brownian motion and prevents sedimentation [18].

Table 2. Statistical model fit summary report.

Parameters	Regression Parameters		Models			Model
			Linear	2FI	Quadratic	
	SD		40.76	45.64	2.71	
	R^2		0.5929	0.6074	0.9990	
	Adjusted R^2		0.4990	−0.3718	0.9978	
Particle size	Predicted R^2		0.3270	−0.2725	0.9941	Quadratic
	%CV		-	-	1.22	
	Ade Precision		35,703.02	67,503.77	313.33	
	Lack of fit	F-value	272.57	394.27	0.61	
			Significant	Significant	Non-Significant	
		p-value	<0.0001	<0.0001	0.6419	
	SD		6.90	7.66	1.43	
	R^2		0.6458	0.6640	0.9918	
	Adjusted R^2		0.5640	0.4623	0.9812	
Entrapment efficiency	Predicted R^2		0.4364	0.0131	0.9013	Quadratic
	%CV		-	-	2.14	
	Adeq. Precision		−983.49	1722.20	172.20	
	Lack of fit	F-value	68.79	97.85	3.49	
			Significant	Significant	Non-Significant	
		p-value	0.0005	0.0003	0.1295	

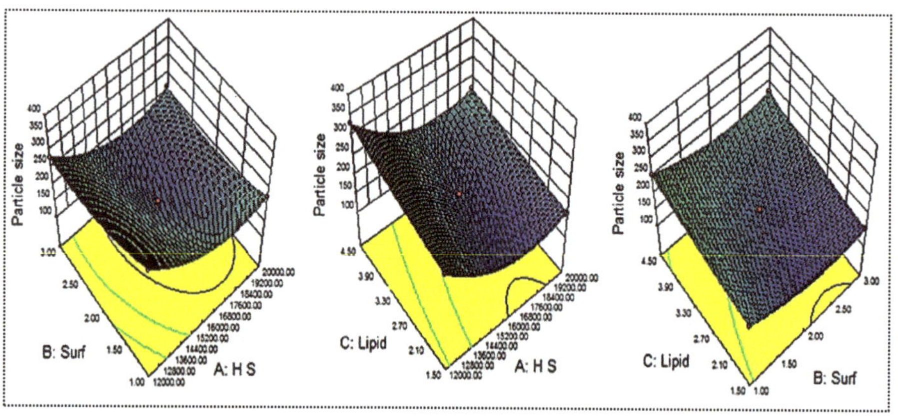

Figure 2. Effect of independent variables (Homogenization speed—A, Surfactant—B, Lipid—C) on the particle size (Y_1).

The effect of independent variables was further statistically analyzed by the polynomial Equation (1).

$$\text{Particle size } (Y_1) = +178.2 - 50.27A - 21.8B + 30.48C + 12.75AB - 3.90AC + 3.80BC + 62.24A^2 + 27.99B^2 + 3.94C^2 \quad (1)$$

The above equation states that homogenization speed (A, coefficient -50.27) and surfactant (B, coefficient -21.80) showed a negative effect. On the other hand, the third variable, lipid concentration (C, coefficient $+30.48$), depicted a positive effect. The quadratic coefficient of A^2, B^2, and C^2 showed the positive values of $+62.24$, 27.99, and $+3.94$, respectively. The interpretation explains that the variables alone and in combination had a positive effect on the particle size. Based on insignificant lack of fit (F = 0.61, P = 0.6419) and the lowest precision (313.33) value, the quadratic model was chosen as the best fitting model to describe the influence of independent variables on particle size.

2.5. Influence of Independent Variables on Entrapment Efficiency (Y_2)

The mean entrapment efficiency of the prepared AM-NLs was found between 48.8 ± 1.1 (AM-NL9) and 80.9 ± 2.9 (AM-NL12), as shown in Table 1. The minimum entrapment efficiency (Y_2) was found from the formulation (AM-NL9) prepared with homogenization speed of 16,000 rpm, surfactant concentration of 1% *w/v*, and lipid concentration of 1.5% *w/v*. The maximum particle size (Y_1) was found from the formulation (AM-NL12) prepared with composition homogenization speed of 16,000 rpm, surfactant concentration of 3% *w/v*, and lipid concentration of 4.5% *w/v*. From the results, a significant difference in the entrapment efficiency was observed by changing the composition of independent variables. The effects of independent variables (homogenization speed, surfactant concentration, and lipid concentration) were evaluated on the 3D response surface plot (Figure 3). As shown in Figure 3, homogenization speed (A) showed a synergistic effect on entrapment efficiency. With the increase in the homogenization speed from a low level (12,000 rpm) to an intermediate level (16,000 rpm), AM entrapment increased. After reaching the intermediate level, the further increase in the homogenization speed decreased the entrapment efficiency. The decrease in entrapment efficiency may be due to the leaching of drugs at a high homogenization speed. The surfactant (B) also showed a dual effect on the entrapment efficiency. The increase in surfactant concentration led to a decrease in interfacial tension and a greater amount of solubilized and entrapped AM. The presence of a sufficient surfactant concentration prevents the drug expulsion and ultimately forms stable nanoparticles. Contrarily, if sufficient lipids are not available for drug entrapment, the reverse results will occur [32]. The third variable, lipid concentration (C), depicted a positive effect on entrapment efficiency. The increase in lipid content resulted in a greater amount of AM entrapped inside the NLs. The possible reason for this is the availability of more space for the accommodation of drugs within the NL vesicles [33]. A low concentration of surfactants and high concentration of lipids gives lesser entrapment due to incomplete emulsification.

The effect of independent variables was further evaluated by the polynomial equation given below:

$$\text{Entrapment efficiency} = +74.71 - 1.62A + 5.98B + 10.13C + 0.70AB - 0.40AC + 2.70BC - 4.42A^2 - 10.02B^2 - 2.51C^2 \quad (2)$$

As depicted in Equation (2), homogenization speed with the magnitudes -1.62 (A) and -4.42 (A^2) exhibited a negative effect on entrapment efficiency. On the other hand, upon increasing the surfactant concentration (magnitude $+5.98$), the entrapment efficiency was increased if sufficient lipids were present in the dispersion. This might be due to the solubility of the drug in the presence of sufficient surfactants. The quadratic model was considered as the best fitting model based on non-significant lack of fit (F = 3.49, *p*-value 0.1295) and lowest precision (172.2) value (Table 2). It described the influence of independent variables on entrapment efficiency.

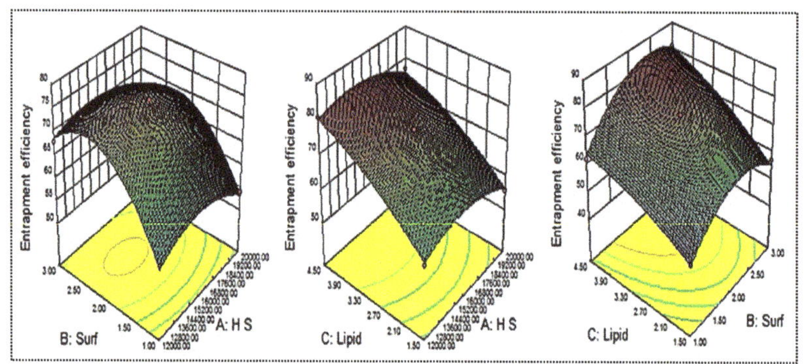

Figure 3. Effects of independent variables (Homogenization speed—A, Surfactant—B, Lipid—C) on the encapsulation efficiency (Y_2).

2.6. Optimized Composition

Table 2 and Figures 2 and 3 indicate that all the process and formulation variables were well controlled, and the obtained values were in an acceptable range. Based on the above results, it was concluded that all three variables had a significant effect on the responses. Among the 17 prepared formulations, AM-NL13 was selected for further point prediction optimization by slightly changing the composition and further evaluated for particle size and entrapment efficiency (Table 3). The practical results of particle size and entrapment efficiency were found to be closer to each other, and the overall desirability was found to be 0.992, which confirms the validity of the model.

Table 3. Point prediction optimization by Box Behnken design.

Code	Homogenization Speed: Surfactant: Lipid	Actual Value		Predicted Value	
		Y_1 (nm)	Y_2 (%)	Y_1 (nm)	Y_2 (%)
AM-NL13	16,000:2:3	176.1 ± 12.4	75.4 ± 2.6	178.20	74.7
AM-NLop	16,500:2.25:3	154.1 ± 6.35	78.15 ± 1.9	161.3	79.8

The selected optimized formulation (AM-NLop) was the composition prepared with homogenization speed of 16,500 rpm, surfactant concentration of 2.25% *w/v*, and lipid concentration of 3% *w/v*. It showed particle size and entrapment efficiency of 154.1 ± 6.35 nm (Figure 4) and 78.15 ± 1.9%. It depicted a PDI value of 0.18 and negative zeta potential (−34.24 mV), indicating higher stability. The low PDI (less than 0.5) and high negative zeta potential (±30 mV) revealed the greater homogeneity of particles. The predicted value of particle size and entrapment efficiency was found to be 161.3 nm and 79.8%, and the value was found to be closer to the predicted values.

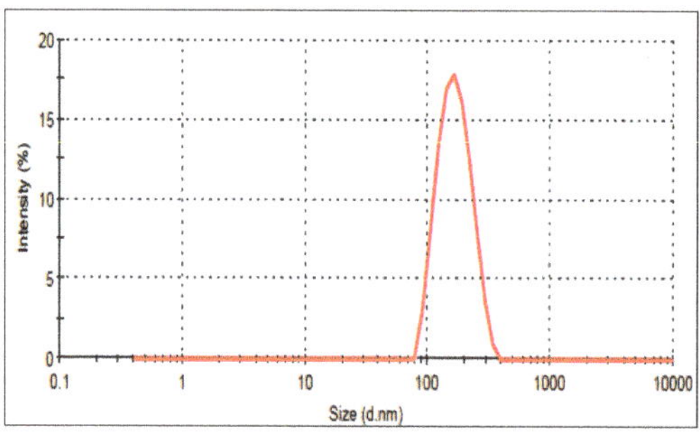

Figure 4. Particle size distribution image of Azithromycin-loaded nano lipid carrier (AM-NLop).

2.7. X-ray Diffraction (XRD) Analysis

Figure 5 shows the diffractogram of AM and AM-NLop. AM exhibited a characteristic highly intense peak at 2-theta levels of 8.2°, 10.2°, 13°, 16.2°, 17.2°, 19.1°, and 29.6°, assuring its crystallinity. The diffractogram of AM-NLop did not exhibit any characteristic sharp peak of AM after encapsulating into NLs. The low intensity and broad AM peaks indicate that AM was encapsulated or solubilized in a lipid matrix. The reduction in the intensities takes place due to some modification in the crystallinity in the NLs, attributed to the disordering of the solid lipid crystalline structure due to the presence of liquid lipids [34].

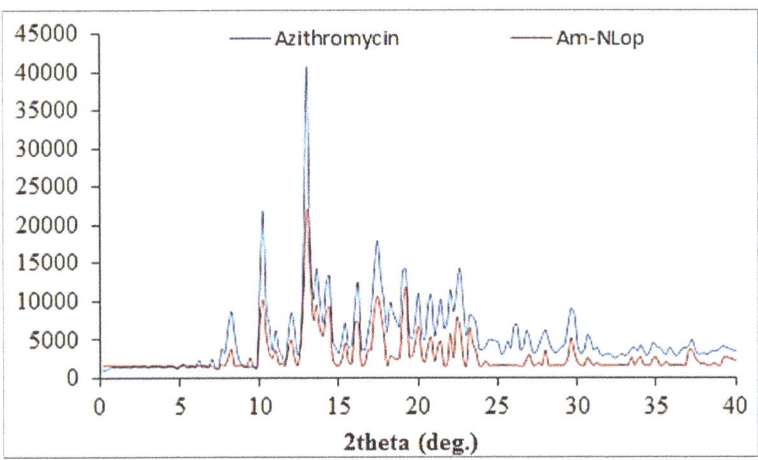

Figure 5. XRD image of Azithromycin and Azithromycin-loaded nano lipid carrier (AM-NLop).

2.8. Gel Evaluation

2.8.1. Clarity and Gelling Ability

The formulation was found to be clear upon observation with a black and white background. The presence of foreign particles may produce irritation to the soft tissue of the ocular region. It was also evaluated for sol to gel properties after gradually converting into a gel after the addition of the sol form into STF. This gelling is due to the replacement of sodium from sodium alginate with Ca^{+2} in tear fluid, forming calcium alginate. Table 4 shows the viscosity results of AM-NLopG and AM-G formulations. The viscosity of AM-NLopG in the solution state was found to be very low, whereas after conversion into gel form, a significant enhancement in the viscosity was observed. The viscosity was also evaluated for the control formulation AM-IG. In solution form, it shows the viscosity of 235.34 ± 7.21 cps, and after conversion into gel state, it showed 495.18 ± 8.72 cps. AM-NLopG showed higher viscosity in the solution state due to the presence of different ingredients used to prepare NLs, and after converting into gel state, greater viscosity was also observed due to higher gelling capacity.

Table 4. Physicochemical characterization of in situ gel.

Physicochemical Parameters	AM-G		AM-NLopG	
	Sol State	Gel State	Sol State	Gel State
Viscosity (cps)	131.25 ± 6.23	402.12 ± 5.52	235.34 ± 7.2	495.18 ± 8.7
Cohesiveness (g)	0.8 ± 0.02	0.94 ± 0.01	0.78 ± 0.09	0.86 ± 0.05
Adhesiveness (N-mm)	13.34 ± 0.92	26.23 ± 1.52	14.24 ± 1.1	29.43 ± 1.1
Hardness (N)	3.54 ± 0.32	8.34 ± 0.38	3.21 ± 0.3	8.76 ± 0.5

2.8.2. Texture Analysis

Texture analysis of AM-NLopIG and AM-IG (control) formulation was performed to determine the mechanical properties, and the results are expressed in Table 4. The data show that cohesiveness, adhesiveness, and hardness of AM-NLopIG were found to be significantly changed ($p < 0.05$) in sol and gel states. It showed cohesiveness of 0.78 ± 0.12 (sol) to 0.86 ± 0.09 (gel), adhesiveness of 14.24 ± 1.03 (sol) to 29.43 ± 1.12 (gel), and hardness of 3.21 ± 0.24 (sol) to 8.76 ± 0.51 (gel). The results of the mechanical properties of AM-NLopG and AM-IG did not show significant changes ($p > 0.05$). Interaction of Ca^{+2} with the gelling polymer in situ gel system directly affects mechanical properties. The higher value of adhesiveness indicates more adhesion to the corneal surface and increases the residence time. However, the cohesiveness of the formulation reduces the irritation and makes it easy to spread on the ocular surface [35].

2.8.3. In Vitro Drug Release

Figure 6 depicts the comparative release profile of AM-NLopIG, AM-IG, and eye drop using the dialysis bag method. The release of AM from AM-NLopIG and AM-IG was found to be $98.65 \pm 4.65\%$ and $49.48 \pm 4.23\%$, respectively, in 24 h. The eye drop depicted $99.45 \pm 5.3\%$ release in 3 h. AM-NLopIG exhibited biphasic release behavior, with initial fast release and later slow release. The initial fast release may be due to the nano-sized particles of NLs and a higher effective surface area. The drug particles adsorbed on the surface of NLs enter the dissolution media. Later, the slower drug release was found to be due to the formation of gel matrix, and the drug needs to cross it as well as the dialysis membrane. The eye drop showed quicker release in 3 h due to the lesser viscosity, and the drug is available for release. AM-IG showed significantly less release than AM-NLopIG due to the poor solubility and non-availability of surfactants. The surfactant helps to increase the solubility and lowers the interfacial tension between the two phases.

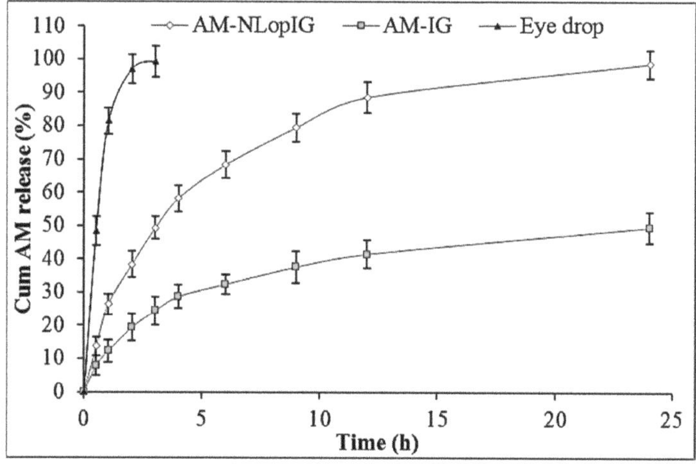

Figure 6. In vitro release data of the Azithromycin nano lipid carrier (AM-NLop), Azithromycin nano lipid carrier-based in situ gel (AM-NLopIG), and AM eye drop. The tests were performed in triplicate and data are shown as mean $\pm$ SD.

2.8.4. Ex Vivo Permeation Study

The comparative ex vivo permeation results of AM-NLopIG, AM-IG, and eye drop showed significant differences in permeation (%) and flux. AM-NLopIG exhibited significantly ($p < 0.05$) higher permeation ($60.76 \pm 4.12\%$) than AM-IG ($33.46 \pm 3.04\%$) and eye drop ($23.31 \pm 3.76\%$). The flux of AM-NLopIG was found to be $153.43\ \mu g/h.cm^2$, which is 1.82-fold higher than AM-IG and 2.6-fold higher than the eye drop. The higher permeation of AM from AM-NLopIG was found due to the increase in AM solubility after nano-sizing. The presence of lipids and surfactants of NLs increases the endocytosis through the corneal epithelium and leads to greater permeation across the membrane [21,36]. The sol form of a prepared formulation is converted to gel form and the contact time increases. The increase in contact time and the used polymer helps to open the tight junction of the membrane and gives greater permeation. The apparent permeability coefficients of AM-NLopIG, AM-IG, and eye drop were calculated as 2.4×10^{-1} cm/s, 1.3×10^{-1} cm/s, and 9.3×10^{-2} cm/s, respectively. AM-NLopIG showed an enhancement ratio 1.82-fold higher than AM-IG and 2.61-fold higher than the eye drop.

2.8.5. Corneal Hydration

The ocular hydration study was performed to evaluate the ocular toxicity after keeping the formulation with the cornea. After treatment, the corneal hydration was found to be 76.45%, which is within the standard limit of 76–80% [37]. The results of the study revealed that AM-NLopIG did not produce any toxicity and did not alter the structure of the cornea. There was no damage to the cornea observed externally and internally, which was further confirmed by histopathology examination.

2.8.6. Histopathology

The cornea was collected after the permeation study and assessed for internal damage in the structure. The histopathology of the treated cornea with AM-NLopIG and control (NaCl, 0.9%) was compared to changes in the structure, as depicted in Figure 7A,B. AM-NLopIG-treated cornea revealed no marked change in the internal structure. No damage in cellular structure or histological alterations were observed in the treated cornea, matching with the normal saline-treated cornea. The epithelium, Bowman's membrane, and stroma of the cornea and the cells were found intact, similar to the control sample. The morphology of the cornea was also well maintained [38]. From the results, we can presume that there is no alteration in the corneal structure, and that AM-NLopIG is nontoxic and compatible with the ocular structure.

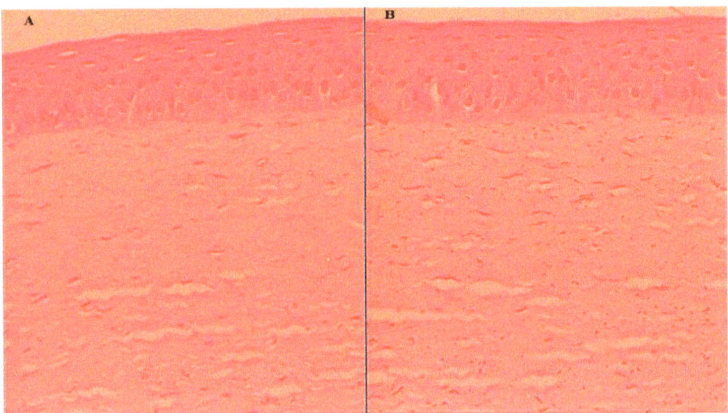

Figure 7. Histopathology image of AM-NLopIG-treated cornea (**A**). NaCl-treated cornea (**B**) taken by light microscope at image scale 40×.

2.8.7. HET-CAM Irritation Study

HET-CAM (hen's egg chorioallantoic membrane) is an in vitro evaluation method. It is helps to determine the irritation potential of formulations and also used as an alternative to the Draize test in rabbits. The results of CAM treated with AM-NLopIG, a negative control (0.9% NaCl), and a positive control (1% SLS) are expressed in Figure 8A–C. AM-NLopIG and the negative control did not exhibit any damage to CAM (blood capillaries). The score was found to be closer to zero and considered as non-irritant. It revealed no damage to the vein or artery, no bleeding and hemorrhage observed. The positive control depicted a high irritant cumulative score of 22.33 (severe irritant) with bleeding and excessive hemorrhage. From the results, it can be concluded that the prepared formulation AM-NLopIG is safe and non-irritant for ocular administration.

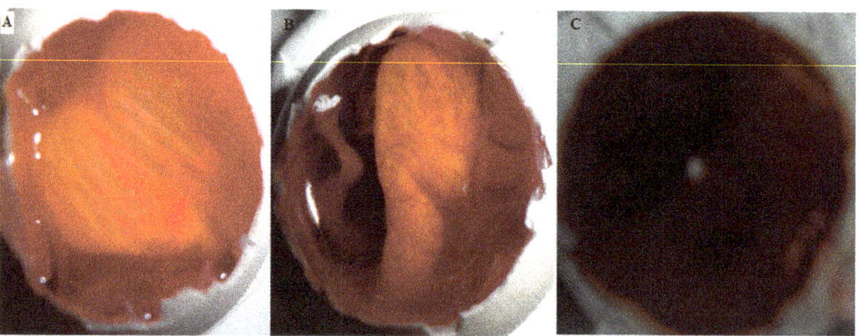

Figure 8. HET-CAM irritation test image treated with AM-NLopIG (**A**), 0.9% NaCl (**B**), and 1% SLS (**C**).

2.8.8. Isotonicity Study

The study was performed to evaluate the damage to the blood cells after treatment with AM-NLopIG. It was evaluated by treating with goat blood and observed under a microscope for any damage to RBCs. AM-NLopIG- and control (0.9% NaCl)-treated blood samples were evaluated for shrinkage and swelling to the blood cells and did not show any damage in RBC, revealing that the formulation is isotonic (Figure 9A,B). The results of the study are supported by the findings of the HET-CAM results. It also revealed no damage to the CAM.

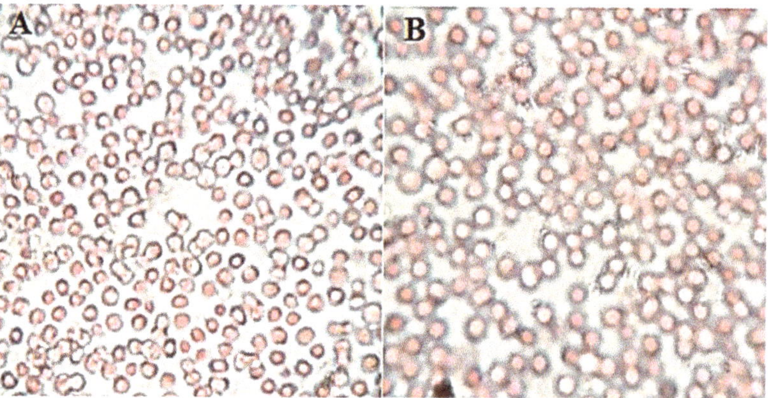

Figure 9. Isotonicity evaluation image of (**A**) AM-NLopIG and 0.9% (**B**) NaCl.

2.8.9. Antimicrobial Study

The antimicrobial activity of AM-NLopIG and eye drop was determined by the cup plate method against *S. aureus* and *E. coli* (Figure 10). AM-NLopIG depicted ZOI of 17.5 ± 1.7 mm and 20.4 ± 2.1 mm against *S. aureus* and *E. coli* at 24 h. The same formulation was further evaluated at 48 h and there was a significant ($p < 0.05$) enhancement in ZOI. It showed a ZOI of 22.8 ± 2.2 mm and 26.1 ± 1.9 mm against *S. aureus* and *E. coli* at 48 h. The eye drop showed less antibacterial activity against *S. aureus* and *E. coli* as 19.1 ± 1.4 mm and 21.1 ± 2.2 mm, respectively, at 24 h. Significantly less activity was observed at 48 h.

It showed ZOI of 13.3 ± 1.2 mm and 15.8 ± 1.7 mm. The prepared formulation AM-NLopIG showed significantly ($p < 0.05$) higher susceptibility (1.4-fold and 1.5-fold) than the eye drop. The enhanced antibacterial activity is due to the nano-sized particles and high surface energy. It will have a large surface area to act on the cell wall of microorganisms and increase the membrane permeability of the bacteria due to more enhanced mucoadhesive properties than the eye drop.

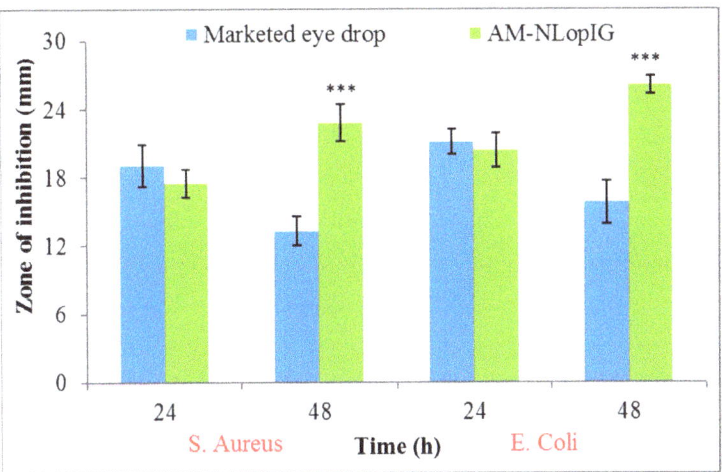

Figure 10. Antimicrobial graph of AM-NLopIG and eye drop evaluated against *S. aureus* and *E. coli*. Results shown as mean ± SD. *** indicates significant difference to the marketed eye drop.

3. Conclusions

An Azithromycin-loaded nano lipid carrier was prepared by the emulsification–homogenization method. The formulations were optimized by BBD design using homogenization speed (A), surfactant concentration (B), and lipid concentration (C) as independent variables. The prepared formulations showed nano-metric size and high encapsulation efficiency. The optimized formulation (AM-NLop) showed a nanoparticle size of 154.1 ± 6.35 nm, entrapment efficiency of 78.15 ± 1.9%, a PDI value of 0.18, and negative zeta potential (−34.24 mV). It was further converted into the sol-gel system to enhance the mucoadhesion. It showed optimal physicochemical properties in the solution as well as in gel form. The release and permeation study results depicted prolonged drug release as well as enhanced permeation (1.82-fold higher than AM-IG and 2.61-fold higher than eye drop). AM-NLopIG was found to be isotonic as well as non-irritant (HET CAM and histopathology). The antibacterial study results revealed greater activity (1.4-fold and 1.5-fold) against both organisms. From the results, it can be concluded that the prepared AM-NLs and AM-NL-based sol-gel system is an ideal delivery system to treat ocular diseases.

4. Materials

Azithromycin (AM) was procured from Alembic Pharmaceutical Ltd. (Ahmedabad, India). Glyceryl behenate (GB, melting point ~70 °C), Labrasol, and Precirol ATO-5 were procured from Gattifosse (Mumbai, India). Tripalmitins, Stearic acid, Myristic acid, Glycerol monostearate, Glyceryl monooleate, methanol, and acetonitrile were procured from Sigma-Aldrich (St. Louis, MO, USA). Isopropyl myristate, Miglyol, Coconut oil, Sunflower oil, Labrasol, and Sesame oil were procured from Loba Chemie (Mumbai, India). The Span 60, Span 20, Kolliphor EL, and Cremophor RH60 were procured from Acros organic (Somerset, NJ, USA). Zaha eye drops (Azithromycin eye solution, 1% *w/v*, Ajanta Pharma, India) were purchased from a local pharmacy.

5. Experimental

5.1. Screening of Lipids and Surfactant

The screening of lipids was performed by the solubility of AM in the different solid lipids for the formulation of NLs. The excess amount of AM was added to each melted solid lipid (Glyceryl behenate, Tripalmitin (~64 °C), Stearic acid (~69 °C), Myristic acid (~52 °C), Glycerol monostearate (~74 °C), Glyceryl monooleate (~40 °C), Precirol ATO-5 (~55 °C)) in a glass vial. Similarly, the excess amount of AM was added to liquid lipids (Isopropyl myristate, Miglyol, Coconut oil, Sunflower oil, Labrasol, Sesame oil) and surfactants (Span 60, Span 20, Kolliphor EL, Cremophor RH60) in a glass vial. The samples were vortexed and kept in a water bath shaker (JULABO, Izumi Osaka, Japan) for 72 h. The mixture was centrifuged at 5000 rpm for 30 min, and then AM in each solid lipid was determined by a UV spectrophotometer (Genesys 10S UV-Vis, Thermo Scientific, Waltham, MA, USA) after appropriate dilution.

5.2. Selection of Solid and Liquid Lipid Ratio

The melted solid and liquid lipids in different ratios (9:1, 8:2, 7:3, 6:4, 5:5, 2:8) were mixed with continuous stirring. The mixture was cooled, and the smear of lipids was prepared over filter paper. The separation of oil droplets over the filter paper was visually observed.

5.3. Optimization

Design Expert software (Design Expert, version 8.0.6, State-Ease Inc., Minneapolis, MN, USA) was used for the optimization [39]. The purpose of this design was to examine the level of independent constraints, namely, homogenization speed (A, 12,000–20,000 rpm), surfactant concentration (B, 1.5–4.5%), and lipid concentration (1.5–4.5%), to achieve the desired size and entrapment efficiency. The polynomial Equation (3) was used to evaluate the statistical assessment of different independent variables on dependent constraints and can be given as:

$$Y = b_0 + b_1 A + b_2 B + b_3 C + b_{12}AB + b_{13}AC + b_{23}BC + b_{11}A^2 + b_{22}B^2 + b_{33}C^2 \qquad (3)$$

where Y is the response related to each independent constraint; b_0 is the intercept; b_1–b_{33} are the regression coefficients obtained for observed experimental values; and A, B, and C are the coded values of the independent factors. The coefficients b_1, b_2, and b_3 show the individual parameter effect on the responses when two other parameters remain constant. The interaction terms (like b_{12}–b_{23}) display the response modification when the two factors are simultaneously altered. Based on the preliminary screening, a range of various independent constraints were decided. These values were fitted in the Design Expert software to obtain the composition of 17 formulations including five center points (common composition). The formulations were prepared, and their particle size and encapsulation efficiency data were placed in BBD to derive the predicted values, different polynomial equations, and model graphs. The obtained results were evaluated to examine the effects of independent constraints on dependent factors.

5.4. Development of Azithromycin-Loaded Nanostructure Lipid Carrier

AM-NLs were developed by the emulsification–homogenization technique as per a previously reported method with slight modifications [20]. The various batches (AM-NL1–AM-NL17) were prepared as per the given composition of Table 1. The solid lipid was melted at above 5 °C of its melting point and the liquid lipid and was added with continuous stirring to make the homogeneous mixture. The aqueous surfactant solution was prepared in distilled water and heated at the same temperature. The hot aqueous surfactant solution was added to the melted lipid mixture at 15,000 rpm for 15 min to form a coarse primary emulsion. The primary coarse emulsion was further subjected to a homogenizer

(T25 digital Ultra-Turrax IKA, Staufen, Germany) for 5 min. The prepared emulsion was cooled to room temperature to form AM-NLs and further stored for evaluation.

6. Characterization

6.1. Size, PDI, and Zeta Potential

The size, PDI, and zeta potential were analyzed by a zeta sizer (NanoZS90, Malvern Instrument Ltd., Malvern, UK). The diluted NLs (100-fold) were placed in cuvettes, and size and PDI were determined. For zeta potential, the same diluted sample was placed in electrode-containing cuvettes, and then zeta potential was determined.

6.2. Entrapment Efficiency (% EE)

The ultracentrifugation method was employed for the determination of EE from AM-NLs. AM-NLs were placed in a centrifugation tube and centrifuged at 15,000 rpm (Remi, cooling centrifuge, Mumbai, India) for 30 min. The supernatant was collected and further diluted to evaluate the AM content by a UV spectrophotometer, and % EE was calculated by the following equation:

$$\% \text{ EE} = \frac{\text{Total AM} - \text{Free AM}}{\text{Total AM}} \times 100 \tag{4}$$

6.3. X-ray Diffraction (XRD) Analysis

An XRD instrument (Ultima IV, Rigaku Inc., Tokyo, Japan) was used to determine the crystallinity of AM after encapsulating into AM-NLop. Each sample was filled in an XRD sample holder and scanned between 5° and 60° at the 2-theta level. Diffractograms of both samples were recorded to compare the change in peak height and width.

6.4. Formulation of AM-NLop In Situ Gel

The optimized formulation (AM-NL13) based on particle size and encapsulation efficiency was further converted into in situ gel by using mucoadhesive polymers. Weighed quantities of sodium alginate (1.5%, *w/v*) and hydroxyl propyl methyl cellulose (0.5%, *w/v*) were soaked overnight in distilled water. AM-NL13 was dispersed in polymeric solution to form in situ gel dispersion equivalent to 1% AM and pH was adjusted to 6.5.

6.5. Clarity and Gelling Ability

The clarity and gelling ability of AM-NLs in situ gel (AM-NLopIG) were checked visually. For gelling strength, AM-NLopIG (50 µL) was added to simulated tear fluid (2 mL) in a vial and gelling time was noted [40].

6.6. Viscosity Determination

The viscosity of AM-NLopIG formulation in solution and gel form was measured by a Brookfield viscometer (Fungi Lab, Madrid, Spain) using spindle number LV3 at 50 rpm angular speed. The temperature was fixed at 37 ± 0.5 °C for the study.

6.7. Texture Analysis

This study was performed using a texture analyzer (TA.XTplus100C, stable Microsystem, Ltd., Surrey GU7 1YL, Warrington, UK) as per the previous prescribed method [41]. AM-NLopIG was placed into the sample holder and pressed by a 20 mm diameter cylindrical probe (20 s interval, 2 mm depth). The hardness, cohesiveness, and adhesiveness were analyzed, and the results were compared with the eye drop.

6.8. In Vitro Drug Release

The comparative release study of AM-NLopIG, AM-IG (control), and eye drop was performed using a dialysis bag (MWCO 12 kDa). The pretreated dialysis bag was filled with each formulation and dipped into STF (250 mL) at a temperature of 37 ± 0.5 °C. At a fixed time, 2 mL release content was collected and the same volume of fresh STF was added to maintain the uniform release media volume. The released content was filtered and further diluted to measure the absorbance using a UV spectrophotometer.

6.9. Trans-Corneal Permeation Study

The study was performed using goat cornea collected from the slaughterhouse. The whole eyeball was collected and stored in NaCl (0.9%) at 4 °C. The cornea was removed from the eyeball with the sclera and washed. Simulated tear fluid (STF) was used as permeation media and filled into the receptor compartment of the diffusion cell (area 0.6 cm^2, volume 10 mL). The cornea was mounted between the donor and acceptor compartment and the temperature was maintained at 37 ± 0.5 °C. AM-NLopIG, AM-IG, and eye drop (2 mg of the AM, in percentage) were filled into the donor compartment, and at a fixed time point, the released content (1 mL) was removed. The blank fresh STF was replaced into a diffusion cell to keep a uniform volume. The collected release content was filtered and further diluted for evaluation by the previously validated HPLC method [42]. The HPLC method was performed using mobile phase composition ammonium acetate buffer (0.05M and pH 8) and acetonitrile (60:40 v/v). The study was performed at a flow rate of 1 mL/min with an injection volume of 20 µL. The % permeation and flux were calculated.

$$\text{Flux} = \frac{\text{Concentration}}{\text{Permeation area} \times \text{ime}} \tag{5}$$

$$\text{Drug Permeation} = \frac{\text{Concentration}}{\text{applied area}} \times \text{dilution} \tag{6}$$

6.10. Corneal Hydration

The corneal hydration test was performed to evaluate the hydration of the cornea after the treatment with AM-NLopIG. The cornea was collected after the permeation study and the initial weight was noted (wet weight). The cornea was placed into the oven (Thermo Scientific, Osterode, Germany) at 80 °C for drying. The cornea was removed from the oven and reweighed to note the dry weight. The % corneal hydration (CH) was calculated using the formula as reported by Mudgil et al. [43].

$$\text{CH (\%)} = \frac{\text{Wet weight} - \text{Dry weight}}{\text{Dry weight}} \times 100 \tag{7}$$

6.11. Histopathology Study

The histopathology of the cornea was evaluated to check the internal damage after treatment with a prepared formulation. The cornea was collected after the permeation study, washed with STF, and stored in formalin solution (10% v/v.) The cornea was treated with 0.9% NaCl solution taken as a control. Each treated cornea was cleaned with alcohol and mounted with molten paraffin. The cross-section of the cornea was cut and stained with hematoxylin and eosin to evaluate under a high-resolution microscope (BA210m Motic microscope, Selangor Darul Ehsan, Malaysia).

6.12. HET-CAM Irritation Study

The in vitro irritation AM-NLopIG, negative control (0.9% NaCl), and positive control (sodium lauryl sulphate, 1% w/v) were evaluated by using hen egg chorioallantoic membrane (HET-CAM). The fertilized hen eggs were procured from a poultry farm and incubated for 10 days in a humidity-controlled incubator at 37 ± 0.5 °C/5% RH. On the 10th day, eggs were collected and shells were removed from the air chamber side. The inner membrane was wetted by adding a drop of 0.9% NaCl and then carefully removed by using forceps to visualize developed CAM. Then, 2–3 drops of the AM-NLopIG, NaCl (0.9% w/v), and sodium lauryl sulphate (1% w/v) were added over CAM and observed for 5 min to check for damage. The scores were given as per the standard scoring scale at different time points, i.e., 0–0.9 for non-irritants, 1–4.9 for weak irritants, 5–8.9 for moderate irritants, and 9–21 for severe irritants [44].

6.13. Antimicrobial Study

The activity of AM-NLopIG and eye drop was evaluated by the cup plate method against *Staphylococcus aureus* (*S. aureus*) and *Escherichia coli* (*E. coli*). The required quantity of nutrient agar media was weighed, transferred into a conical flask, and dissolved in distilled water. The media was sterilized by autoclave at 121 °C for 15 min. The micro-organisms *S. aureus* and *E. coli* (0.1 mL) were mixed with nutrient agar media and transferred to a sterilized Petri plate under aseptic conditions. The media was kept for solidification in an aseptic area. Using a sterilized borer, 5 mm wells were created and the samples (AM-NLopIG and eye drop) were added to evaluate their efficacy. The Petri plates were kept for 1 h and further incubated at 37 °C for 24 h and 48 h to evaluate the zone of inhibition (ZOI).

Author Contributions: S.J.G. and M.N.b.J.—Conceptulization, Resources and Funding; A.Z. and S.S.I.—Experi-mental work and project administration; M.Y. and M.K.—software and data curation; S.A.—Supervision, review and editing; M.M.G. and F.M.A.—Data analysis and writing original draft. All authors have read and agreed to the published version of the manuscript.

Funding: Princess Nourah bint Abdulrahman University Researchers Supporting Project number (PNURSP2022R108), Princess Nourah bint Abdulrahman University, Riyadh, Saudi Arabia.

Institutional Review Board Statement: Not Applicable.

Informed Consent Statement: Not Applicable.

Data Availability Statement: Not Applicable.

Acknowledgments: This research project is supported by Princess Nourah bint Abdulrahman University Researchers Supporting Project number (PNURSP2022R108), Princess Nourah bint Abdulrahman University, Riyadh, Saudi Arabia.

Conflicts of Interest: The authors declare no conflict of interest.

References

1. Tangri, P.; Khurana, S. Basics of ocular drug delivery systems. *Int. J. Res. Pharm. Biomed. Sci.* **2011**, *2*, 1541–1552.
2. Wafa, H.G.; Essa, E.A.; El-Sisi, A.E.; El Maghraby, G.M. Ocular films versus film-forming liquid systems for enhanced ocular drug delivery. *Drug Deliv. Transl. Res.* **2021**, *11*, 1084–1095. [CrossRef] [PubMed]
3. Gilhotra, R.M.; Nagpal, K.; Mishra, D.N. Azithromycin novel drug delivery system for ocular application. *Int. J. Pharm. Investig.* **2011**, *1*, 22–28. [CrossRef] [PubMed]
4. Landucci, E.; Bonomolo, F.; De Stefani, C.; Mazzantini, C.; Pellegrini-Giampietro, D.E.; Bilia, A.R.; Bergonzi, M.C. Preparation of Liposomal Formulations for Ocular Delivery of Thymoquinone: In Vitro Evaluation in HCEC-2 e HConEC Cells. *Pharmaceutics* **2021**, *13*, 2093. [CrossRef]
5. Owodeha-Ashaka, K.; Ilomuanya, M.O.; Iyire, A. Evaluation of sonication on stability-indicating properties of optimized pilocarpine hydrochloride-loaded niosomes in ocular drug delivery. *Prog. Biomater.* **2021**, *10*, 207–220. [CrossRef]
6. Dandamudi, M.; McLoughlin, P.; Behl, G.; Rani, S.; Coffey, L.; Chauhan, A.; Kent, D.; Fitzhenry, L. Chitosan-Coated PLGA Nanoparticles Encapsulating Triamcinolone Acetonide as a Potential Candidate for Sustained Ocular Drug Delivery. *Pharmaceutics* **2021**, *13*, 1590. [CrossRef]
7. Nair, A.B.; Shah, J.; Al-Dhubiab, B.E.; Jacob, S.; Patel, S.S.; Venugopala, K.N.; Morsy, M.A.; Gupta, S.; Attimarad, M.; Sreeharsha, N.; et al. Clarithromycin Solid Lipid Nanoparticles for Topical Ocular Therapy: Optimization, Evaluation and In Vivo Studies. *Pharmaceutics* **2021**, *13*, 523. [CrossRef]
8. Youshia, J.; Kamel, A.O.; El Shamy, A.; Mansour, S. Gamma sterilization and in vivo evaluation of cationic nanostructured lipid carriers as potential ocular delivery systems for antiglaucoma drugs. *Eur. J. Pharm. Sci.* **2021**, *163*, 105887. [CrossRef]
9. Kassaee, S.N.; Mahboobian, M.M. Besifloxacin-loaded ocular nanoemulsions: Design, formulation and efficacy evaluation. *Drug Deliv. Transl. Res.* **2022**, *12*, 229–239. [CrossRef]
10. Sun, K.; Hu, K. Preparation and Characterization of Tacrolimus-Loaded SLNs in situ Gel for Ocular Drug Delivery for the Treatment of Immune Conjunctivitis. *Drug Des. Dev. Ther.* **2021**, *15*, 141–150. [CrossRef]
11. Elmowafy, M.; Al-Sanea, M.M. Nanostructured lipid carriers (NLCs) as drug delivery platform: Advances in formulation and delivery strategies. *Saudi Pharm. J.* **2021**, *29*, 999–1012. [CrossRef] [PubMed]

12. Gan, L.; Wang, J.; Jiang, M.; Bartlett, H.; Ouyang, D.; Eperjesi, F.; Liu, J.; Gan, Y. Recent advances in topical ophthalmic drug delivery with lipid-based nanocarriers. *Drug Discov. Today* **2013**, *18*, 290–297. [CrossRef] [PubMed]
13. Tian, C.; Zeng, L.; Tang, L.; Yu, J.; Ren, M. Sustained Delivery of Timolol Using Nanostructured Lipid Carriers-Laden Soft Contact Lenses. *AAPS PharmSciTechnol* **2021**, *22*, 212. [CrossRef] [PubMed]
14. Lakhani, P.; Patil, A.; Taskar, P.; Ashour, E.; Majumdar, S. Curcumin-loaded Nanostructured Lipid Carriers for Ocular Drug Delivery: Design Optimization and Characterization. *J. Drug Deliv. Sci. Technol.* **2018**, *47*, 159–166. [CrossRef] [PubMed]
15. Lakhani, P.; Patil, A.; Wu, K.W.; Sweeney, C.; Tripathi, S.; Avula, B.; Taskar, P.; Khan, S.; Majumdar, S. Optimization, stabilization, and characterization of amphotericin B loaded nanostructured lipid carriers for ocular drug delivery. *Int. J. Pharm.* **2019**, *572*, 118771. [CrossRef] [PubMed]
16. EL Kiss Berko, S.; Gácsi, A.; Kovacs, A.; Katona, G.; Soós, J.; Csányi, E.; Grof, I.; Harazin, A.; Deli, M.A.; Budai-Szűcs, M. Design and Optimization of Nanostructured Lipid Carrier Containing Dexamethasone for Ophthalmic Use. *Pharmaceutics* **2019**, *11*, 679.
17. Seyfoddin, A.; Al-Kassas, R. Development of solid lipid nanoparticles and nanostructured lipid carriers for improving ocular delivery of acyclovir. *Drug Dev. Ind. Pharm.* **2013**, *39*, 508–519. [CrossRef]
18. Baig, M.S.; Owida, H.; Njoroge, W.; Siddiqui, A.R.; Yang, Y. Development and evaluation of cationic nanostructured lipid carriers for ophthalmic drug delivery of besifloxacin. *J. Drug Deliv. Sci. Techol.* **2020**, *55*, 101496. [CrossRef]
19. Ustundag Okur, N.; Yozgatlı, V.; Okur, M.E.; Yolta¸s, A.; Siafaka, P.I. Improving therapeutic efficacy of voriconazole against fungal keratitis: Thermo-sensitive in situ gels as ophthalmic drug carriers. *J. Drug Deliv. Sci. Technol.* **2019**, *49*, 323–333. [CrossRef]
20. Youssef, A.; Dudhipala, N.; Majumdar, S. Ciprofloxacin Loaded Nanostructured Lipid Carriers Incorporated into In-Situ Gels to Improve Management of Bacterial Endophthalmitis. *Pharmaceutics* **2020**, *12*, 572. [CrossRef]
21. Gade, S.; Patel, K.K.; Gupta, C.; Anjum, M.M.; Deepika, D.; Agrawal, A.K.; Singh, S. An Ex Vivo Evaluation of Moxifloxacin Nanostructured Lipid Carrier Enriched In Situ Gel for Transcorneal Permeation on Goat Cornea. *J. Pharm. Sci.* **2019**, *108*, 2905–2916. [CrossRef] [PubMed]
22. Patil, A.; Lakhani, P.; Taskar, P.; Avula, B.; Majumdar, S. Carboxyvinyl Polymer and Guar-Borate Gelling System Containing Natamycin Loaded PEGylated Nanolipid Carriers Exhibit Improved Ocular Pharmacokinetic Parameters. *J. Ocul. Pharmacol. Ther.* **2020**, *36*, 410–420. [CrossRef] [PubMed]
23. Shelley, H.; Rodriguez-Galarza, R.M.; Duran, S.H.; Abarca, E.M.; Babu, R.J. In Situ Gel Formulation for Enhanced Ocular Delivery of Nepafenac. *J. Pharm. Sci.* **2018**, *107*, 3089–3097. [CrossRef] [PubMed]
24. Noreen, S.; Ghumman, S.A.; Batool, F.; Ijaz, B.; Basharat, M.; Noureen, S.; Kausar, T.; Iqbal, S. Terminalia arjuna gum/alginate in situ gel system with prolonged retention time for ophthalmic drug delivery. *Int. J. Biol. Macromol.* **2020**, *152*, 1056–1067. [CrossRef]
25. Makwana, S.B.; Patel, V.A.; Parmar, S.J. Development and characterization of in-situ gel for ophthalmic formulation containing ciprofloxacin hydrochloride. *Results Pharma Sci.* **2015**, *6*, 1–6. [CrossRef]
26. Srividya, B.; Cardoza, R.M.; Amin, P.D. Sustained ophthalmic delivery of ofloxacin from a pH triggered in situ gelling system. *J. Control. Release* **2001**, *73*, 205–211. [CrossRef]
27. Liu, Z.D.; Li, J.W.; Nie, S.F.; Liu, H.; Ding, P.T.; Pan, W.S. Study of an alginate/HPMC-based in situ gelling ophthalmic delivery system for gatifloxacin. *Int. J. Pharm.* **2006**, *315*, 12–17. [CrossRef]
28. Szekalska, M.; Pucilowska, A.; Szymanska, E.; Ciosek, P.; Winnicka, K. Alginate: Current use and future perspectives in pharmaceutical and biomedical applications. *Int. J. Polym. Sci.* **2016**, *2016*, 7697031. [CrossRef]
29. Mandal, S.; Thimmasetty, M.K.; Prabhushankar, G.; Geetha, M. Formulation and evaluation of an in situ gel-forming ophthalmic formulation of moxifloxacin hydrochloride. *Int. J. Pharm. Investig.* **2012**, *2*, 78–82. [CrossRef]
30. Tundisi, L.L.; Mostaço, G.B.; Carricondo, P.C.; Petri, D.F.S. Hydroxypropyl methylcellulose: Physicochemical properties and ocular drug delivery formulations. *Eur. J. Pharm. Sci.* **2021**, *159*, 105736. [CrossRef]
31. Jazuli, I.; Annu Nabi, B.; Moolakkadath, T.; Alam, T.; Baboota, S.; Ali, J. Optimization of Nanostructured Lipid Carriers of Lurasidone Hydrochloride Using Box-Behnken Design for Brain Targeting: In Vitro and In Vivo Studies. *J. Pharm. Sci.* **2019**, *108*, 3082–3090. [CrossRef] [PubMed]
32. Yang, G.; Wu, F.; Chen, M.; Jin, J.; Wang, R.; Yuan, Y. Formulation design, characterization, and in vitro and in vivo evaluation of nanostructured lipid carriers containing a bile salt for oral delivery of gypenosides. *Int. J. Nanomed.* **2019**, *14*, 2267–2280. [CrossRef] [PubMed]
33. Pandey, S.S.; Patel, M.A.; Desai, D.T.; Patel, H.P.; Gupta, A.R.; Joshi, S.V.; Shah, D.O.; Maulvi, F.A. Bioavailability enhancement of repaglinide from transdermally applied nanostructured lipid carrier gel: Optimization, in vitro and in vivo studies. *J. Drug Deliv. Sci. Technol.* **2020**, *57*, 101731. [CrossRef]
34. Shamma, R.N.; Aburahma, M.H. Follicular delivery of spironolactone via nanostructured lipid carriers for management of alopecia. *Int. J. Nanomed.* **2014**, *9*, 5449–5460. [CrossRef]
35. Singh, Y.; Vuddanda, P.R.; Jain, A.; Parihar, S.; Chaturvedi, T.P.; Singh, S. Mucoadhesive gel containing immunotherapeutic nanoparticulate satranidazole for treatment of periodontitis: Development and its clinical implications. *RSC Adv.* **2015**, *5*, 47659–47670. [CrossRef]
36. Balguri, S.P.; Adelli, G.R.; Janga, K.Y.; Bhagav, P.; Majumdar, S. Ocular disposition of ciprofloxacin from topical, PEGylated nanostructured lipid carriers: Effect of molecular weight and density of poly (ethylene) glycol. *Int. J. Pharm.* **2017**, *529*, 32–43. [CrossRef]

37. Nagarwal, R.C.; Kumar, R.; Pandit, J.K. Chitosan coated sodium alginate-chitosan nanoparticles loaded with 5-FU for ocular delivery: In vitro characterization and in vivo study in rabbit eye. *Eur. J. Pharm. Sci.* **2012**, *47*, 678–685. [CrossRef]
38. Mittal, N.; Kaur, G. Leucaena leucocephala (Lam.) galactomannan nanoparticles: Optimization and characterization for ocular delivery in glaucoma treatment. *Int. J. Biol. Macromol.* **2019**, *139*, 1252–1262. [CrossRef]
39. Fasolo, D.; Pippi, B.; Meirelles, G.; Zorzi, G.; Fuentefria, A.M.; Poser, G.V.; Teixeira, H.F. Topical delivery of antifungal Brazilian red propolis benzophenones-rich extract by means of cationic lipid nanoemulsions optimized by means of Box-Behnken Design. *J. Drug Deliv. Sci. Technol.* **2020**, *56*, 101573. [CrossRef]
40. Kesarla, R.; Tank, T.; Vora, P.A.; Shah, T.; Parmar, S.; Omri, A. Preparation and evaluation of nanoparticles loaded ophthalmic in situ gel. *Drug Deliv.* **2016**, *23*, 2363–2370. [CrossRef]
41. Harish, N.M.; Prabhu, P.; Charyulu, R.N.; Gulzar, M.A.; Subrahmanyam, E.V. Formulation and Evaluation of in situ Gels Containing Clotrimazole for Oral Candidiasis. *Indian J. Pharm. Sci.* **2009**, *71*, 421–427. [CrossRef] [PubMed]
42. Zeng, A.; Liu, X.; Zhang, S.; Zheng, Y.; Huang, P.; Du, K.; Fu, Q. Determination of azithromycin in raw materials and pharmaceutical formulations by HPLC coupled with an evaporative light scattering detector. *Asian J. Pharm. Sci.* **2014**, *9*, 107–116. [CrossRef]
43. Mudgil, M.; Pawar, P.K. Preparation and in vitro/ex vivo evaluation of moxifloxacin-loaded PLGA nanosuspensions for ophthalmic application. *Sci. Pharm.* **2013**, *81*, 591–606. [PubMed]
44. Budai, P.; Lehel, J.; Tavaszi, J.; Kormos, E. HET-CAM test for determining the possible eye irritancy of pesticides. *Acta Vet. Hung.* **2010**, *58*, 369–377. [CrossRef]

 *gels*

Article

Preparation and Characterization of a Novel Mucoadhesive Carvedilol Nanosponge: A Promising Platform for Buccal Anti-Hypertensive Delivery

El-Sayed Khafagy [1,2,*], Amr S. Abu Lila [3,4], Nahed Mohamed Sallam [5], Rania Abdel-Basset Sanad [5], Mahgoub Mohamed Ahmed [6], Mamdouh Mostafa Ghorab [2], Hadil Faris Alotaibi [7], Ahmed Alalaiwe [1], Mohammed F. Aldawsari [1], Saad M. Alshahrani [1], Abdullah Alshetaili [1], Bjad K. Almutairy [1], Ahmed Al Saqr [1] and Shadeed Gad [2]

[1] Department of Pharmaceutics, College of Pharmacy, Prince Sattam Bin Abdulaziz University, Al-Kharj 11942, Saudi Arabia; a.alalaiwe@psau.edu.sa (A.A.); moh.aldawsari@psau.edu.sa (M.F.A.); sm.alshahrani@psau.edu.sa (S.M.A.); a.alshetaili@psau.edu.sa (A.A.); b.almutairy@psau.edu.sa (B.K.A.); a.alsaqr@psau.edu.sa (A.A.S.)

[2] Department of Pharmaceutics and Industrial Pharmacy, Faculty of Pharmacy, Suez Canal University, Ismailia 41522, Egypt; mghorab@hotmail.com (M.M.G.); shaded_abdelrahman@pharm.suez.edu.eg (S.G.)

[3] Department of Pharmaceutics and Industrial Pharmacy, Faculty of Pharmacy, Zagazig University, Zagazig 44519, Egypt; a.abulila@uoh.edu.sa

[4] Department of Pharmaceutics, College of Pharmacy, University of Hail, Hail 81442, Saudi Arabia

[5] Department of Pharmaceutics, National Organization for Drug Control and Research (NODCAR), Giza 12553, Egypt; n_mohamed_80@yahoo.com (N.M.S.); sanadrania3@yahoo.com (R.A.-B.S.)

[6] Department of Molecular Drug Evaluation, National Organization for Drug Control and Research (NODCAR), Giza 12553, Egypt; mahgouba3@gmail.com

[7] Department of Pharmaceutical Sciences, College of Pharmacy, Princess Nourah Bint Abdul Rahman University, P.O. Box 84428, Riyadh 11671, Saudi Arabia; hfalotaibi@pnu.edu.sa

[*] Correspondence: e.khafagy@psau.edu.sa; Tel.: +966-533-564-286

Citation: Khafagy, E.-S.; Abu Lila, A.S.; Sallam, N.M.; Sanad, R.A.-B.; Ahmed, M.M.; Ghorab, M.M.; Alotaibi, H.F.; Alalaiwe, A.; Aldawsari, M.F.; Alshahrani, S.M.; et al. Preparation and Characterization of a Novel Mucoadhesive Carvedilol Nanosponge: A Promising Platform for Buccal Anti-Hypertensive Delivery. *Gels* 2022, 8, 235. https://doi.org/10.3390/gels8040235

Academic Editors: Maddalena Sguizzato, Rita Cortesi and Rachel Yoon Chang

Received: 3 March 2022
Accepted: 9 April 2022
Published: 11 April 2022

Publisher's Note: MDPI stays neutral with regard to jurisdictional claims in published maps and institutional affiliations.

Abstract: Carvedilol (CRV) is a non-selective third generation beta-blocker used to treat hypertension, congestive heart failure and angina pectoris. Oral administration of CRV showed poor bioavailability (25%), which might be ascribed to its extensive first-pass metabolism. Buccal delivery is known to boost drugs bioavailability. The aim of this study is to investigate the efficacy of bilosomes-based mucoadhesive carvedilol nanosponge for enhancing the oral bioavailability of CRV. The bilosomes were prepared, optimized and characterized for particle size, surface morphology, encapsulation efficiency and ex-vivo permeation studies. Then, the optimized formula was incorporated into a carboxymethyl cellulose/hydroxypropyl cellulose (CMC/HPC) composite mixture to obtain buccal nanosponge enriched with CRV bilosomes. The optimized bilosome formula (BLS9), showing minimum vesicle size, maximum entrapment, and highest cumulative in vitro release, exhibited a spherical shape with 217.2 nm in diameter, 87.13% entrapment efficiency, and sustained drug release for up to 24 h. In addition, ex-vivo drug permeation across sheep buccal mucosa revealed enhanced drug permeation with bilosomal formulations, compared to aqueous drug suspension. Consequently, BLS9 was incorporated in a CMC/HPC gel and lyophilized for 24 h to obtain bilosomal nanosponge to enhance CRV buccal delivery. Morphological analysis of the prepared nanosponge revealed improved swelling with a porosity of 67.58%. The in vivo assessment of rats indicated that CRV-loaded nanosponge efficiently enhanced systolic/diastolic blood pressure, decreased elevated oxidative stress, improved lipid profile and exhibited a potent cardio-protective effect. Collectively, bilosomal nanosponge might represent a plausible nanovehicle for buccal delivery of CRV for effective management of hypertension.

Keywords: bilosomes; buccal mucosa; carvedilol; heart biomarkers; mucoadhesive nanosponge

1. Introduction

Compared to other routes of administration, the oral route represents the most favored, convenient, and extensively used route of administration since it provides benefits such as patient compliance, painless administration, etc. Nevertheless, the limited oral absorption, and consequently, the low oral bioavailability, due to weak water solubility, low membrane permeability, and/or P-glycoprotein efflux, pose a hurdle against the development of new chemical entities for oral administration [1].

Carvedilol (CAR) is a non-selective third generation beta-blocker used to treat hypertension, congestive heart failure and angina pectoris [2]. Carvedilol is practically water insoluble; with a solubility value of 4.4 μg/mL in water, and it has a weak basic character (pKa value 6.8) [3]. Oral administration of CRV showed poor oral bioavailability ($\approx$25%) due to high exposure to first-pass metabolism [4]. Because of its high lipophilicity and limited water solubility, it is classified as a class II in the biopharmaceutical classification system. Several solubility-improving approaches, including micronization, inclusion complexation and solid dispersion, have been adopted to improve the aqueous solubility of CRV and thereby enhance its oral bioavailability [5–8]. Nevertheless, because of the limitation of these conventional dosage forms, these approaches have shown modest improvement in therapeutic efficacy.

Nanotechnology-based drug delivery systems are considered to be a viable strategy for improving the therapeutic efficacy of such lipophilic drugs [9,10]. Recently, various nanosystems have been reported to enhance lipophilic drugs' solubility, bioavailability, and stability through lipid-based formulations [11–13]. Among them, liposomes and niosomes, showed the efficacy to entrap both lipophilic and hydrophilic drugs [14], and to protect the encapsulated drug from degradation by the gastrointestinal tract (GIT) enzymes to some extent [15,16]. However, drug leakage and storage instability are identified as the primary disadvantages of conventional nano-vesicular carriers. Bile salt stabilized nano-vesicular systems, known as bilosomes, have been recently introduced as an alternative to the commonly used conventional nano-vesicular carriers (liposomes and niosomes). Bilosomes are novel vesicular nanocarriers in which bile salts are incorporated into the lipid bilayers of conventional nano-vesicular systems. Because of their configuration, they show a higher resistance to disruption by gastric secretion along with higher storage stability, compared to conventional nano-vesicular systems [17–19]. Accordingly, encapsulating a lipophilic drug such as CRV into bilosomes might offer a viable mean for enhancing drug aqueous solubility and thereby its systemic bioavailability.

Buccal cavity has been widely tested as an alternative to the oral route as a site for drug administration [20]. Systemic drug delivery via the buccal route has several advantages over oral administration, such as avoiding the first-pass metabolism and avoiding pre-systemic elimination within the gastrointestinal tract [21]. Recently, a large variety of dosage forms have been developed and commercialized (tablets, gels, films, etc.) for buccal administration [22–24]. Among these systems, mucoadhesive sponges have shown the advantages of keeping their swollen gel structure for a prolonged time, allowing for a longer residence time and more efficient medication absorption. In addition, by incorporating a nano-vesicular system, such as bilosomes, in a sponge to form a nanosponge, the developed system can combine the advantages of both platforms and, thus, opens the door to more enhanced buccal drug delivery with unique advances such as improved localization in the buccal mucosa, minimized burst release, and controlled drug release [25,26].

The aim of this study was to explore the efficacy of mucoadhesive cellulosic derivative sponges loaded with CRV bilosomes (CRV BL-SP) as an innovative anti-hypertensive dosage form. The in vivo efficacy of such an innovative formula was assessed via pharmacodynamic evaluation of specific biochemical parameters as lipid profile (cholesterol, triglycerides, LDL, and HDL), oxidative stress parameters (ROS, NO, TAC) in the heart, compared to the commercially available drug.

2. Results and Discussion

2.1. Effect of Formulation Variables on Physicochemical Characteristics of CRV-Loaded Bilosomes

2.1.1. Effect on Vesicle Size, Polydispersity and Zeta Potential

The mean size, PDI, and ZP of the prepared CRV-loaded bilosomes are summarized in Table 1. The vesicular size of different CRV-loaded bilosomes ranged from 217.2 ± 2.0 nm (BLS9) to 311.5 ± 4.3 (BLS2), depending on the effect of both lipid concentration and cholesterol concentration used in preparing different bilosome formulations. At fixed cholesterol concentrations, increasing lipid concentration from 1 to 3% significantly produced smaller bilosomes ($p < 0.0001$) (Table 1). The mean particle size of BLS9 prepared at 3% lipid concentration (217.2 ± 2 nm) was significantly lower than that of BLS6 (235.1 ± 5.1 nm) or BLS3 (252.8 ± 5.1 nm), prepared at 2 and 1% lipid concentration, respectively. This finding might be attributed to the availability of a greater surface area at a higher lipid level, leading to the incorporation of more CRV into the vesicle bilayer, which in turn could improve vesicular membrane packing, and consequently result in a significant decrease in vesicles size [27]. Similar findings were reported by El Menshawe et al., who demonstrated the negative impact of lipid concentration on the vesicle size of terbutaline-loaded bilosomes [28].

Table 1. Composition and physicochemical characteristics of CRV-loaded bilosomes.

Formulation Code	SPC Conc. (% *w/w*)	Chol Conc. (% *w/w*)	EE (%)	DL	PS (nm)	PDI	ZP (mV)
BLS1	1	0	51.03 ± 0.5	1.86 ± 0.04	311.5 ± 4.3	0.41 ± 0.005	-28.5 ± 3.0
BLS2	1	10	59.81 ± 1.4	2.15 ± 0.04	271.2 ± 5.8	0.54 ± 0.003	-26.1 ± 3.7
BLS3	1	30	77.05 ± 0.9	2.26 ± 0.03	252.8 ± 5.1	0.67 ± 0.004	-33.6 ± 1.0
BLS4	2	0	78.81 ± 1.5	1.10 ± 0.00	251.4 ± 3.4	0.36 ± 0.015	-34.8 ± 2.1
BLS5	2	10	82.8 ± 2.8	1.12 ± 0.04	241.3 ± 5.8	0.37 ± 0.006	-34.7 ± 2.9
BLS6	2	30	85.72 ± 1.4	1.15 ± 0.02	235.1 ± 5.1	0.21 ± 0.004	-38.1 ± 1.0
BLS7	3	0	79.57 ± 1.2	0.73 ± 0.01	231.3 ± 7.1	0.43 ± 0.003	-37.6 ± 1.9
BLS8	3	10	82.4 ± 1.8	0.75 ± 0.01	229.5 ± 6.0	0.43 ± 0.002	-43.3 ± 1.1
BLS9	3	30	87.13 ± 0.5	0.78 ± 0.02	217.2 ± 2.0	0.18 ± 0.002	-46.1 ± 1.1

Each formula contains 62.5 mg CRV, DCP (5% *w/w* of total lipid), and 25 mg SDC. EE, entrapment efficiency; DL, drug loading; PS, particle size, PDI, polydispersity index; ZP, zeta potential. Data represents mean $\pm$ SD of three independent experiments.

Similarly, incorporation of cholesterol within bilosomal membrane was found to significantly affect the vesicular size of CRV-loaded bilosomes. At fixed lipid concentrations, incorporating cholesterol within the bilosomal membrane resulted in a significant decrease in particle size. The particle size of BLS3, prepared at cholesterol concentration of 30%, was significantly lower than that of BLS1 lacking cholesterol (252.8 ± 5.1 vs. 311.5 ± 4.3 nm, respectively). Furthermore, increasing cholesterol concentration from 10 to 30% was found to reduce the particle size of CRV-loaded bilosomes. The particle size of BLS3, prepared with 30% cholesterol (252.8 ± 5.1 nm), was smaller than that of BLS2 (271.2 ± 5.8 nm), prepared with 10% cholesterol (Table 1). This effect might be ascribed to the membrane rigidizing effect of cholesterol, which makes the vesicle bilayer more compact. These results are inconsistent with those of Abdel Rahman et al. [29] who reported that cholesterol concentration had a significant impact on reducing the particle size of tretinoin-loaded phospholipid vesicles.

The polydispersity index (PDI) of any colloidal system is a measure of its size distribution. It is regarded as an index that might indicate the homogeneity or heterogeneity of the produced colloidal system [30]. Generally, low PDI values refer to a homogenous system. In this study, all CRV-loaded bilosomes had PDI values ranging from 0.18 to 0.67 (Table 1), which may be interpreted as an acceptable mid-range showing a good size distribution and assuming homogeneity of the formulations.

The zeta potential is a key indicator of the colloidal stability of bilosomal formulations. Generally, a zeta potential value ± 20 mV is crucial for colloidal stability due to electrical repulsion between particles [30]. As depicted in Table 1, the zeta potential of all bilosomal formulations was in negative range (-26.1 ± 3.7 mV to -46.1 ± 1.1 mV) (Table 1). In addition, at a fixed cholesterol concentration, a significant increase ($p < 0.05$) in the zeta potential values was observed upon by increasing phospholipid concentration from 1 to 3%.

2.1.2. Effect on Percentage Entrapment Efficiency

Entrapment efficiency is regarded as a crucial criterion for assessing the quality of the fabricated bilosomes. As depicted in Table 1, the percentage entrapment efficiency of the formulated bilosomes fluctuated from 51.03 ± 0.5 to 87.13 ± 0.5%, depending on different formulation variables. Increasing lipid concentration from 1 to 3% remarkably elevated entrapment efficiency of bilosomes from 51.03 ± 0.5 (BLS1) to 79.57 ± 1.2 (BLS7) (Table 1). Such synergistic effect might be attributed to the higher surface area of lipid bilayer vesicles obtained at higher lipid concentration, which could provide more lipophilic area for the lipophilic drug (carvedilol) to be entrapped, giving rise to increased entrapment efficiency. These findings are consistent with those of Hasan et al. [31], who highlighted the synergistic impact of increasing phospholipid content from 1 to 3% on mebeverine hydrochloride entrapment efficiency inside ethosomes.

In the same context, the entrapment efficiency of the prepared bilosomes increased as cholesterol concentration increased from 0 to 30%. At fixed lipid concentration, the entrapment efficiency of BLS1 and BLS3 formulations were 51.03 ± 0.5% and 77.05 ± 0.9%, respectively (Table 1). The positive effect of increasing cholesterol concentration on the entrapment efficiency might be ascribed, on the one hand, to cholesterol's membrane stabilizing effect, which prevents the leakage of an entrapped drug outside the vesicles, and, on the other hand, to its ability to increase the hydrophobicity of bilosomal phospholipid bilayer, and thus, allowing higher entrapment of the lipophilic drug, carvedilol, within bilosomal vesicles [29].

2.1.3. Effect on In Vitro Drug Release

In vitro drug release pattern of CRV-loaded bilosomes, as well as free drug, are depicted in Figure 1. As shown in Figure 1, the rate of drug released from free CRV suspension was significantly higher than the investigated bilosomal formulations ($p < 0.05$). In addition, all bilosomal formulations exhibited biphasic release profiles with a rapid initial drug release in the first 3 h (ranging from 20.03 to 44.67%), followed by a sustained release profile till 24 h. The cumulative percentage release of CRV from bilosomes ranged from 60.12% (BLS3) to 89.3% (BLS9) after 24 h. The initial fast release of CRV from bilosomal formulation might be attributed to the presence of drug in the outer shell of the bilosome vesicles [32]. The impact of different formulation variables—namely, lipid concentration and cholesterol concentration—on the in vitro drug release from formulated bilosomes was also investigated. It was obvious that increasing lipid concentrations had a positive effect on the percentage of drug release from BLS. This effect might be ascribed, on the one hand, to the amphiphilic properties of the soya bean phosphatidylcholine, and on the other hand, to the negative effect of increasing lipid concentration on the vesicle size. In contrast, cholesterol concentration did not significantly affect the in vitro drug release pattern from the prepared bilosomal formulations ($p = 0.894$).

Analysis of release data, fitted into different release kinetic models (Table S1), revealed that CRV release from different bilosomal formulations followed the Higuchi kinetics, illustrating a diffusion-controlled mechanism [33], whilst free CRV suspension followed first-order kinetics.

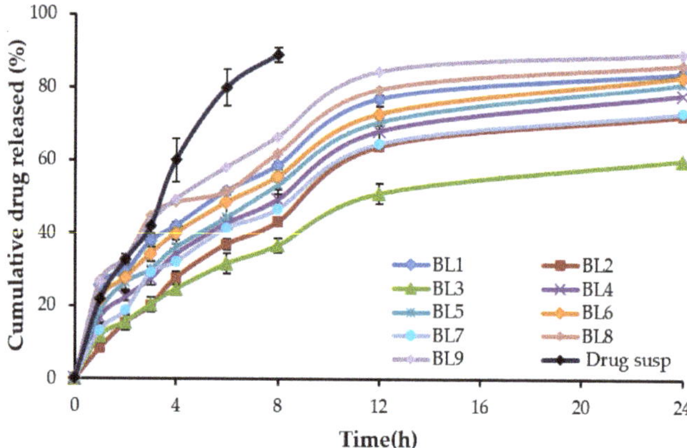

Figure 1. In vitro release profile of different CRV-loaded bilosomal formulations and CRV aqueous suspension. Data represents mean ± SD.

2.2. Selection of Optimized Formula

Particle size, entrapment efficiency percentage and in vitro drug release from a vesicular carrier represent key parameters that might affect the in vivo behavior of the entrapped drug. In this study, therefore, the optimized formula was selected to have the least PS with maximum entrapment efficiency and maximum cumulative percent of CRV released after 24 h. Under such constrains, BLS9, containing lipid concentration (3%) and cholesterol (30% *w/w*), was chosen as the optimum formula among all other formulations for further investigations.

2.3. Characterization of the Optimized CRV-Loaded Bilosomes

2.3.1. Transmission Electron Microscopic (TEM) Analysis

The TEM micrograph of the optimized bilosome formulation (BLS9) is shown in Figure 2. The optimized bilosome was depicted in well-identified small unilamellar vesicles with a spherical shape. No accumulation or drug crystals were detected.

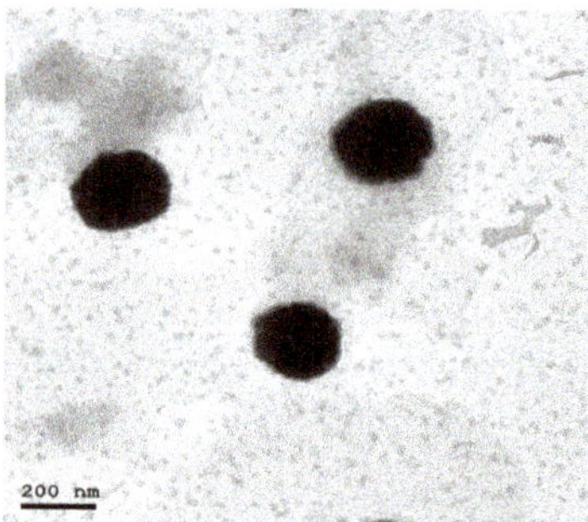

Figure 2. Transmission electron image of optimized bilosome formulation (BLS9).

2.3.2. Differential Scanning Calorimetry

The DSC thermograms of pure CRV, soybean phosphatidylcholine (SPC), cholesterol, and lyophilized optimized bilosomes (BLS9) formula are represented in Figure 3. The DSC thermogram of pure CRV unveiled a melting endothermic peak at 116 °C, corresponding to its melting point. Soybean phosphatidylcholine (SPC) and cholesterol exhibited endothermic peaks at 173.4 °C and 148.08 °C, respectively. Of interest, no endothermal peaks were observed in the thermogram of the lyophilized optimized bilosome formulation (BLS9), signifying that the drug is completely entrapped within bilosomal vesicles.

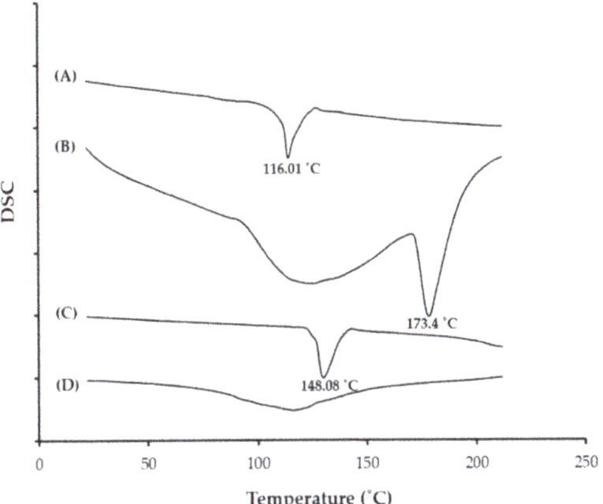

Figure 3. DSC thermograms of (A) pure carvedilol; (B) Soybean phosphatidylcholine; (C) Cholesterol and (D) optimized bliosomal formulation (BLS9).

2.4. Ex-Vivo Release Study

2.4.1. Ex-Vivo Permeation Studies

The ex-vivo permeation study of CRV from optimized bilosomal vesicles and CRV aqueous suspension was conducted using sheep buccal mucosa. As shown in Figure 4, encapsulation of CRV in bilosome vesicles leads to a significant permeation improvement ($p < 0.05$) across the mucosal membrane relative to the control CRV suspension with a three-fold increase in the transdermal flux (J_{max}). J_{max} of optimized bilosomal formulation was 22.82 ± 0.41 µg/cm^2/h across sheep buccal mucosa, as compared to CRV aqueous suspension (7.83 ± 0.11 µg/cm^2/h). This significant increase in transdermal flux of bilosomes might be attributed, on the one hand, to the small vesiclular size of CRV-loaded bilosomes, along with high phospholipids content, which increases the affinity of bilosomes to the biological membranes, thus increasing its permeation [34], and on the other hand, to the existence of bile salt that might help the permeation of the drug by opening the tight junction of the membrane [35]. Various permeation parameters through sheep buccal mucosa, were calculated and summarized in Table 2.

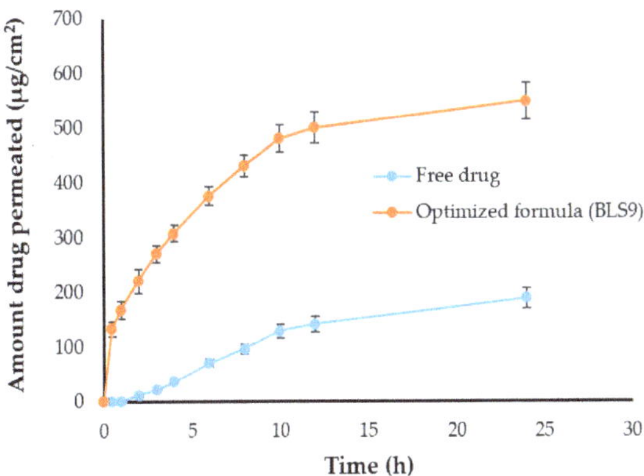

Figure 4. Ex-vivo permeation of CRV-loaded liposomes from sheep buccal mucosa. The data are represented as mean $\pm$ SD ($n = 3$).

Table 2. Ex-vivo permeation parameters for CRV aqueous suspension and CRV-loaded bilosomes.

Permeation Parameters	CRV Aqueous Suspension	CRV-Loaded Bilosome
Total amount of drug permeated (μg/cm^2)	188.0 ± 1.9	548.4 ± 6.9
J_{max} (μg/cm^2/h)	7.83 ± 0.11	22.82 ± 0.41
ER (cm^{-2}·h^{-1})	1	2.91
C_p (cm/h)	0.0017 ± 0.001	0.0071 ± 0.003

2.4.2. Confocal Laser Scanning Microscopy (CLSM) Study

Confocal laser scanning microscopy (CLSM) is a valuable image tool to study the localization of bilosomes in the mucosal membrane. As illustrated in Figure 5, compared to rhodamine B solution, Rh-B loaded bilosomes showed deep penetration across the mucosal membrane, as manifested by higher great fluorescence intensity and homogeneous distribution. This deep penetration of bilosomes might be ascribed to the small vesicle size, which bestow a remarkably higher surface area. In addition, the presence of bile salts might help in enhancing drug penetration through the mucosal membrane [36]. Similar findings were reported by Saifi et al. [33] who demonstrated that entrapment of the antiviral, acyclovir, within bilosomal vesicles, could efficiently enhance drug penetration through intestinal mucosa, resulting in improved drug oral bioavailability.

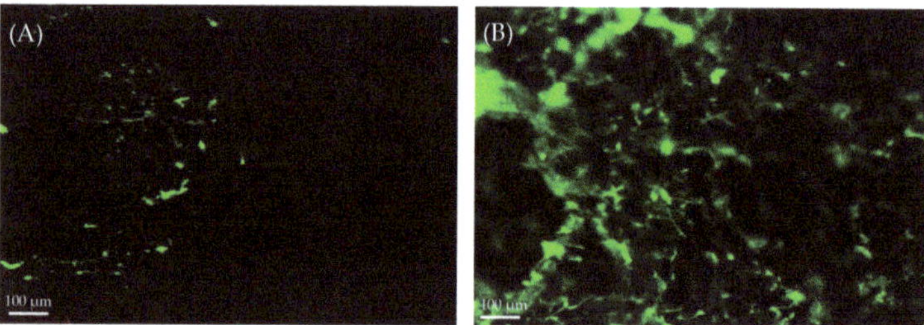

Figure 5. Confocal images of cross-sections of buccal mucosa after application of (**A**) rhodamine solution and (**B**) rhodamine-loaded bilosomes.

2.5. Stability Study

After three months of storage at 4 ± 1 °C, there was no apparent alteration in the appearance of the optimized bilosome dispersion (BLS9). In addition, no remarkable differences were observed ($p > 0.05$) in the EE% and PS of the stored bilosomes compared to the fresh ones. The stored samples showed EE% of 84.5 ± 1.25% and particle size of 229.4 ± 2.65 nm; compared to EE% of 87.13 ± 0.5% and particle size of 217.2 ± 2 nm for fresh samples. These results clearly indicate the physical stability of bilosomal dispersion under the storage condition.

2.6. Development of CRV-Loaded Bilosomal Sponge (CRV Nanosponge)

CRV bilosomes loaded sponge (CRV nanosponge) and non-vesicular CRV sponge were successfully prepared in a matrix of carboxymethyl cellulose/hydroxypropyl cellulose (CMC/HPC) by a solvent casting method. The lyophilized sponge had a smooth surface texture with a porous structure. The porous structure of the sponge is created after water removal during lyophilization.

2.7. Characterization of CRV Nanosponge

2.7.1. Sponge Morphology

Scanning electron microscopy (SEM) was adopted to examine the surface morphology of the prepared CRV nanosponge. SEM micrographs of CRV nanosponge illustrate a porous structure (Figure S1). This porous structure has resulted from the lyophilization process. The removal of water from bilosomal dispersion incorporated into the sponge by lyophilization created the vacant spaces, contributing to the recognized large number of uniform pores [37].

2.7.2. pH, Drug Content and Porosity

CRV nanosponge was evaluated for pH, drug content and porosity. The pH of the formulated CRV nanosponge was 6.5 ± 0.22, which is considered compatible with buccal mucosa [38], nullifying the possibility of buccal mucosal irritation upon administration. Drug content of the prepared CRV nanosponge was found to be $97.75 \pm 0.92\%$, indicating efficient drug incorporation. The porosity of the sponges is a critical factor in determining fluid absorption capability, and mechanical performance. In this study, CRV nanosponge showed improved porosity ($67.58\% \pm 1.64$), compared to non-vesicular CRV sponge ($47.49\% \pm 1.32$). The higher porosity of bilosomal nanosponge might be attributed to the incorporation of drug-loaded bilosome vesicles in the dispersion form, which would produce a highly porous structure upon lyophilization [37].

2.7.3. Swelling Ratio and Mucoadehison Time

The degree of swelling of bioadhesive polymers is an essential factor that significantly affects adhesion power. Generally, adhesion occurs shortly after the beginning of swelling, whereas uptake of water results in relaxation of the originally entangled or twisted polymer chains, resulting in exposure of polymer bioadhesive sites for bonding to occur. The higher the polymer's swelling, the faster the formation of adhesive bonds, and the better the adhesion is, which in turn would allow sufficient time for the drug entrapped in the bilosomes to permeate the buccal mucosa, making buccal delivery ideal. The swelling ratio of CRV nanosponge in PBS solutions as a function of time is given in Figure S2. It is obvious that CRV nanosponge showed an increasing trend in the swelling ratio with prolonging the immersing time. The swelling ratio of the prepared sponge enriched with bilosomes increased from 1.45 to 3.68 upon immersion in PBS for 1 and 4 h, respectively. This swelling ratio range is considered adequate for buccal delivery, based on previous findings of Hazzah et al. [26] who formulated curcumin nanoparticle-loaded sponge for buccal delivery with swelling ratios fluctuating from 1 to 3.5.

2.7.4. In Vitro Mucoadhesion Time

Mucoadhesion time is considered a key determinant of the efficacy of drugs intended for buccal administration. Prolonging the residence time of mucoadhesive drug delivery systems in the buccal cavity allows intensified contact with the epithelial barrier, resulting in enhanced drug absorption, and thereby improves drug bioavailability. In this study, no displacement or detachment of the nanosponge was observed over 3 h. This relatively long mucoadhesion time is considered suitable for contacting with buccal mucosa, helping to wet the dosage form with the mucosal substrate, and thus allowing efficient drug release from the formulation.

2.7.5. In Vitro Release of CRV from CRV Nanosponge

In vitro drug release pattern of CRV from either bilosomes or bilosomal sponge (nanosponge) are depicted in Figure 6. As shown in Figure 6, the in vitro release of CRV from BLS9 sponge after 24 h was much lower than that from bilosomes; the percent cumulative release of CRV from nanosponges and bilosomes were $69.77 \pm 4.9\%$ and $89.33\% \pm 2.5\%$, respectively. The slower drug release from nanosponge might be attributed to the fact that, upon sponge hydration, the polymer regains its gel structure, leading to an increase in diffusional path length for the drug, which consequently may delay the release [37].

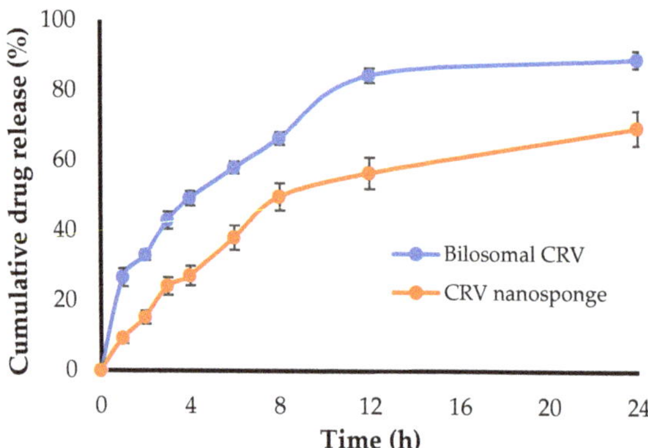

Figure 6. In vitro release profile of CRV from CRV-loaded bilosomes and CRV-loaded nanosponge. Data represents mean ± SD.

2.8. In Vivo Studies of CRV Nanosponge

2.8.1. Effect on Systolic and Diastolic Blood Pressure

In order to assess the pharmacological efficacy of the CRV nanosponge, the hypotensive potential of CRV was challenged against $CdCl_2$-induced hypertension in rats. Hypertension was induced by intraperitoneal injection $CdCl_2$ into rats for 14 days. Then, animals were treated with either CRV nanosponge or a commercially available marketed product (Carvid®, MULTI-Apex Pharma, Alexandria, Egypt) for two weeks. Finally, the systolic and diastolic blood pressures were recorded by the tail cuff method. As illustrated in Figure 7, intraperitoneal injection of $CDCl_2$ for 14 consecutive days resulted in a significant increase in both systolic and diastolic blood pressure, compared to the negative control group. Treatment with either CRV nanosponge or commercially available marketed product (Carvid®) significantly lowered both systolic and diastolic blood pressure, compared to $CdCl_2$-treated animals. Of interest, superior to Carvid®, CRV nanosponge was efficient in reducing both systolic and diastolic blood pressures to the normal control level. These results suggest the efficacy of CRV nanosponge in mitigating the blood pressure in hypertensive rats, to the near-normal control level.

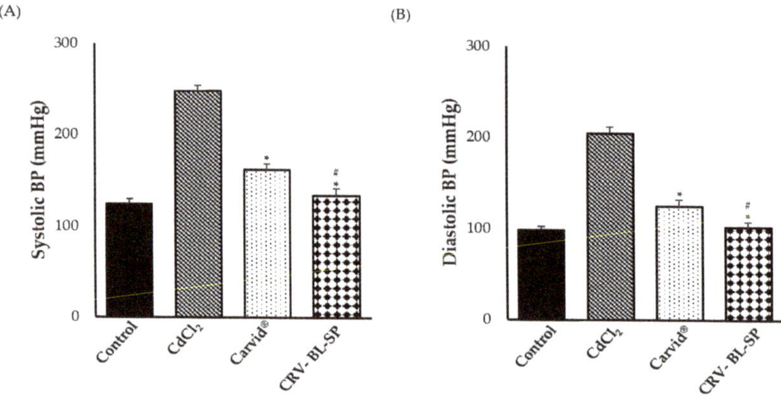

Figure 7. Effect of cadmium exposure and carvedilol bilosomes loaded sponge on (A) Systolic blood pressure and (B) Diastolic blood pressure. * $p < 0.05$ vs. $CdCl_2$-intoxicated rats, # $p < 0.05$ vs. Carvid®-treated group.

2.8.2. Effect on Cardiac Biomarkers

Cadmium (Cd) is a potent cardiotoxic heavy metal that is reported to induce oxidative stress and membrane disturbances in cardiac myocytes [39]. Serum lactate dehydrogenase (LDH) and creatine kinase (CK) are widely used markers of tissue damage. These enzymes are released into the bloodstream from the heart, and they could reflect the alterations in myocardial membrane permeability. Accordingly, in the present study, we followed the changes of such markers in $CdCl_2$-intoxicated rats and evaluated the protective effect of carvedilol. As shown in Figure 8, there was a marked elevation in serum LDH and CK activities for $CdCl_2$-intoxicated rats, as compared to the control animals, indicating possible myocardial damage. In contrast, pretreatment with either CRV nanosponge or commercially available marketed product (Carvid®) could induce a significant ($p < 0.05$) reduction in the activities of the tested enzymes, compared to the $CdCl_2$-intoxicated rats. Most importantly, CRV nanosponge could efficiently restore cardiac enzyme activities to their normal values as in the negative control group, which in turn, was beneficial in maintaining normal cardiac function. The plausible mechanism underlying the efficacy of CRV nanosponge on $CdCl_2$-induced hypertensive rats might be linked to the powerful antioxidant activity of CRV, which restricts the leak of CK and LDH enzymes from myocardium tissue.

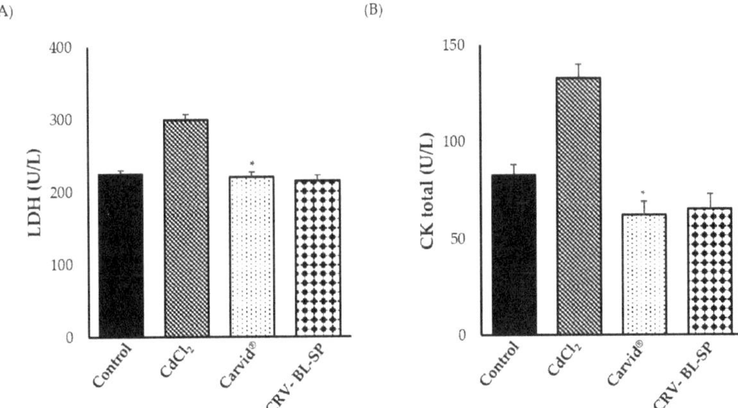

Figure 8. Effect of cadmium exposure and carvedilol bilosomes loaded sponge on cardiac biomarkers; (**A**) Serum lactate dehydrogenase (LDH); and (**B**) Serum creatine kinase (CK). * $p < 0.05$ vs. $CdCl_2$-intoxicated rats.

2.8.3. Effect of CRV Nanosponge on Oxidative Stress Markers

Oxidative stress is a pivotal factor and key promoter of a number of cardiovascular diseases [40]. Accordingly, in this study we evaluated the efficacy of CRV nanosponge to alleviate oxidative stress in $CdCl_2$-intoxicated rats. As shown in Figure 9, $CdCl_2$ significantly elevated the levels of oxidative stress markers in the heart such as reactive oxygen species (ROS) and lipid peroxidation (LPO), and decreased nitric oxide (NO) and total antioxidant capacity (TAC) levels, compared to the control group, suggesting vascular endothelial cells dysfunction [41]. On the other hand, CRV nanosponge and commercially available marketed product (Carvid®) significantly ($p < 0.05$) decreased the heart contents of ROS and LPO while increasing the content of NO and TCA, compared to the $CdCl_2$-treated animals (Figure 9). Cadmium has been reported to induce endothelial dysfunction via reducing NO• bioavailability [42]. Exposure to Cd causes an increase in O_2^-• production in the thoracic aorta, which could effectively scavenge NO to form strong oxidant peroxynitrite ($ONOO^-$). Besides its action as a β-blocker, carvedilol has been demonstrated to have a potent antioxidant capacity. CRV exerts both ROS-scavenging and ROS-suppressive properties. In addition, it by preserves endothelium-derived nitric oxide activity. Lopez et al. [43] have emphasized the potential of carvedilol to protect against free-radical-induced endothelial dysfunction via by scavenging free radicals and enhancing the effect of NO. Accordingly, in this study, the efficient antioxidant activity of bliosomal CRV nanosponge might be ascribed to the reduced peroxynitrite formation in the endothelial cells and increase NO's bioavailability through scavenging superoxide ions and inhibiting the production of superoxide radicals [44].

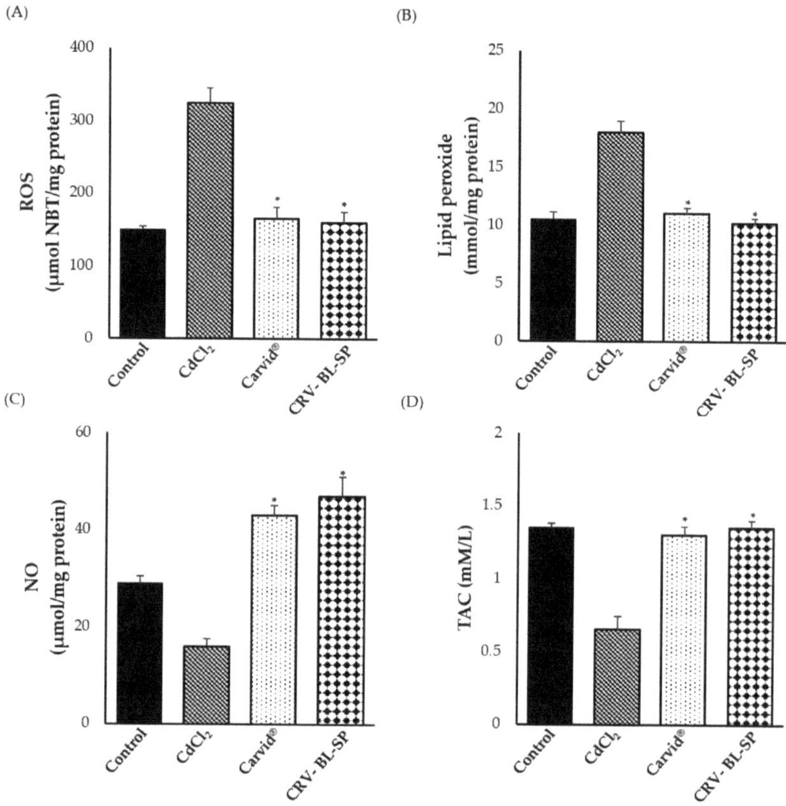

Figure 9. Effect of cadmium exposure and carvedilol bilosomes loaded sponge on oxidative stress markers levels; (**A**) reactive oxygen species (ROS); (**B**) lipid peroxidation (LPO); (**C**) nitric oxide (NO) and (**D**) total antioxidant capacity (TAC). * $p < 0.05$ vs. CdCl$_2$-intoxicated rats.

2.8.4. Effect on the Serum Lipid Profile Parameters

Recent studies have revealed novel properties of carvedilol, which may function to protect the vasculature and heart chronic pathological conditions such as atherosclerosis. In the current study, therefore, the lipid profile (cholesterol, low density lipoprotein (LDL), high density lipoprotein (HDL) and triglycerides (TG)) was assessed in the serum of different treated groups. As shown in Figure 10, cholesterol, LDL, HDL and TG levels in serum showed significant elevations ($p < 0.05$) in the CdCl$_2$-treated rats, compared to normal ones. On the contrary, treatment with either CRV nanosponge or commercially available marketed product (Carvid®) significantly reduced Cholesterol, LDL, and TG levels, while increasing the levels of HDL, compared to the CDCl2-treated group. The efficacy of CRV nanosponge in lessening the levels of cholesterol, TG, and LDL while increasing the levels of HDL, might be ascribed to the ability of carvedilol to inhibit the oxidation of LDL, preventing the creation of oxidized-LDL, which is thought to promote the development of atherosclerotic plaque [45], and thus protects vascular endothelial from damage.

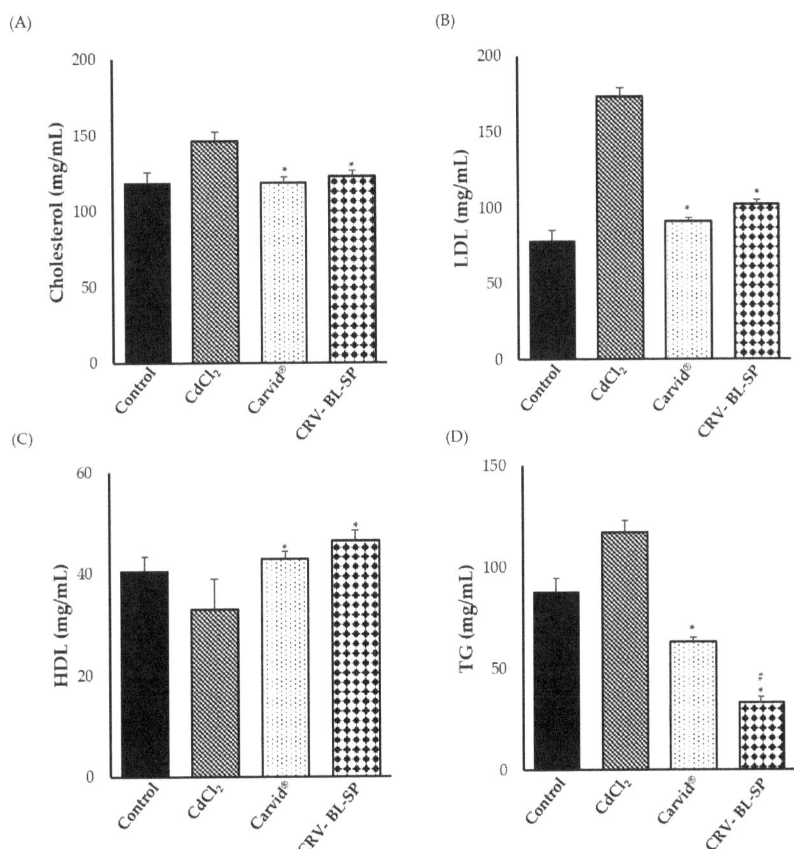

Figure 10. Effect of cadmium exposure and carvedilol bilosomes loaded sponge on lipid profile levels; (**A**) Cholesterol; (**B**) low density lipoproteins (LDL); (**C**) High density lipoproteins (HDL) and (**D**) triglycerides (TG). * $p < 0.05$ vs. CdCl$_2$-intoxicated rats, # $p < 0.05$ vs. Carvid®-treated group.

2.9. Histopathological Examination

To gain an insight into the cardio-protective effect of different CRV formulations, heart tissues were collected from animals at the end of the experiment, and histological analysis was conducted (Figure 11). Obvious gross morphological anomalies were observed in myocardial cells of CdCl$_2$-treated rats. Histological section of CdCl$_2$-treated rats showed localized fatty changes in some myocardial cells and the presence of vascular blood gaps lined by damaged endothelium. In addition, there was edema with inflammatory cells in the pericardium. On the other hand, a micrograph of heart tissue of Carvid®-treated rats showed cardiac blood capillaries exhibiting moderate dilation. Of interest, histological sections in the heart tissues of animals treated with CRV nanosponge showed no remarkable histopathological alterations in myocardial cells. Collectively, histopathological studies indicated a superior cardio-protective effect of CRV nanosponge, compared to commercially available drugs (Carvid®).

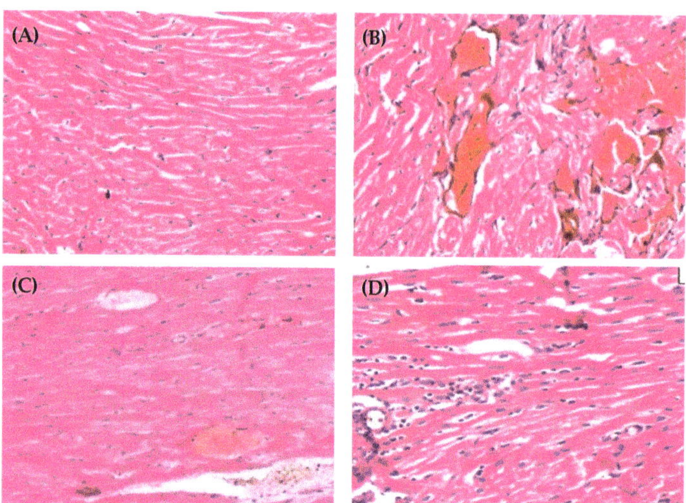

Figure 11. Histopathological sections of heart tissue of (**A**) control group, (**B**) CdCl2-intoxicated group, (**C**) Carvid®-treated group and (**D**) CRV nanosponge-treated group.

3. Conclusions

In this study, CRV-bilosomes were developed and optimized. The optimized CRV-loaded bilosomes significantly increased buccal permeability as manifested by deeper penetration of drug-loaded bilosomes through sheep buccal mucosa. In addition, the optimized bilosomes formulation was successfully incorporated into a CMC/HPC sponge. The designed CMC/HPC bilosomal nanosponge showed acceptable characteristics such as pH and swelling ratio and a good residence time that allowed enough contact time for penetration. In addition, in vivo studies revealed that a CRV nanosponge could efficiently enhance systolic and diastolic blood pressure, decrease elevated oxidative stress, improve lipid profile and exhibit a potent cardio-protective effect. To sum up, bilosomal nanosponge might represent a promising carrier for buccal delivery of carvedilol for the management of hypertension with a superior cardio-protective effect, compared to the commercially available product (Carvid®).

4. Materials and Methods

4.1. Materials

Carvedilol (CRV) was provided by SAGA Pharmaceutical Company (Cairo, Egypt). Soybean phosphatidylcholine (SPC), cholesterol (CH), sodium deoxycholate (SDC), dialysis membrane (12,000–14,000 molecular weight cut-off), Griess reagent, thiobarbituric acid, carboxymethyl cellulose sodium salt (Na CMC; 50–200 cps) were purchased from Sigma Aldrich (St. Louis, MO, USA). Hydroxypropyl cellulose (HPC; 100,000 cps) was obtained from Fluka (Buchs, Switzerland). All other chemicals and reagents used were of analytical grade.

4.2. Preparation of Bilosomes (BLS)

CRV-loaded bilosomes were prepared by the thin-film hydration/sonication technique [46]. Briefly, accurately weighted amounts of Soybean phosphatidylcholine (SPC), cholesterol (Chol), diacetyl phosphate (DCP), and carvedilol (CRV) were completely dissolved in 5 mL dichloromethane. The resultant lipid solution was then evaporated at 60 °C under reduced pressure, using a rotary evaporator, to produce a thin lipid film. The lipid film was then hydrated using 25 mL distilled water containing 25 mg sodium deoxycholate (SDC). The formed CRV-loaded bilosomes were size-reduced by sonication

(Crest UltraSonicator 575DAE, Trenton, NJ, USA) for 3 cycles of 5 min with an interval of 5 min. The obtained bilosome dispersions were stored in a refrigerator at 4 °C till further investigations. The composition of the prepared bilosomes is given in Table 1.

4.3. Characterization of CRV-Loaded Bilosomes

4.3.1. Determination of Particle Size

A laser scattering particle size analyzer (Malvern Instrument Ltd., Worcestershire, UK) was used to determine particle size, polydispersity index, and zeta potential of the bilosomes. Before measurement, 100 µL of CRV-loaded bilosomes were suitably diluted (100 times) with distilled water. All measurements were conducted at 25 °C in triplicates.

4.3.2. Transmission Electron Microscope (TEM)

The morphology of CRV-loaded bilosomes was detected by transmission electron microscope (Joel JEM 1230, Tokyo, Japan). Briefly, CRV-loaded bilosomes were diluted with distilled water and stained with phosphotungstic acid. The samples were then applied onto a carbon-coated copper gride and allowed to dry before being visualized by TEM at an accelerating voltage of 100 kV [47].

4.3.3. Entrapment Efficiency (EE %) and Drug Loading (DL)

CRV entrapped in bilosomes was separated from the un-entrapped drug by centrifugation of bilosomal dispersion at 15,000 rpm for 0.5 h at 4 °C using a cooling centrifuge (Beckman, Fullerton, Canada). The collected supernatants were then analyzed spectrophotometrically for CRV at λmax of 242 nm (Shimadzu, UV-1601 PC model, Kyoto, Japan) [48]. EE (%) and DL were calculated using the following equations [29].

$$EE\ (\%) = \frac{(Di - Dn)}{Di} \times 100$$

$$DL = \frac{(Di - Dn)}{L} \times 100$$

where, Di is initial drug amount, Dn is the unentrapped drug, and L is total lipid concentration.

4.3.4. Differential Scanning Calorimetry (DSC)

DSC thermograms of pure CRV, SPC, cholesterol, and the optimized lyophilized CRV-loaded bliosome formulation were recorded using Shimadzu differential scanning calorimeter (DSC-50, Kyoto, Japan). Samples (2 mg) were subjected to heat in the range of 10 to 300 °C at a constant heating rate of 10 °C/min in a standard aluminum pot under an inert nitrogen flow of 25 mL/min.

4.4. In Vitro Drug Release Study

The in vitro release study was conducted via a dialysis bag method using cellophane membrane (MW cut-off 12–14 KDa), which permits the diffusion of free drug but not lipid vesicles [49]. Briefly, adequate volume of CRV-loaded bilosomes, containing CRV equivalent to 6.25 mg, were placed in a cellophane bag suspended into 250 mL phosphate-buffered saline (PBS; pH 6.8) containing 0.5% sodium lauryl sulfate, and was maintained at 37 ± 1 °C and stirred at 50 rpm for 24 h. 1 mL samples were withdrawn at 1, 2, 3, 4, 5, 6, 8, 12 and 24 h, and was replenished with an equal volume of fresh buffer to maintain the sink conditions. The release of CRV from bilosomes was compared with CRV suspension. Drug concentrations in the samples were determined spectrophotometrically at 242 nm and cumulative drug release was estimated using the following equation:

$$Cumulative\ drug\ release\ (\%) = \frac{Mt}{Mi} \times 100$$

where Mi is the initial amount of the CRV in bilosomes and Mt is the amount of drug released at time t.

To gain an insight into the mechanism of drug release from bilosomes, the results obtained from the release studies were kinetically analyzed for the order of drug release. The formulae were fitted to zero-order, first-order, and the Higuchi diffusion models, and various correlation coefficients (R2) were determined.

4.5. Ex-Vivo Permeation Study

In order to assess the ability of bilosomes to enhance drug permeation in vivo, the ex-vivo permeation of CRV-loaded bilosomes was investigated using sheep buccal mucosa. Briefly, sheep buccal mucosa was fixed at one end of a glass cylinder, while the other end of the cylinder was connected to the shafts of dissolution apparatus I (Figure S3). The permeation medium consisted of 250 mL phosphate buffer saline (PBS; pH 6.8) containing 0.5% sodium lauryl sulfate, and was maintained at 37 ± 1 °C and stirred at 50 rpm for 24 h. Adequate volume of CRV-loaded bilosomes, equivalent to 6.25 mg drug, was placed in the glass cylinders. At definite time points (0.5, 1, 2, 3, 4, 6, 8, 12, and 24 h), 1 mL aliquot samples were withdrawn from the permeation medium, followed by compensation with equal volumes of fresh medium. A similar experiment was conducted for CRV suspension, containing 6.25 mg drug. The amount of drug permeated was analyzed using a UV spectrophotometer (Shimadzu, Tokyo, Japan) at 242 nm. The cumulative amounts of drug permeated per unit area ($\mu g/cm^2$) was graphically displayed as a function time, and the permeation parameters were determined. Drug flux at 24 h (Jmax) and enhancement ratio (ER) were calculated according to the following equations [31]:

$$\text{Jmax} = \frac{\text{Amount of drug permeated per unit area}}{\text{Time}}$$

$$\text{ER} = \frac{\text{Jmax}_{\text{Bilosomal formulation}}}{\text{Jmax}_{\text{CRV suspension}}}$$

Besides, the permeability coefficient C_p (cm/h) of CRV from bilosomes was calculated by dividing the slope of the curve by initial drug concentration.

4.6. Confocal Laser Scanning Microscopy (CLSM) Study

Confocal laser scanning microscopy (CLSM) was adopted to study the depth of penetration of the optimized bilosomes into the sheep buccal mucosa. Briefly, rhodamine B solution and rhodamine B-loaded bilosome formulation were applied to sheep buccal mucosa, and was then removed and washed with distilled water. The involved area was frozen at −20 °C and was then sectioned with a cryostat into 20 μm slices and put on slides and covered by glass coverslips. The slides were then microscopically inspected using an inverted laser scanning confocal microscope LSM 710 (CRVl Zeiss, Germany) equipped with an He/Ne laser operating at 524 nm for fluorescence excitation [33].

4.7. Stability Studies for Carvedilol-Loaded Bilosomes

The optimized CRV-loaded bilosomes preparation was tested for stability by keeping them in glass vials at 4 ± 1 °C for three months. Physical stability was evaluated by comparing the results of particle size and entrapment efficiency before and after storage time [46].

4.8. Preparation of Carvedilol Nanosponge

The optimized carvedilol-loaded bilosomal dispersion was added to carboxymethyl cellulose/hydroxypropyl cellulose (CMC/HPC) aqueous blend (1:1 ratio) and then stirred for 4 h to obtain a homogenous mixture. The resultant mixture was subsequently poured into a mold (1 cm in diameter), and lyophilized for 24 h.

4.9. Characterization of Carvedilol Nanosponge

4.9.1. Sponge Morphology

The external surface of CRV-nanopsonge was observed by scanning electron microscopy (SEM, Quanta 250 FEG, FEI Co., Tokyo, Japan). The sponge was placed on double-sided adhesive carbon tape. The sponges were then sputter-coated with gold and set in a chamber of SEM and examined under a low vacuum at 200× magnification [26].

4.9.2. Sponge pH

The surface pH of CRV-nanosponge was determined by a pH-meter (Jenway 3510, Barloworld Scientific, Stone, Staffordshire, UK). Briefly, one unit sponge (1 cm diameter) of CRV-nanosponge was allowed to swell in contact with 2 mL of simulated saliva fluid (pH 6.8) for 2 h at room temperature. Then the surface pH of the nanosponge was determined by bringing the electrode of the pH-meter into contact with the sponge surface and left to equilibrate for 1 min [50].

4.9.3. Ex-Vivo Mucoadhesion Time

Sheep buccal mucosa was attached to the inner side of a beaker containing 100 mL simulated saliva fluid (SSF; pH 6.8), as depicted in Figure S4. One unit sponge (1 cm diameter) was moistened with 50 µL of SSF and then fixed to sheep buccal mucosa after applying a slight force. The simulated saliva fluid was stirred at a rate of 50 rpm and mucoadhesive time was recorded when a complete detachment of nanosponge occurred [50].

4.9.4. Porosity Determination

Samples of known volume (V) and weight (Wi) were immersed in a graduated cylinder containing ethanol at 25 °C and soaked for 24 h [37]. The final wet sponge weight was recorded as Wf. Porosity (%) was calculated as follows:

$$\text{Porosity (\%)} = ((\text{Wf} - \text{Wi})/(\rho\ \text{ethanol}) \div \text{V}) \times 100$$

where ρ ethanol: Density of ethanol

4.9.5. Swelling Ratio

The sponge was initially weighed, then immersed in phosphate-buffered saline (PBS; pH 6.8). The sponge was them incubated at 37 °C for 1, 2, 3, and 4 h and reweighed after each time of incubation. The swelling ratio was estimated as follows [26]:

$$\text{Swelling ratio} = (\text{Ww} - \text{Wi})/\text{Wi}$$

Wi is the sponge's initial weight, and Ww is the sponge's weight after hydration and removing the excess buffer.

4.9.6. In Vitro Release of CRV Nanosponge

CRV-loaded nanosponge and bilosomal CRV, containing CRV equivalent to 6.25 mg, were placed in cellophane bag (cut-off 12–14 KDa) suspended into 250 mL phosphate buffer saline (PBS; pH 6.8) containing 0.5% sodium lauryl sulfate, and maintained at 37 ± 1 °C and stirred at 50 rpm. The cumulative % of CRV released was conducted for up to 24 h as previously discussed in Section 2.4.

4.10. In Vivo Studies

4.10.1. Experimental Animals

Male Albino rats (180–200 g) were used for in vivo experiments. The study protocol was approved by the research ethics committee, Suez Canal University, Egypt (201804RA2). Rats were freely allowed to have food and water at a 12 h light/12 h dark cycle at room temperature (25 ± 1 °C).

4.10.2. Experimental Protocol

Twenty-four male Albino rats (180–200 g) were categorized into 4 groups. The first group received 0.5% CMC orally and served as a normotensive control group. To induce hypertension in the remaining three groups, animals were intraperitoneally injected with cadmium chloride dissolved in 0.9% saline (1 mg/kg/day) for 14 days [51]. After the induction of hypertension, the animals were treated with either 0.5% CMC orally for 14 days (Group II), buccal administration of CRV-nanosponge (6.25 mg CRV/kg) for 14 days (Group III), or orally treated with a commercially available CRV product (6.25 mg CRV/kg) suspended in 0.5% CMC vehicle daily for another 14 days (Group IV). At the end of the experiment, rats were euthanized, and blood samples were collected without an anticoagulant. The separated serum was used for the assessment of lactate dehydrogenase [52] and creatine phosphokinase [53] activities using commercial enzymatic kits (Reactivos GPL, Barcelona, Spain). Cholesterol, LDL, HDL, and TG were also assessed. Total antioxidant capacity (TAC) was evaluated using a commercial kit (Biodiagnostic, Cairo, Egypt). All measurements were performed according to the manufacturer instructions.

4.10.3. Heart Homogenate Isolation

The heart of each animal was excised instantly, washed with saline, and stored at $-80\ ^\circ$C. A definite weight of the heart was then homogenized in chilled 50 mM phosphate-buffered saline (PBS; pH 7.4), centrifuged for 15 min at 1200 and 4 $^\circ$C using a cooling centrifuge (Sigma 30K, Osterode, Germany). The supernatants were used for the detection of the oxidative stress parameters.

4.10.4. Lipid Peroxidation Determination (LPO)

Lipid peroxidation (LPO) measurement of the heart was done by a colorimetric reaction with thiobarbituric acid as previousely described [54].

4.10.5. Determination of Nitric Oxide (NO)

Nitric oxide was estimated as nitrite concentration. The technique used depends on the Griess reaction, which converts nitrite to a deep purple azo compound measured photometrically at 540 nm [55].

4.10.6. Reactive Oxygen Species Determination (ROS)

This test measures the intracellular conversion of nitro blue tetrazolium (NBT) to formazan by superoxide anion, which was used to measure the production of reactive oxygen species (ROS) as previously described [56].

4.10.7. Histopathological Examination

Autopsy heart samples were taken from different treated animals and fixed for 24 h in 10% formal saline. The samples were then washed with distilled water, followed by treatment with serial alcohol dilutions (methyl, ethyl, and absolute ethyl) for dehydration. Specimens were cleared in xylene and submerged for 24 h in hot air ovens at 56 $^\circ$C. The tissues were examined by the electric light microscope.

4.11. Statistical Analysis

All experimental data were expressed as mean values $\pm$ S.D. Statistical analysis was performed using GraphPad Prism 8 (GraphPad Software, San Diego, CA, USA) adopting one-way analysis of variance (ANOVA). Post hoc testing was done for inter-group comparisons using Tukey's multiple comparisons test. Differences were considered statistically significant at $p \leq 0.05$.

Supplementary Materials: The following supporting information can be downloaded at: https://www.mdpi.com/article/10.3390/gels8040235/s1, Figure S1: Transmission electron image of CRV nanosponge; Figure S2: Swelling of CRV nanosponge; Figure S3: Cartoon depicting the apparatus utilized in ex-vivo permeation study of the optimized CRV-loaded bilosomes from Sheep buccal mucosa; Figure S4: Cartoon depicting the apparatus utilized in measuring the ex-vivo mucoadhesion time of the formulated BLS nanosponge; Table S1: Kinetic analysis of release data of carvedilol from different bilosomal formulations and aqueous CRV suspension.

Author Contributions: E.-S.K., A.S.A.L., N.M.S., R.A.-B.S. and S.G.; conceptualization; E.-S.K., N.M.S., R.A.-B.S., M.M.G. and S.G.; methodology, E.-S.K., A.S.A.L., N.M.S., R.A.-B.S., M.M.A., M.M.G. and S.G.; data curation, E.-S.K., A.S.A.L., N.M.S., R.A.-B.S., M.M.G., M.M.A. and S.G.; formal analysis, E.-S.K., A.S.A.L., N.M.S., R.A.-B.S. and S.G.; investigation, E.-S.K., A.S.A.L., N.M.S., R.A.-B.S., M.M.A., M.M.G. and S.G.; validation, E.-S.K., A.S.A.L., N.M.S., R.A.-B.S., M.M.A., M.M.G., A.A. (Ahmed Alalaiwe), M.F.A., S.M.A., A.A. (Abdullah Alshetaili), A.A.S. and S.G.; visualization, E.-S.K., N.M.S., R.A.-B.S., M.M.G., H.F.A., A.A. (Ahmed Alalaiwe), M.F.A., S.M.A., A.A. (Abdullah Alshetaili), B.K.A., A.A.S. and S.G.; project administration, E.-S.K., R.A.-B.S., M.M.A., M.M.G., H.F.A., A.A. (Ahmed Alalaiwe), M.F.A., S.M.A., A.A. (Ahmed Alalaiwe), B.K.A., A.A.S. and S.G.; supervision, E.-S.K., A.A.S., N.M.S., R.A.-B.S., M.M.A., M.M.G., H.F.A., A.A. (Ahmed Alalaiwe), M.F.A., S.M.A., A.A. (Abdullah Alshetaili), B.K.A., A.A.S. and S.G.; resources, E.-S.K., A.A.S., N.M.S., R.A.-B.S., M.M.A., M.M.G. and S.G.; writing—original draft preparation, E.-S.K., A.S.A.L., N.M.S., R.A.-B.S., M.M.A., M.M.G., H.F.A., A.A. (Ahmed Alalaiwe), M.F.A., S.M.A., A.A. (Abdullah Alshetaili), B.K.A., A.A.S. and S.G.; writing—review and editing. All authors have read and agreed to the published version of the manuscript.

Funding: This work was supported by Princess Nourah bint Abdulrahman University researchers supporting project number (PNURSP2022R205), Princess Nourah bint Abdulrahman University, Riyadh, Saudi Arabia.

Institutional Review Board Statement: The animal study protocol was approved by the Research Ethics Committee, Suez Canal Univer-sity, Egypt (201804RA2).

Informed Consent Statement: Not applicable.

Data Availability Statement: Not applicable.

Acknowledgments: The authors extend their appreciation to Princess Nourah bint Abdulrahman University, Riyadh, Saudi Arabia for supporting this work under researcher supporting project number (PNURSP2022R205).

Conflicts of Interest: The authors declare no conflict of interest.

References

1. Agrawal, U.; Sharma, R.; Gupta, M.; Vyas, S.P. Is nanotechnology a boon for oral drug delivery? *Drug Discov. Today* **2014**, *19*, 1530–1546. [CrossRef] [PubMed]
2. Stafylas, P.C.; Sarafidis, P.A. Carvedilol in hypertension treatment. *Vasc. Health Risk Manag.* **2008**, *4*, 23–30. [CrossRef] [PubMed]
3. Halder, S.; Ahmed, F.; Shuma, M.L.; Azad, M.A.K.; Kabir, E.R. Impact of drying on dissolution behavior of carvedilol-loaded sustained release solid dispersion: Development and characterization. *Heliyon* **2020**, *6*, e05026. [CrossRef] [PubMed]
4. Neugebauer, G.; Akpan, W.; Kaufmann, B.; Reiff, K. Stereoselective disposition of carvedilol in man after intravenous and oral administration of the racemic compound. *Eur. J. Clin. Pharmacol.* **1990**, *38*, 108–111. [CrossRef]
5. Rasenack, N.; Müller, B.W. Dissolution rate enhancement by in situ micronization of poorly water-soluble drugs. *Pharm. Res.* **2002**, *19*, 1894–1900. [CrossRef]
6. Yuvaraja, K.; Khanam, J. Enhancement of carvedilol solubility by solid dispersion technique using cyclodextrins, water soluble polymers and hydroxyl acid. *J. Pharm. Biomed. Anal.* **2014**, *96*, 10–20. [CrossRef]
7. Potluri, R.H.; Bandari, S.; Jukanti, R.; Veerareddy, P.R. Solubility enhancement and physicochemical characterization of carvedilol solid dispersion with Gelucire 50/13. *Arch. Pharm. Res.* **2011**, *34*, 51–57. [CrossRef]
8. Sid, D.; Baitiche, M.; Elbahri, Z.; Djerboua, F.; Boutahala, M.; Bouaziz, Z.; Le Borgne, M. Solubility enhancement of mefenamic acid by inclusion complex with β-cyclodextrin: In silico modelling, formulation, characterisation, and in vitro studies. *J. Enzym. Inhib. Med. Chem.* **2021**, *36*, 605–617. [CrossRef]
9. Venishetty, V.K.; Chede, R.; Komuravelli, R.; Adepu, L.; Sistla, R.; Diwan, P.V. Design and evaluation of polymer coated carvedilol loaded solid lipid nanoparticles to improve the oral bioavailability: A novel strategy to avoid intraduodenal administration. *Colloids Surf. B Biointerfaces* **2012**, *95*, 1–9. [CrossRef]
10. Sharma, M.; Sharma, R.; Jain, D.K.; Saraf, A. Enhancement of oral bioavailability of poorly water soluble carvedilol by chitosan nanoparticles: Optimization and pharmacokinetic study. *Int. J. Biol. Macromol.* **2019**, *135*, 246–260. [CrossRef]
11. Pham, M.N.; Van Vo, T.; Tran, V.T.; Tran, P.H.; Tran, T.T. Microemulsion-Based Mucoadhesive Buccal Wafers: Wafer Formation, In Vitro Release, and Ex Vivo Evaluation. *AAPS PharmSciTech* **2017**, *18*, 2727–2736. [CrossRef] [PubMed]

12. Mohsin, K.; Long, M.A.; Pouton, C.W. Design of lipid-based formulations for oral administration of poorly water-soluble drugs: Precipitation of drug after dispersion of formulations in aqueous solution. *J. Pharm. Sci* **2009**, *98*, 3582–3595. [CrossRef] [PubMed]

13. Kalepu, S.; Manthina, M.; Padavala, V. Oral lipid-based drug delivery systems—An overview. *Acta Pharmaceutica Sinica B* **2013**, *3*, 361–372. [CrossRef]

14. Abd El Azim, H.; Nafee, N.; Ramadan, A.; Khalafallah, N. Liposomal buccal mucoadhesive film for improved delivery and permeation of water-soluble vitamins. *Int. J. Pharm.* **2015**, *488*, 78–85. [CrossRef] [PubMed]

15. Abdelbary, A.A.; Abd-Elsalam, W.H.; Al-Mahallawi, A.M. Fabrication of novel ultradeformable bilosomes for enhanced ocular delivery of terconazole: In vitro characterization, ex vivo permeation and in vivo safety assessment. *Int. J. Pharm.* **2016**, *513*, 688–696. [CrossRef] [PubMed]

16. Shukla, A.; Mishra, V.; Kesharwani, P. Bilosomes in the context of oral immunization: Development, challenges and opportunities. *Drug Discov. Today* **2016**, *21*, 888–899. [CrossRef]

17. Aburahma, M.H. Bile salts-containing vesicles: Promising pharmaceutical carriers for oral delivery of poorly water-soluble drugs and peptide/protein-based therapeutics or vaccines. *Drug Deliv.* **2016**, *23*, 1847–1867. [CrossRef]

18. Jain, S.; Harde, H.; Indulkar, A.; Agrawal, A.K. Improved stability and immunological potential of tetanus toxoid containing surface engineered bilosomes following oral administration. *Nanomedicine* **2014**, *10*, 431–440. [CrossRef]

19. Wilkhu, J.S.; McNeil, S.E.; Anderson, D.E.; Perrie, Y. Characterization and optimization of bilosomes for oral vaccine delivery. *J. Drug Target.* **2013**, *21*, 291–299. [CrossRef]

20. Abd-Elbary, A.; Makky, A.M.; Tadros, M.I.; Alaa-Eldin, A.A. Laminated sponges as challenging solid hydrophilic matrices for the buccal delivery of carvedilol microemulsion systems: Development and proof of concept via mucoadhesion and pharmacokinetic assessments in healthy human volunteers. *Eur. J. Pharm. Sci.* **2016**, *82*, 31–44. [CrossRef]

21. Patel, V.F.; Liu, F.; Brown, M.B. Advances in oral transmucosal drug delivery. *J. Control. Release* **2011**, *153*, 106–116. [CrossRef] [PubMed]

22. Shergill, M.; Patel, M.; Khan, S.; Bashir, A.; McConville, C. Development and characterisation of sustained release solid dispersion oral tablets containing the poorly water soluble drug disulfiram. *Int. J. Pharm.* **2016**, *497*, 3–11. [CrossRef] [PubMed]

23. Chou, W.H.; Galaz, A.; Jara, M.O.; Gamboa, A.; Morales, J.O. Drug-Loaded Lipid-Core Micelles in Mucoadhesive Films as a Novel Dosage Form for Buccal Administration of Poorly Water-Soluble and Biological Drugs. *Pharmaceutics* **2020**, *12*, 1168. [CrossRef] [PubMed]

24. Patel, D.; Prajapati, J.; Patel, U.; Patel, V. Formulation of thermoresponsive and buccal adhesive in situ gel for treatment of oral thrush containing poorly water soluble drug bifonazole. *J. Pharm. Bioallied Sci.* **2012**, *4*, S116–S117. [CrossRef] [PubMed]

25. Portero, A.; Teijeiro-Osorio, D.; Alonso, M.J.; Remuñán-López, C. Development of chitosan sponges for buccal administration of insulin. *Carbohydrate Polymers* **2007**, *68*, 617–625. [CrossRef]

26. Hazzah, H.A.; Farid, R.M.; Nasra, M.M.; El-Massik, M.A.; Abdallah, O.Y. Lyophilized sponges loaded with curcumin solid lipid nanoparticles for buccal delivery: Development and characterization. *Int. J. Pharm.* **2015**, *492*, 248–257. [CrossRef]

27. Manca, M.L.; Zaru, M.; Manconi, M.; Lai, F.; Valenti, D.; Sinico, C.; Fadda, A.M. Glycerosomes: A new tool for effective dermal and transdermal drug delivery. *Int. J. Pharm.* **2013**, *455*, 66–74. [CrossRef]

28. El Menshawe, S.F.; Aboud, H.M.; Elkomy, M.H.; Kharshoum, R.M.; Abdeltwab, A.M. A novel nanogel loaded with chitosan decorated bilosomes for transdermal delivery of terbutaline sulfate: Artificial neural network optimization, in vitro characterization and in vivo evaluation. *Drug Deliv. Transl. Res.* **2020**, *10*, 471–485. [CrossRef]

29. Rahman, S.A.; Abdelmalak, N.S.; Badawi, A.; Elbayoumy, T.; Sabry, N.; El Ramly, A. Tretinoin-loaded liposomal formulations: From lab to comparative clinical study in acne patients. *Drug Deliv* **2016**, *23*, 1184–1193. [CrossRef]

30. Al Saqr, A.; Khafagy, E.-S.; Alalaiwe, A.; Aldawsari, M.F.; Alshahrani, S.M.; Anwer, M.K.; Khan, S.; Lila, A.S.A.; Arab, H.H.; Hegazy, W.A.H. Synthesis of Gold Nanoparticles by Using Green Machinery: Characterization and In Vitro Toxicity. *Nanomaterials* **2021**, *11*, 808. [CrossRef]

31. Hasan, A.A.; Samir, R.M.; Abu-Zaid, S.S.; Abu Lila, A.S. Revitalizing the local anesthetic effect of Mebeverine hydrochloride via encapsulation within ethosomal vesicular system. *Colloids Surf. B Biointerfaces* **2020**, *194*, 111208. [CrossRef] [PubMed]

32. Ramalingam, N.; Natesan, G.; Dhandayuthapani, B.; Perumal, P.; Balasundaram, J.; Natesan, S. Design and characterization of ofloxacin niosomes. *Pak. J. Pharm. Sci.* **2013**, *26*, 1089–1096. [PubMed]

33. Aboud, H.M.; Ali, A.A.; El-Menshawe, S.F.; Elbary, A.A. Nanotransfersomes of carvedilol for intranasal delivery: Formulation, characterization and in vivo evaluation. *Drug Deliv.* **2016**, *23*, 2471–2481. [CrossRef] [PubMed]

34. Tayel, S.A.; El-Nabarawi, M.A.; Tadros, M.I.; Abd-Elsalam, W.H. Duodenum-triggered delivery of pravastatin sodium via enteric surface-coated nanovesicular spanlastic dispersions: Development, characterization and pharmacokinetic assessments. *Int. J. Pharm.* **2015**, *483*, 77–88. [CrossRef]

35. Zafar, A.; Imam, S.S.; Alruwaili, N.K.; Yasir, M.; Alsaidan, O.A.; Alshehri, S.; Ghoneim, M.M.; Khalid, M.; Alquraini, A.; Alharthi, S.S. Formulation and Evaluation of Topical Nano-Lipid-Based Delivery of Butenafine: In Vitro Characterization and Antifungal Activity. *Gels* **2022**, *8*, 133. [CrossRef]

36. Moghimipour, E.; Ameri, A.; Handali, S. Absorption-Enhancing Effects of Bile Salts. *Molecules* **2015**, *20*, 14451–14473. [CrossRef]

37. Sanad, R.A.; Abdel-Bar, H.M. Chitosan-hyaluronic acid composite sponge scaffold enriched with Andrographolide-loaded lipid nanoparticles for enhanced wound healing. *Carbohydr. Polym.* **2017**, *173*, 441–450. [CrossRef]

38. Anisha, B.S.; Sankar, D.; Mohandas, A.; Chennazhi, K.P.; Nair, S.V.; Jayakumar, R. Chitosan-hyaluronan/nano chondroitin sulfate ternary composite sponges for medical use. *Carbohydr. Polym.* **2013**, *92*, 1470–1476. [CrossRef]
39. Ansari, M.N.; Ganaie, M.A.; Rehman, N.U.; Alharthy, K.M.; Khan, T.H.; Imam, F.; Ansari, M.A.; Al-Harbi, N.O.; Jan, B.L.; Sheikh, I.A.; et al. Protective role of Roflumilast against cadmium-induced cardiotoxicity through inhibition of oxidative stress and NF-κB signaling in rats. *Saudi Pharm. J.* **2019**, *27*, 673–681. [CrossRef]
40. Xu, C.; Hu, Y.; Hou, L.; Ju, J.; Li, X.; Du, N.; Guan, X.; Liu, Z.; Zhang, T.; Qin, W.; et al. β-Blocker carvedilol protects cardiomyocytes against oxidative stress-induced apoptosis by up-regulating miR-133 expression. *J. Mol. Cell Cardiol.* **2014**, *75*, 111–121. [CrossRef]
41. Urbański, K.; Nowak, M.; Guzik, T.J. Oxidative stress and vascular function. *Postepy Biochem.* **2013**, *59*, 424–431. [PubMed]
42. Angeli, J.K.; Cruz Pereira, C.A.; de Oliveira Faria, T.; Stefanon, I.; Padilha, A.S.; Vassallo, D.V. Cadmium exposure induces vascular injury due to endothelial oxidative stress: The role of local angiotensin II and COX-2. *Free. Radic. Biol. Med.* **2013**, *65*, 838–848. [CrossRef]
43. Lopez, B.L.; Christopher, T.A.; Yue, T.L.; Ruffolo, R.; Feuerstein, G.Z.; Ma, X.L. Carvedilol, a new beta-adrenoreceptor blocker antihypertensive drug, protects against free-radical-induced endothelial dysfunction. *Pharmacology* **1995**, *51*, 165–173. [CrossRef]
44. Yue, T.-L.; Ruffolo, R.R.; Feuerstein, G. Antioxidant Action of Carvedilol: A Potential Role in Treatment of Heart Failure. *Heart Fail. Rev.* **1999**, *4*, 39–52. [CrossRef]
45. Feuerstein, G.Z.; Ruffolo, R.R., Jr. Carvedilol, a novel vasodilating beta-blocker with the potential for cardiovascular organ protection. *Eur. Heart J.* **1996**, *17* (Suppl. B), 24–29. [CrossRef] [PubMed]
46. El-Nabarawi, M.A.; Shamma, R.N.; Farouk, F.; Nasralla, S.M. Bilosomes as a novel carrier for the cutaneous delivery for dapsone as a potential treatment of acne: Preparation, characterization and in vivo skin deposition assay. *J. Liposome Res.* **2020**, *30*, 1–11. [CrossRef]
47. Saifi, Z.; Rizwanullah, M.; Mir, S.R.; Amin, S. Bilosomes nanocarriers for improved oral bioavailability of acyclovir: A complete characterization through in vitro, ex-vivo and in vivo assessment. *J. Drug Deliv. Sci. Technol.* **2020**, *57*, 101634. [CrossRef]
48. Aboelwafa, A.A.; El-Setouhy, D.A.; Elmeshad, A.N. Comparative study on the effects of some polyoxyethylene alkyl ether and sorbitan fatty acid ester surfactants on the performance of transdermal carvedilol proniosomal gel using experimental design. *AAPS PharmSciTech* **2010**, *11*, 1591–1602. [CrossRef]
49. Moghimipour, E.; Tafaghodi, M.; Balouchi, A.; Handali, S. Formulation and in vitro Evaluation of Topical Liposomal Gel of Triamcinolone Acetonide. *Res. J. Pharm. Biol. Chem. Sci.* **2013**, *4*, 101–107.
50. Kassem, M.A.; ElMeshad, A.N.; Fares, A.R. Lyophilized sustained release mucoadhesive chitosan sponges for buccal buspirone hydrochloride delivery: Formulation and in vitro evaluation. *AAPS PharmSciTech* **2015**, *16*, 537–547. [CrossRef]
51. Akhtar, M.S.; Jabeen, Q.; Akram, M.; Khan, H.U.; Karim, S.; Malik, M.N.; Muhammad, N.; Salma, U. Antihypertensive Activity of Aqueous-Methanol Extract of Berberis Orthobotrys Bien Ex Aitch in Rats. *Trop. J. Pharm. Res.* **2013**, *12*, 393–399.
52. Buhl, S.N.; Jackson, K.Y. Optimal conditions and comparison of lactate dehydrogenase catalysis of the lactate-to-pyruvate and pyruvate-to-lactate reactions in human serum at 25, 30, and 37 degrees C. *Clin. Chem.* **1978**, *24*, 828–831. [CrossRef] [PubMed]
53. Szasz, G.; Gruber, W.; Bernt, E. Creatine kinase in serum: 1. Determination of optimum reaction conditions. *Clin. Chem.* **1976**, *22*, 650–656. [CrossRef] [PubMed]
54. Mihara, M.; Uchiyama, M. Determination of malonaldehyde precursor in tissues by thiobarbituric acid test. *Anal. Biochem.* **1978**, *86*, 271–278. [CrossRef]
55. Wang, C.C.; Huang, Y.J.; Chen, L.G.; Lee, L.T.; Yang, L.L. Inducible nitric oxide synthase inhibitors of Chinese herbs III. Rheum palmatum. *Planta Med.* **2002**, *68*, 869–874. [CrossRef] [PubMed]
56. Vrablic, A.S.; Albright, C.D.; Craciunescu, C.N.; Salganik, R.I.; Zeisel, S.H. Altered mitochondrial function and overgeneration of reactive oxygen species precede the induction of apoptosis by 1-O-octadecyl-2-methyl-rac-glycero-3-phosphocholine in p53-defective hepatocytes. *FASEB J.* **2001**, *15*, 1739–1744. [CrossRef]

Article

Tacrolimus-Loaded Solid Lipid Nanoparticle Gel: Formulation Development and In Vitro Assessment for Topical Applications

Abdul Shakur Khan [1], Kifayat Ullah Shah [1,*], Mohammed Al Mohaini [2,3], Abdulkhaliq J. Alsalman [4], Maitham A. Al Hawaj [5], Yousef N. Alhashem [6], Shakira Ghazanfar [7], Kamran Ahmad Khan [1], Zahid Rasul Niazi [1] and Arshad Farid [8,*]

[1] Faculty of Pharmacy, Gomal University, Dera Ismail Khan 29050, Pakistan; abdulshakurkhan01@gmail.com (A.S.K.); dr.kamrangu@gmail.com (K.A.K.); zahidscholar1@gmail.com (Z.R.N.)

[2] Basic Sciences Department, College of Applied Medical Sciences, King Saud bin Abdulaziz University for Health Sciences, Alahsa 31982, Saudi Arabia; mohainim@ksau-hs.edu.sa

[3] King Abdullah International Medical Research Center, Alahsa 31982, Saudi Arabia

[4] Department of Clinical Pharmacy, Faculty of Pharmacy, Northern Border University, Rafha 91911, Saudi Arabia; kaliqs@gmail.com

[5] Department of Pharmacy Practice, College of Clinical Pharmacy, King Faisal University, Ahsa 31982, Saudi Arabia; hawaj@kfu.edu.sa

[6] Clinical Laboratory Sciences Department, Mohammed Al-Mana College for Medical Sciences, Dammam 34222, Saudi Arabia; yousefa@machs.edu.sa

[7] National Institute for Genomics Advanced Biotechnology, National Agricultural Research Centre, Park Road, Islamabad 45500, Pakistan; shakira_akmal@yahoo.com

[8] Gomal Center of Biochemistry and Biotechnology, Gomal University, Dera Ismail Khan 29050, Pakistan

* Correspondence: kifayatrph@gmail.com (K.U.S.); arshadfarid@gu.edu.pk (A.F.)

Citation: Khan, A.S.; Shah, K.U.; Mohaini, M.A.; Alsalman, A.J.; Hawaj, M.A.A.; Alhashem, Y.N.; Ghazanfar, S.; Khan, K.A.; Niazi, Z.R.; Farid, A. Tacrolimus-Loaded Solid Lipid Nanoparticle Gel: Formulation Development and In Vitro Assessment for Topical Applications. *Gels* **2022**, *8*, 129. https://doi.org/10.3390/gels8020129

Academic Editors: Maddalena Sguizzato, Rita Cortesi and Rachel Yoon Chang

Received: 20 January 2022
Accepted: 14 February 2022
Published: 18 February 2022

Publisher's Note: MDPI stays neutral with regard to jurisdictional claims in published maps and institutional affiliations.

Abstract: The currently available topical formulations of tacrolimus have minimal and variable absorption, elevated mean disposition half-life, and skin irritation effects resulting in patient non-compliance. In our study, we fabricated tacrolimus-loaded solid lipid nanoparticles (SLNs) that were converted into a gel for improved topical applications. The SLNs were prepared using a solvent evaporation method and characterized for their physicochemical properties. The particle size of the SLNs was in the range of 439 nm to 669 nm with a PDI of $\leq$0.4, indicating a monodispersed system. The Zeta potential of uncoated SLNs (F1–F5) ranged from -25.80 to -15.40 mV. Those values reverted to positive values for chitosan-decorated formulation (F6). The drug content and entrapment efficiency ranged between 0.86 ± 0.03 and 0.91 ± 0.03 mg/mL and 68.95 ± 0.03 and $83.68 \pm 0.04\%$, respectively. The pH values of 5.45 to 5.53 depict their compatibility for skin application. The surface tension of the SLNs decreased with increasing surfactant concentration that could increase the adherence of the SLNs to the skin. The release of drug from gel formulations was significantly retarded in comparison to their corresponding SLN counterparts ($p \leq 0.05$). Both SLNs and their corresponding gel achieved the same level of drug permeation, but the retention of the drug was significantly improved with the conversion of SLNs into their corresponding gel formulation ($p \leq 0.05$) due to its higher bioadhesive properties.

Keywords: tacrolimus; chitosan; solid lipid nanoparticles; gel; topical drug delivery

1. Introduction

Human skin provides the most comfortably accessible route of drug administration. Topical administration is the route of choice for cutaneous pathologies like atopic dermatitis since it rarely presents systemic adverse effects when compared to other routes of drug administration [1]. The stratum corneum of the skin, composed of flat dead cells encompassing high keratin filaments surrounded by a lipophilic matrix consisting of keratin, ceramides, cholesterol, cholesterol esters, and various other fatty acids, provides a natural physical barrier against particle penetration [2,3]. Many transdermal methods have been tried to overcome the barrier function of the stratum corneum and achieve required transdermal permeability, but nanotechnology has developed an attractive niche in transdermal drug delivery. The physicochemical properties of nanoparticles such as size, shape, viscosity, and surface tension have a significant effect on dermal drug delivery [4], but the importance of nanoparticle composition cannot be underestimated [5].

The SLNs composed of physiological lipids that are solid at room temperature have a generally recognized as safe (GRAS) status for topical application. These carriers combine the advantages of emulsions, liposomes, and polymeric nanoparticles [6]. The SLNs can be used on inflamed skin due to the nonirritating and nontoxic properties of lipid content [7]. Due to their small particle size and lipoidal occlusive nature, these carriers amplify the concentration of lipophilic agents on the skin surface [8]. These carriers can encapsulate both lipophilic and hydrophilic drug molecules and provide controlled drug release and targeted drug delivery to specific cells, enhanced physical stability, good tolerability, and ease of scale-up [9]. The SLNs reduce water loss from the skin surface by forming a thin film that enhances the appearance of healthy human skin and reduces atopic eczema symptoms [10]. Chitosan, a natural polymer, is widely used in food, cosmetic, and drug delivery research due to its biocompatible and biodegradable nature and its antibacterial and wound-healing activities. When the SLNs are decorated with chitosan by physical adsorption, they can modify the physicochemical properties of the carriers by imparting a positive charge, which is derived from the protonated amino groups in chitosan, to their surface [11]. The use of chitosan in modification of nanocarriers can improve the stability of SLN dispersions and the degree of cellular interaction in biological models due to chitosan's muco- and bioadhesive properties [12]. The SLN dispersions could be easily incorporated into their corresponding gel to achieve a semisolid consistency for skin application [13].

Solid lipid nanoparticles (SLNs) have proven efficacy in maximizing dermal drug concentration due to their lipoidal nature, capacity for higher drug encapsulation of class 11 drugs, biocompatibility, and safety when delivered topically [14]. These carriers are considered superior to other nanoparticulate carriers; they are physiological, able to hydrate the skin, and can be easily decorated with positively charged ligands like chitosan for improving the retention of drugs in the inflamed area of the skin [15,16]. Due to the potential advantages of SLNs in topical drug delivery, we formulated an immune suppressant drug, tacrolimus, as SLNs for potential application in atopic dermatitis.

Atopic dermatitis is a common inflammatory skin disorder with limited treatment options, affecting 10% of adults and 20% of children worldwide [17]. Tacrolimus is an immunosuppressant drug that belongs to the macrolide family, possessing prominent therapeutic efficacy in inflammatory conditions like atopic dermatitis [18]. Tacrolimus belongs to the BCS-II drug group due to its low solubility (4–12 µg/mL) and high permeability [19]. Tacrolimus cannot readily cross the stratum corneum due to its high molecular weight (822.95 g/mol) and strong lipophilicity (partition coefficient log P = 3.96 ± 0.83) [20]. The US-FDA approved ointment formulation (Protopic®; Astellas Pharma, Tokyo, Japan) for the treatment of atopic dermatitis has limitations, e.g., a sticky sensation due to its greasy nature. In previous studies, tacrolimus has been encapsulated in mesoporous silica nanoparticles functionalized with amino and phosphonate groups that improve its solubility sevenfold in comparison to free tacrolimus [21]. Significantly higher drug retention in the skin was achieved when the mesoporous silica nanoparticles were converted into Carbopol gel, suggesting that nanoparticle-loaded gel formulation could be a promising strategy for the topical delivery of tacrolimus. The goal of our study was to formulate tacrolimus-loaded SLNs that were coated with chitosan and loaded into gel formulations to prevent lipid degradation and drug leaching and provide controlled release of the drug for potential application in the treatment of atopic dermatitis.

2. Results and Discussion

2.1. Preparation of SLNs

The SLNs were fabricated using a solvent evaporation method from generally recognized as safe ingredients [22] as mentioned in Table 1. The components of the formulation include stearic acid, a low-cost, easily available solid lipid with high drug loading efficiency [23]. Polysorbate 80 and sorbitan monooleate were used as potent surfactants with desired HLB values for promoting colloidal stability [24]. Ethanol overall contributes to the homogeneity of SLNs. Chitosan, a natural coating polymer, was used as a decorating ligand for increasing drug penetration and promoting drug retention in inflamed tissue. Chitosan has controlled release properties with added drug loading and entrapment efficiency [25].

Table 1. (**A**): Composition of drug-loaded SLNs formulations; (**B**): Composition of optimized SLN gel formulation.

Sample	Tacrolimus (gm)	Tween 80 (gm)	Span 80 (gm)	Stearic Acid (gm)	Ethanol (gm)	Chitosan (gm)	Distill Water (gm)
F1	0.1	1	—	1	10	—	87.90
F2	0.1	1.25	—	1	10	—	87.65
F3	0.1	1.50	—	1	10	—	87.40
F4	0.1	1.25	0.25	1	10	—	87.40
F5	0.1	1.25	0.50	1	10	—	87.15
F6	0.1	1.25	0.50	1	10	0.0015	87.14

Sample	Sodium Alginate (g)	Glycerol (g)	Trietanolamine (g)
Gel (F2)	1	5	1
Gel (F5)	1	5	1
Gel (F6)	1	5	1

2.2. Characterization of Tacrolimus-Loaded SLNs Formulation

The physicochemical properties of SLNs were shown to have a significant impact on topical drug delivery [26] and are mentioned in Table 2. The particle size of uncoated SLNs was in the range of 439 to 669 nm, while coated SLNs had a droplet size of 523 ± 3.79 nm (Figures 1 and 2). The particle size decreased from 669 ± 5.06 (F1) to 489 ± 6.81 (F2) with increasing concentration of Tween 80 due to lower interfacial tension, thereby improving miscibility between the layers of SLN dispersions. Any further increase in the concentration of Tween 80 in formulation resulted in an increasing trend of particle size from F2 to F3 due to the formation of micelles. At the critical micelle concentration, the surface becomes fully loaded with Tween 80 molecules, and thus beyond the critical micelle concentration, the interfacial tension change is nearly negligible resulting in the increased particle sizes [27]. Among Tween-80-based formulations, F2 was selected and further added with Span 80 for improved stabilization of dispersion. Addition of Span 80 in SLN formulations (F4 & F5) reduced the particle size significantly (Student t-test; $p \leq 0.05$). The F5 was then coated with chitosan (F6), thereby increasing its particle size [28]. The PDI values of both coated and uncoated tacrolimus-loaded SLNs were in the acceptable range of <0.4, presenting the homogeneous dispersion system (Table 2). The PDI value of F6 was significantly lower than other formulations, which could be due to repulsive forces of the positive charges ($-NH_2$) and additional surfactant action of chitosan [29]. The Zeta potential values ranged from -25.80 (F1) to -15.40 (F5) and reverted to a positive value in the case of chitosan-coated formulation (F6) due to the protonated amino group of chitosan, thereby confirming the successful decoration of SLN droplets [28]. The zeta potential values decreased from F1 to F5 due to increased nonionic surfactant concentration. The potential charges of all the tacrolimus-loaded SLNs showed good electrochemical stability due to electrostatic repulsion amongst particles [24]. The TEM analyses support the finding of photon correlation spectroscopy with respect to particle size and homogeneity. The particles were spherical and uniformly dispersed (Figure 1).

Table 2. Physicochemical characteristics of the SLNs.

Formulation Code	Size (nm)	PDI	Zeta Potential (mV)	Drug Content (mg/mL)	Entrapment Efficiency (%)
F1	669 ± 5.06	0.302	−25.80 ± 0.05	0.86 ± 0.03	68.95 ± 0.03
F2	489 ± 6.81	0.318	−23.10 ± 0.02	0.89 ± 0.02	72.26 ± 0.05
F3	639 ± 8.43	0.342	−20.20 ± 0.04	0.87 ± 0.04	69.71 ± 0.02
F4	578 ± 4.12	0.358	−19.30 ± 0.02	0.88 ± 0.03	78.38 ± 0.04
F5	439 ± 4.44	0.372	−15.70 ± 0.02	0.89 ± 0.02	80.45 ± 0.05
F6	523 ± 3.79	0.292	17.40 ± 0.07	0.91 ± 0.03	83.68 ± 0.04

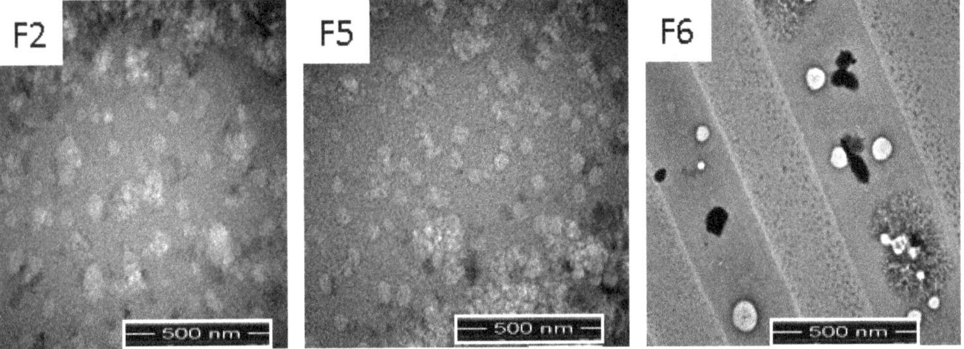

Figure 1. Morphology of optimized SLN formulations using TEM analysis.

Lipid components and the nature of surfactants are the two main components of SLNs that can affect drug encapsulation efficiency [30]. At constant lipid concentration, both drug content and encapsulation efficiency increased with the increasing concentration of surfactant (Table 2). The lower drug entrapment efficiency values in the case of the first three formulations (F1, F2, and F3) were due to the fabrication of SLNs with only one surfactant (Tween 80). The longer alkyl chain of polysorbate 80 reduces the HLB value, which ultimately reduces the encapsulation efficiency [30]. The decreased entrapment efficiency could also be explained by the partition phenomenon, characterized by the increased partitioning of drug from inner phase to outer phase due to the presence of high surfactant concentration in the exterior phase; this supports an increased leakage of drug from internal to external phase [31,32]. The Span 80 with an HLB value of 4.3 has a higher affinity for lipids when compared to the Tween 80 with an HLB value of 15.0. The combination of two surfactants produces nanocarriers with higher stability [23]. The maximum entrapment efficiency of F6 could be attributed to reduced leakage of drug from stable SLN droplets (Table 2; [28]).

Healthy skin with a pH of approximately 4.9 to 5.9 owns acid/base assets, which might alter the mark of ionization of ionizable drug, ultimately affecting the drug permeation through the skin. The pH of the topical formulation is a significant factor regarding solubility and consistency of the drug product throughout the storing period [33]. The pH of prepared SLNs was in the range of 5.45 to 5.53 as tabulated in Table 3, depicting their suitability as topical formulation [34]. The viscosity of tacrolimus-loaded SLNs ranged from 11.56 to 39.87 (Table 3). It has been previously elaborated that the viscosity of SLNs is the function of lipid component, type and quantity of surfactant, and the quantity of aqueous phase [21]. The viscosity increased with the increasing concentration of surfactant. The addition of chitosan has also been shown to have a prominent effect on the viscosity of the formulation, thereby increasing the contact of nanoparticles with the skin surface and ultimately affecting the skin retention and permeation properties of the topical formulation [24,25]. The relations between two immiscible ingredients are governed by surface tension [35]. The surface tension of the tacrolimus-loaded SLNs decreased with increasing surfactant concentration (Table 3). The increased surfactant concentration minimizes the superficial tension and thereby reduces the surface area of each particle in the formulation. The minimum surface tension observed for F6 was due to the presence of double surfactant with additional surfactant properties of chitosan that could probably increase the retention and permeation of the tacrolimus. The specific gravity and density of all tacrolimus-loaded SLNs were close to that of deionized water.

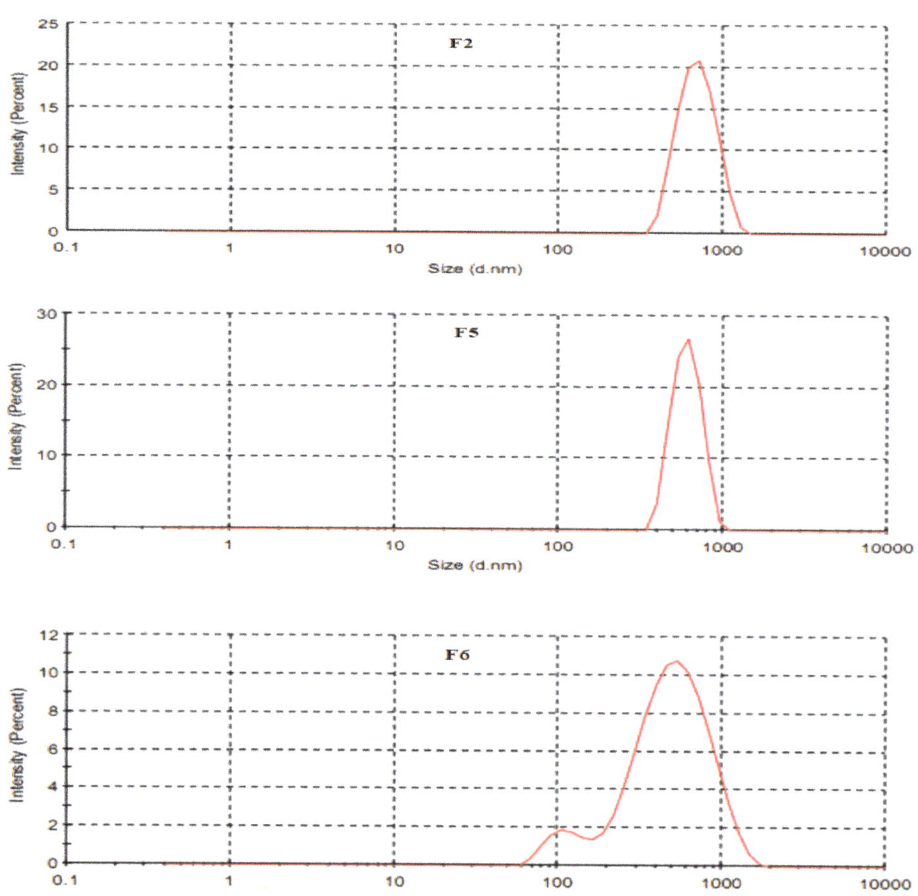

Figure 2. Particle size distribution of optimized SLN formulations using DLS analysis.

Table 3. Viscosity, specific gravity, Surface tension, pH, and density of prepared SLN formulations.

Formulation Code	Viscosity	Surface Tension (Dynes/cm^2)	pH	Specific Gravity	Density
F1	11.56 ± 0.43	30.82 ± 0.81	5.45 ± 0.14	0.977 ± 0.09	0.947 ± 0.02
F2	14.22 ± 0.22	28.33 ± 0.78	5.36 ± 0.05	0.984 ± 0.04	0.952 ± 0.04
F3	23.11 ± 0.26	27.66 ± 0.49	5.44 ± 0.07	0.991 ± 0.06	0.961 ± 0.04
F4	25.87 ± 0.55	22.79 ± 0.63	5.11 ± 0.12	0.992 ± 0.01	0.979 ± 0.03
F5	36.95 ± 0.54	21.43 ± 0.52	5.29 ± 0.14	0.993 ± 0.02	0.988 ± 0.03
F6	39.87 ± 0.43	19.81 ± 0.22	5.53 ± 0.07	0.996 ± 0.03	0.997 ± 0.05

2.3. Drug Release

The in vitro drug release, being a valuable indicator of in vivo drug performance, was performed at skin temperature of $32 \pm 2\ ^\circ$C using a modified Franz diffusion cell. The cellulose acetate membrane with a pore size of 0.45 µm was clamped between the donor and recipient compartment of the Franz diffusion cell. The release of drug from the SLNs was biphasic; a burst drug release was followed by controlled release of drug. The initial abrupt release of drug could be due to the availability of drug at the interphase and in the hydrophilic phase of SLNs. The release of drug was comparatively retarded from F1 to F5 due to increased viscosity and higher surfactant concentrations (Figure 3A; Table 3). The release of drug from the chitosan-coated formulation was significantly retarded when compared to uncoated SLNs due to decreased solubility of chitosan at physiological pH and the higher viscosity (ANOVA, $p \leq 0.05$; [36]. Another reason could be the enhanced drug entrapment efficiency of chitosan-coated SLNs [37]. When the SLNs were converted to their corresponding gel formulations using sodium alginate as a gelling agent, the release of drug was significantly retarded due to presence of polymeric matrix as mentioned previously (Figure 3B; [38]).

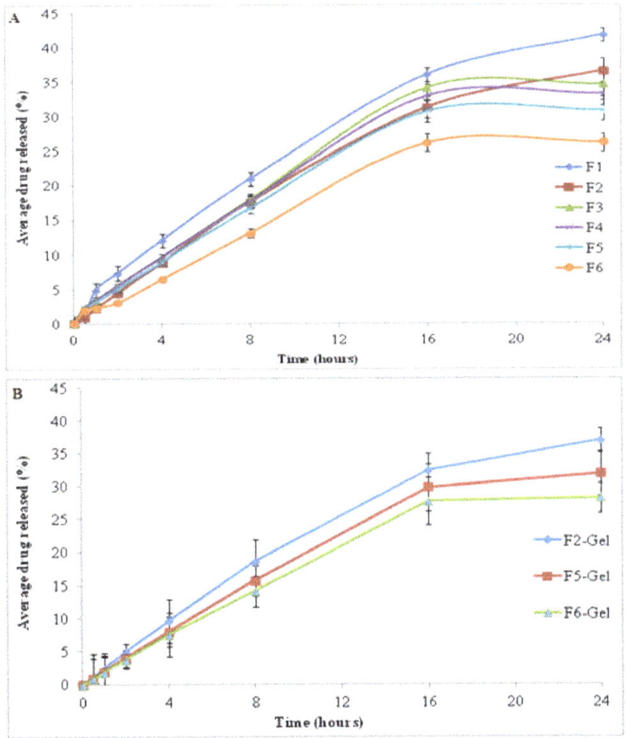

Figure 3. Drug release pattern from (**A**) SLN formulations and (**B**) SLN-loaded gel.

2.4. Drug Permeation through the Rats' Skin

The permeation of tacrolimus through the full thickness skin of Sprague dawley rats from applied SLNs and their corresponding gel formulation is shown in Figure 4A,B, respectively. The total amount of drug permeated linearly increased from F1 to F6 due to decreased surface tension. The increasing concentration of surfactant and decreased Zeta potential values could also result in improved drug permeation from F1 to F5. The innate features of SLNs, including compatibility with the dermal lipids, the occlusive effect, and their lesser particulate size may also be held responsible to improve the capacity of SLNs to permeate the skin [39,40]. The minor change among the ordinary particle dimensions of the prepared formulations was not the main aspect that inclined the drug permeation into the skin; it was the thermodynamic properties that caused the discharge at a topical temperature [41]. The higher drug permeation in the case of chitosan-coated SLNs (F6) could be due to the interaction of positively charged chitosan and negatively charged stratum corneum of the skin [37]. In response to this interaction, secondary changes occurred in the skin with the resulting disorganization of lipids and opening of skin pores that facilitate the drug penetration by intercellular and/or transcellular pathways [42]. Tween 80, a hydrophilic surfactant, hydrates the stratum corneum via lipid fluidization with water [43]. The permeation of drug from SLN-loaded gel is shown in Figure 4B It can be observed that the permeated percentage was lower than the nongel formulation. This retarding infiltration of the drug from the gel formulation could be due to the higher viscosity thereby decreasing the skin drug diffusion within tested time duration. However, the gel formulation increases the overall retention of drug within the skin layers [37,44].

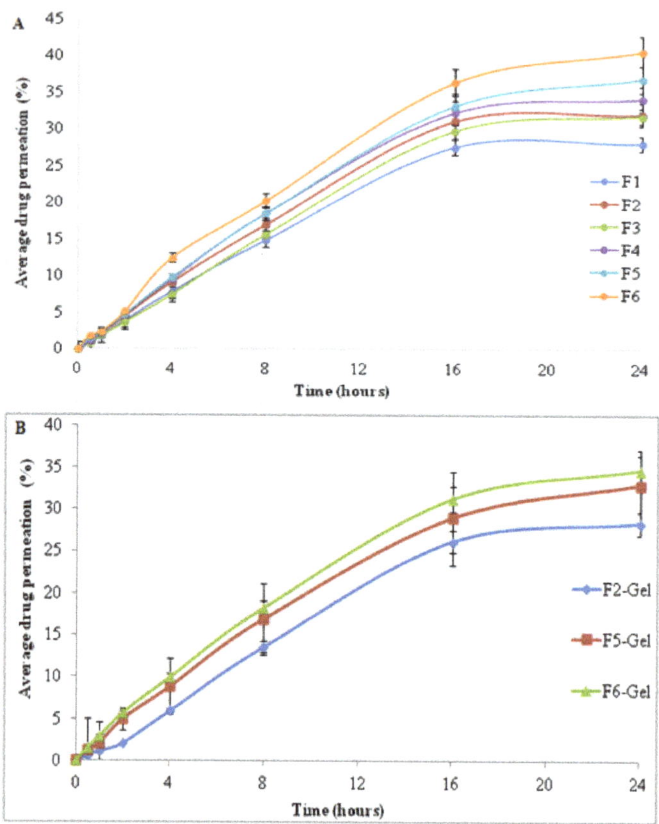

Figure 4. Permeation of drug through the skin from (**A**) SLNs formulation and (**B**) SLN-loaded gel.

2.5. Skin Drug Retention

The amount of drug retained in the skin was estimated by extracting the drug from the skin tissue. The retention of drug within intact skin from SLNs was comparatively higher for the chitosan-coated formulation than noncoated carriers due to its bioadhesive properties. However, the change was not significant (Figure 5A; F2 to F6; $p \geq 0.05$). The higher retention of drug in the case of chitosan-coated SLNs was due to the formation of a dense occlusive layer of solid lipid that melts and penetrates through the skin due to suitable physicochemical properties of SLNs and the penetration enhancement effect of chitosan [45]. When skin drug retention of SLNs was compared to their corresponding gel, the latter achieved significantly higher drug retention ($p \leq 0.05$; Figure 5B). This higher retention of gel formulations could be attributed to the penetration enhancement effect of the gelling agent that reduces water loss from the skin, thereby producing hydration for improved retention of drug within the skin [44,46].

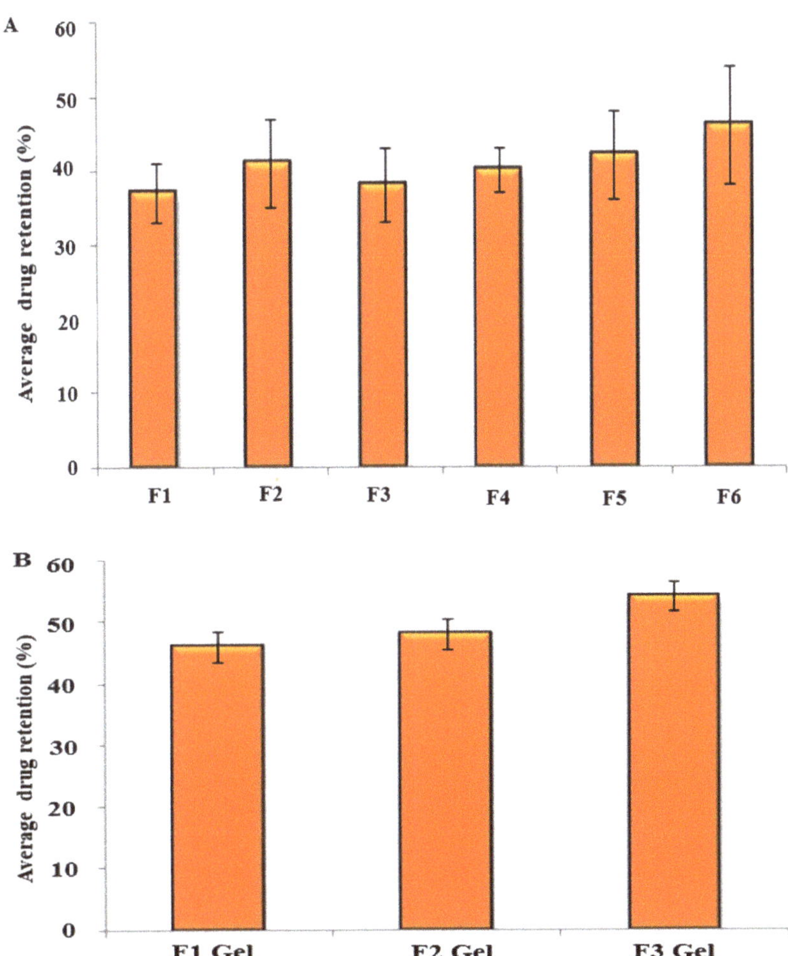

Figure 5. Drug retention profile in the skin from (**A**) SLN formulations and (**B**) SLN-loaded gel.

3. Conclusions

Tacrolimus, a drug of choice in atopic dermatitis, was successfully formulated in SLNs, having suitable physicochemical properties for skin drug application. The SLNs achieved optimum permeation ranging from 25 to 40%. The release of drug from all formulations was controlled; however, the addition of chitosan sustained drug release significantly. The total amount of drug permeated increased linearly from F1 to F6 due to the increasing concentration of surfactants thereby reducing surface tension and promoting skin drug interaction. Conversion of SLNs into their corresponding gel improved skin drug retention, thereby achieving the goal of topical drug application in the management of atopic dermatitis.

4. Materials and Methods

4.1. Materials

The tacrolimus was kindly gifted to us by Siam Pharmaceuticals (Pvt) Ltd., Islamabad, Pakistan. Stearic acid (Sigma Aldrich, St. Louis, MO, USA) was used as a solid lipid while polysorbate 80 and Sorbiton monooleate 80 (Sigma Aldrich, St. Louis, MO, USA) were used as surfactants. Ethanol (Merk, Kenilworth, NJ, USA) was used as solvent, and chitosan (LMW; Sigma Aldrich, St. Louis, MO, USA) was utilized as a penetration enhancer. Sodium alginate and glycerol along with humectant trietanolamine were used as gelling agents. All other chemicals used in the study were of pharmaceutical grades and were used without any purification.

4.2. Preparation of Tacrolimus-Loaded SLNs

The tacrolimus-loaded SLNs were prepared using varying concentrations of surfactants and cosurfactants via a solvent emulsification technique as mentioned previously [25]. Briefly, the stearic acid was melted on a water bath (Memmert WPE 45, Schwabach, Germany) at $70 \pm 2\ °C$ for 90 min. The tacrolimus previously dissolved in ethanol was mixed with melted lipid with constant magnetic stirring for 15 min at $70 \pm 2\ °C$. The aqueous phase of SLNs was prepared by adding Tween 80 to distilled water with continuous magnetic stirring at $70 \pm 2\ °C$ for 60 min. To improve the stability of colloidal dispersion, Span 80 was also added in some formulations (F4–F6). The final dispersion was then achieved by mixing the two phases at constant magnetic stirring for 2.0 h at $70 \pm 2\ °C$. The final formulation was subjected homogenization at 10,000 rpm (Ultra-turrax, DXT45) for 5 min followed by solvent evaporation. The optimized formulation (F5) was coated with 85% deacetylated chitosan and had a molecular weight of 50,000 to get the final formulation (F6) as mentioned in Table 1A.

4.3. SLN-Loaded Gel Preparation

The optimized SLNs formulations (F2, F5 and F6) were converted into a gel by the addition of glycerol and sodium alginate to the dispersion. Trietanolamine was also added as a humectant upon continuous stirring, which yielded the desired gel. The prepared gel incorporated with the drug-loaded SLNs was preserved in the refrigerator at 2–8 °C until used. The composition of prepared SLNs and their respective gel formulations is shown in Table 1A,B respectively.

4.4. Physicochemical Characterization

The SLN colloidal dispersions were evaluated for their size and size distribution (PDI), surface charge, morphology, drug content, surface tension, viscosity, pH, and density to predict their suitability for topical application.

4.4.1. Particle Size and Polydispersity Index

The particle size and particle size distribution were investigated at $25 \pm 1\ °C$ using a Zetasizer (Malvern Instruments, Worcestershire, UK). In brief, 10 µL of the sample was added to 1 mL of deionized water, mixed well, and subjected to vortex mixing for 2 min followed by photon correlation spectroscopic analysis as mentioned previously [47]. The results obtained were recorded in triplicate and the average particle size and PDI values were tabulated.

4.4.2. Zeta Potential

The surface charge on prepared SLN colloidal dispersions was evaluated using Zetasizer (Malvern Instruments, Worcestershire, UK). An aliquot of 700 µL nondiluted colloidal dispersion was added to Zeta potential cell and evaluated for its charge in triplicate as mentioned previously [48].

4.4.3. Structural Analysis

Transmission electron microscopy (TEM) at specified magnification was used to evaluate the facet shape and structural features of the particulates. A double-sided adhesive tape adhered to the aluminium stub was used for the placement of dispersion drops followed by vacuum drying. The drops were then coated with the thin layer of gold and the images were taken at suitable magnification as previously mentioned [49].

4.4.4. Drug Encapsulation Efficiency

The drug content and encapsulation efficiency of the SLN dispersions was investigated spectrophotometrically. Briefly, 1.5 mL of the sample was added to centrifuge tubes and centrifugated at 14,000 rpm for 10 min using a centrifuge machine (D3024, SCILOGEX, Rocky Hill, CT, USA). The supernatant layer was analyzed with a UV spectrometer (UV-160, Shimadzu, Kyoto, Japan) for free drug content after appropriate dilution with phosphate buffer pH 7.4 at λ max 294 nm. The entrapped drug in the SLN droplets was determined by first dissolving the sediment layer in ethanol followed by dilution and vortex mixing for 10 min [50]. The encapsulation efficiency of prepared SLNs was then calculated as mentioned in Equation (1):

$$\text{Encapsulation efficiency} = (\text{Total drug content-free drug content})/(\text{Total drug content}) \times 100 \quad (1)$$

4.4.5. Density

The density of prepared SLNs was determined using an ordinary pycnometer at $25 \pm 1\ ^\circ\text{C}$. Initially the weight of the empty pycnometer of known volume (V) was recorded as W1, followed by the weight of the water-filled pycnometer (W2). The weight of the SLN-filled pycnometer was labeled as W3. The density of SLNs was calculated using the following equation [48].

$$\text{Density of water} = W2/V \quad (2)$$

$$\text{Density of SLNs} = W3/V \quad (3)$$

$$\text{Specific gravity of SLNs} = (\text{SLNs density})/(\text{Water density}) \quad (4)$$

4.4.6. Surface Tension

A laboratory stalagmometer was used for measuring the surface tension of the SLN dispersions. The instrument was cleaned with un-ionized water and filled to the point A with double distilled water. The filled water was then run down drop wise to the point B while recording the number of drops. Similarly, same steps were repeated for the SLN dispersions. The density of SLNs was already calculated as mentioned above in Section 4.4.5. The surface tension of SLNs was then calculated using Equation (5) as mentioned previously [51].

$$\gamma 1 = (\delta 1 * n1 \div \delta 2 * n2) * \gamma 2 \quad (5)$$

where:
$\gamma 1$: Surface tension of dispersion;
$\gamma 2$: Surface tension of water;
$\delta 1$: Density of dispersion;
$\delta 2$: Density of water;
n1: Number of drops of SLNs;
n2: Number of drops of de-ionized water.

4.4.7. Viscosity

The viscosity of SLN dispersions was determined using an Ostwald viscometer at $25 \pm 1\ °C$ (Poulten selfe & Lee Ltd., Essex, UK). The instrument was washed, dried, and fixed vertically on the stand. The apparatus was filled with distilled water up to mark A. The distilled water was allowed to flow down to mark B, and the time of flow between the two points was recorded. Similarly, water was replaced with SLN dispersions, and the time of flow from point A to B was noted. The final viscosity of SLNs was calculated using equation 6 [51]. The readings were recorded for three consecutive experiments and the values were shown as mean $\pm$ SD.

$$\eta1 = (d1 * t1 \_ d2 * t2) * \eta2 \tag{6}$$

where:
$\eta1$: Viscosity of the SLNs sample
$\eta2$: Viscosity of the pure distilled water
$t1$: Flow time of prepared SLNs sample
$t2$: Flow time of pure distilled water
$d1$: Density of SLNs sample
$d2$: Density of pure distilled water

4.4.8. pH

A commonly used pH meter (Accumet meter and Denver instruments) was utilized at $25 \pm 1\ °C$ for predicting the pH of the SLNs. A known buffer solution with a pH of 7 was used to calibrate the instrument. After calibration, the pH of the SLN dispersions was determined and tabulated as mean $\pm$ SD of triplicate readings.

4.5. Ex Vivo Drug Release

The release of drug from designed SLNs and their corresponding gel was determined using Franz diffusion cell (PRIME GEAR Inc. NO 4G 1-22-3-12, Delhi, India). The cellulose acetate membrane (0.45 µm) was fixed between the donor and receptor compartment. The receptor compartment was filled to its capacity with acetate buffer pH 5.5 at $32 \pm 2\ °C$ to mimic the skin conditions. The magnetic stirring was stabilized at 600 rpm to keep sink condition. The donor compartment was fed with 1 gm of the sample. An aliquot of 0.5 mL was withdrawn from the receptor compartment through the sampling port with the help of a calibrated syringe at the time intervals of 0, 0.5, 1, 2, 4, 8, 16 and 24 h. The samples were analyzed using a UV-visible spectrophotometer (UV-1601 Shimadzu, Kyoto, Japan) at λ max of 294 nm [52]. The cumulative release of tacrolimus from the designed carrier system was calculated using the standard calibration curve. All the results were recorded as the mean of triplicate experiments.

4.6. Animal Ethics Approval

The approval of animal use in research and procedures to be carried out for isolation of skin from Sprague dowley rats was granted from Quality Enhancement Cell (QEC), Gomal University, D.I. Khan Khyber Pakhtunkhwa, Pakistan.

4.7. Ex Vivo Drug Permeation and Retention

The permeation of drug through the dead keratinized layer of the skin (stratum corneum) and its potential deposition in desired tissue of the body is a critical factor for the success of topical and transdermal drug delivery. The drug permeation of the designed carriers through the rats' (Sprague dawley) skin was evaluated using Franz diffusion cells (PermeGear, Inc., Hellertown, PA, USA). The Franz diffusion cell consists of 6 cells with a diffusional surface area of $0.654\ cm^2$ and a receptor cell volume of 5 mL, operated at continuous stirring mode (600 rpm) at $37\ °C$. The rat weighing 200 ± 20 g was sacrificed by the cervical dislocation method, followed by complete hair removal from the abdominal area using a razor. The full-thickness skin was surgically excised from Sprague dawley rats as per approved protocols in compliance with the Ethical Review Committee, Gomal University, D.I. Khan, Pakistan. The excess fats were removed, washed with normal saline and stored at -20 °C for future experimental use. The skin was clamped between the donor and recipient compartment of the Franz diffusion cell. To maintain sink conditions, phosphate buffer 7.4 was added to the volume of the receptor compartment. The temperature and magnetic stirring were kept constant at 37 ± 2 °C and 600 rpm, respectively. The SLN's, having a known quantity of the drug, were fed to the contributor chambers of the apparatus. An aliquot of 0.5 mL was withdrawn from the receptor compartment through the sampling port with the help of a calibrated syringe at the time intervals of, 0, 0.5, 1, 2, 4, 8, 16, and finally 24 h. The deficiency of withdrawal sample was top-up using the same quantity of fresh buffer. The drug content in each sample was determined using a UV-visible spectrophotometer at λ max of 294 nm.

Following the permeation experiment, the percentage of drug retained in the skin was determined by isolating the mounted skin on the Franz diffusion cell, carefully weighed and cut into pieces. The cut pieces of the isolated skin were extracted with 10 mL of methanol by mechanically shaking in a water bath (BW-20G, Lab. Companion, Billerica MA, USA) at 37 ± 1 °C overnight. To further improve the extraction efficiencies, tissue cell lysis reagent was introduced at a ratio of 1:20 and sonicated (EUROSTAR digital, IKA®, Kuala Lumpur, Malaysia) for 30 min. All the samples were harvested by ultracentrifugation (12,000 rpm) at 10 °C for 20 min and quantification of tacrolimus in the rats' skin was performed using a UV-visible spectrophotometer as mentioned previously [53].

4.8. Statistical Analysis

All the values were tabulated as a mean $\pm$ SD for triplicate experiments. The data was statistically evaluated using one way ANOVA or paired t-test and a p value of < 0.05 was considered statistically significant.

Author Contributions: Conceptualization, A.S.K. and K.U.S.; Data curation, M.A.M., A.J.A., Y.N.A. and M.A.A.H.; Investigation, K.A.K. and A.S.K.; Methodology, A.F. and A.S.K.; Project administration, K.U.S. and A.F.; Software, Z.R.N.; Supervision, K.U.S. and A.F.; Validation, Y.N.A., S.G. and M.A.M.; Writing—original draft, A.S.K., K.U.S. and A.F.; Writing—review & editing, A.J.A., Z.R.N. and K.A.K. All authors have read and agreed to the published version of the manuscript.

Funding: This research received no external funding.

Institutional Review Board Statement: Not applicable.

Informed Consent Statement: Not applicable.

Data Availability Statement: Requests to access the datasets should be directed to the corresponding author.

Acknowledgments: The authors gratefully acknowledge Gomal Centre of Pharmaceutical Sciences, Faculty of Pharmacy Gomal University D.I. Khan, KPK, Pakistan, for providing financial support and a laboratory facility for this research work.

Conflicts of Interest: The authors declare no conflict of interest.

References

1. Prince, G.T.; Cameron, M.C.; Fathi, R.; Alkousakis, T. Topical 5-fluorouracil in dermatologic disease. *Int. J. Dermatol.* **2018**, *57*, 1259–1264. [CrossRef]
2. Sala, M.; Diab, R.; Elaissari, A.; Fessi, H. Lipid nanocarriers as skin drug delivery systems: Properties, mechanisms of skin interactions and medical applications. *Int. J. Pharm.* **2018**, *535*, 1–17. [CrossRef]
3. Chu, D. Overview of biology, development, and structure of skin. In *Fitzpatrick's Dermatology in General Medicine*, 8th ed.; Wolff, K., Goldsmith, L.A., Katz, S.I., Gilchrest, B.A., Paller, A.S., DJ Leffell, D.J., Eds.; McGraw Hill Medical: New York, NY, USA, 2008.
4. Del Prado-Audelo, M.L.; Caballero-Florán, I.H.; Sharifi-Rad, J.; Mendoza-Muñoz, N.; González-Torres, M.; Urbán-Morlán, Z.; Florán, B.; Cortes, H.; Leyva-Gómez, G. Chitosan-decorated nanoparticles for drug delivery. *J. Drug Deliv. Sci. Technol.* **2020**, *59*, 101896. [CrossRef]

5. Prow, T.W.; Grice, J.E.; Lin, L.L.; Faye, R.; Butler, M.; Becker, W.; Wurm, E.M.; Yoong, C.; Robertson, T.A.; Soyer, H.P. Nanoparticles and microparticles for skin drug delivery. *Adv. Drug Deliv. Rev.* **2011**, *63*, 470–491. [CrossRef]

6. Patel, D.K.; Kesharwani, R.; Kumar, V. Etodolac loaded solid lipid nanoparticle based topical gel for enhanced skin delivery. *Biocatal. Agric. Biotechnol.* **2020**, *29*, 101810. [CrossRef]

7. Chen, X.; Peng, L.; Gao, J. Novel topical drug delivery systems and their potential use in scars treatment. *Asian J. Pharm. Sci.* **2012**, *7*, 155–167.

8. Puglia, C.; Blasi, P.; Rizza, L.; Schoubben, A.; Bonina, F.; Rossi, C.; Ricci, M. Lipid nanoparticles for prolonged topical delivery: An in vitro and in vivo investigation. *Int. J. Pharm.* **2008**, *357*, 295–304. [CrossRef]

9. Kaur, R.; Sharma, N.; Tikoo, K.; Sinha, V. Development of mirtazapine loaded solid lipid nanoparticles for topical delivery: Optimization, characterization and cytotoxicity evaluation. *Int. J. Pharm.* **2020**, *586*, 119439. [CrossRef]

10. Yaghmur, A.; Mu, H. Recent advances in drug delivery applications of cubosomes, hexosomes, and solid lipid nanoparticles. *Acta Pharm. Sin. B* **2021**, *11*, 871–885. [CrossRef]

11. Kang, B.-S.; Choi, J.-S.; Lee, S.-E.; Lee, J.-K.; Kim, T.-H.; Jang, W.S.; Tunsirikongkon, A.; Kim, J.-K.; Park, J.-S. Enhancing the in vitro anticancer activity of albendazole incorporated into chitosan-coated PLGA nanoparticles. *Carbohydr. Polym.* **2017**, *159*, 39–47. [CrossRef]

12. Honary, S.; Zahir, F. Effect of zeta potential on the properties of nano-drug delivery systems-a review (Part 2). *Trop. J. Pharm. Res.* **2013**, *12*, 265–273.

13. Souto, E.; Wissing, S.; Barbosa, C.; Müller, R. Development of a controlled release formulation based on SLN and NLC for topical clotrimazole delivery. *Int. J. Pharm.* **2004**, *278*, 71–77. [CrossRef] [PubMed]

14. Silverberg, J.I.; Hanifin, J.M. Adult eczema prevalence and associations with asthma and other health and demographic factors: A US population–based study. *J. Allergy Clin. Immunol.* **2013**, *132*, 1132–1138. [CrossRef]

15. Chang, K.-T.; Lin, H.Y.-H.; Kuo, C.-H.; Hung, C.-H. Tacrolimus suppresses atopic dermatitis-associated cytokines and chemokines in monocytes. *J. Microbiol. Immunol. Infect.* **2016**, *49*, 409–416. [CrossRef]

16. Zhang, D.; Pan, X.; Wang, S.; Zhai, Y.; Guan, J.; Fu, Q.; Hao, X.; Qi, W.; Wang, Y.; Lian, H. Multifunctional poly (methyl vinyl ether-co-maleic anhydride)-graft-hydroxypropyl-β-cyclodextrin amphiphilic copolymer as an oral high-performance delivery carrier of tacrolimus. *Mol. Pharm.* **2015**, *12*, 2337–2351. [CrossRef]

17. Goebel, A.S.; Neubert, R.H.; Wohlrab, J. Dermal targeting of tacrolimus using colloidal carrier systems. *Int. J. Pharm.* **2011**, *404*, 159–168. [CrossRef]

18. Sezer, A.D.; Cevher, E. Topical drug delivery using chitosan nano-and microparticles. *Expert Opin. Drug Deliv.* **2012**, *9*, 1129–1146. [CrossRef]

19. Dianzani, C.; Foglietta, F.; Ferrara, B.; Rosa, A.C.; Muntoni, E.; Gasco, P.; Della Pepa, C.; Canaparo, R.; Serpe, L. Solid lipid nanoparticles delivering anti-inflammatory drugs to treat inflammatory bowel disease: Effects in an in vivo model. *World J. Gastroenterol.* **2017**, *23*, 4200. [CrossRef]

20. Seo, Y.G.; Kim, D.-W.; Cho, K.H.; Yousaf, A.M.; Kim, D.S.; Kim, J.H.; Kim, J.O.; Yong, C.S.; Choi, H.-G. Preparation and pharmaceutical evaluation of new tacrolimus-loaded solid self-emulsifying drug delivery system. *Arch. Pharmacal. Res.* **2015**, *38*, 223–228. [CrossRef]

21. Shin, S.-B.; Cho, H.-Y.; Kim, D.-D.; Choi, H.-G.; Lee, Y.-B. Preparation and evaluation of tacrolimus-loaded nanoparticles for lymphatic delivery. *Eur. J. Pharm. Biopharm.* **2010**, *74*, 164–171. [CrossRef]

22. Wang, Y.; Sun, J.; Zhang, T.; Liu, H.; He, F.; He, Z. Enhanced oral bioavailability of tacrolimus in rats by self-microemulsifying drug delivery systems. *Drug Dev. Ind. Pharm.* **2011**, *37*, 1225–1230. [CrossRef] [PubMed]

23. Pizzol, C.D.; Filippin-Monteiro, F.B.; Restrepo, J.A.S.; Pittella, F.; Silva, A.H.; Alves de Souza, P.; Machado de Campos, A.; Creczynski-Pasa, T.B. Influence of surfactant and lipid type on the physicochemical properties and biocompatibility of solid lipid nanoparticles. *Int. J. Environ. Res. Public Health* **2014**, *11*, 8581–8596. [CrossRef] [PubMed]

24. Küchler, S.; Herrmann, W.; Panek-Minkin, G.; Blaschke, T.; Zoschke, C.; Kramer, K.D.; Bittl, R.; Schäfer-Korting, M. SLN for topical application in skin diseases—Characterization of drug–carrier and carrier–target interactions. *Int. J. Pharm.* **2010**, *390*, 225–233. [CrossRef] [PubMed]

25. Khan, S.; Shaharyar, M.; Fazil, M.; Baboota, S.; Ali, J. Tacrolimus-loaded nanostructured lipid carriers for oral delivery–optimization of production and characterization. *Eur. J. Pharm. Biopharm.* **2016**, *108*, 277–288. [CrossRef]

26. Danaei, M.; Dehghankhold, M.; Ataei, S.; Hasanzadeh Davarani, F.; Javanmard, R.; Dokhani, A.; Khorasani, S.; Mozafari, M. Impact of particle size and polydispersity index on the clinical applications of lipidic nanocarrier systems. *Pharmaceutics* **2018**, *10*, 57. [CrossRef]

27. Ridolfi, D.M.; Marcato, P.D.; Justo, G.Z.; Cordi, L.; Machado, D.; Durán, N. Chitosan-solid lipid nanoparticles as carriers for topical delivery of tretinoin. *Colloids Surf. B Biointerfaces* **2012**, *93*, 36–40. [CrossRef]

28. Al-Nemrawi, N.; Alsharif, S.; Dave, R. Preparation of chitosan-TPP nanoparticles: The influence of chitosan polymeric properties and formulation variables. *Int. J. Appl. Pharm.* **2018**, *10*, 60–65. [CrossRef]

29. Ruckmani, K.; Sankar, V. Formulation and optimization of zidovudine niosomes. *Aaps Pharmscitech* **2010**, *11*, 1119–1127. [CrossRef]

30. Zirak, M.B.; Pezeshki, A. Effect of surfactant concentration on the particle size, stability and potential zeta of beta carotene nano lipid carrier. *Int. J. Curr. Microbiol. Appl. Sci.* **2015**, *4*, 924–932.

31. Müller, R.H.; Radtke, M.; Wissing, S.A. Solid lipid nanoparticles (SLN) and nanostructured lipid carriers (NLC) in cosmetic and dermatological preparations. *Adv. Drug Deliv. Rev.* **2002**, *54*, S131–S155. [CrossRef]
32. Knudsen, N.Ø.; Pedersen, G.P. pH and drug delivery. In *pH of the Skin: Issues and Challenges*; Karger Publishers: Berlin, Germany, 2018; Volume 54, pp. 143–151.
33. El-Housiny, S.; Shams Eldeen, M.A.; El-Attar, Y.A.; Salem, H.A.; Attia, D.; Bendas, E.R.; El-Nabarawi, M.A. Fluconazole-loaded solid lipid nanoparticles topical gel for treatment of pityriasis versicolor: Formulation and clinical study. *Drug Deliv.* **2018**, *25*, 78–90. [CrossRef] [PubMed]
34. Liu, D.; Jiang, S.; Shen, H.; Qin, S.; Liu, J.; Zhang, Q.; Li, R.; Xu, Q. Diclofenac sodium-loaded solid lipid nanoparticles prepared by emulsion/solvent evaporation method. *J. Nanoparticle Res.* **2011**, *13*, 2375–2386. [CrossRef]
35. Üner, M.; Yener, G. Importance of solid lipid nanoparticles (SLN) in various administration routes and future perspectives. *Int. J. Nanomed.* **2007**, *2*, 289.
36. Bansal, V.; Sharma, P.K.; Sharma, N.; Pal, O.P.; Malviya, R. Applications of chitosan and chitosan derivatives in drug delivery. *Adv. Biol. Res.* **2011**, *5*, 28–37.
37. Pople, P.V.; Singh, K.K. Development and evaluation of topical formulation containing solid lipid nanoparticles of vitamin A. *Aaps Pharmscitech* **2006**, *7*, E63–E69. [CrossRef]
38. Olbrich, C.; Müller, R.H.; Tabatt, K.; Kayser, O.; Schulze, C.; Schade, R. Stable biocompatible adjuvants—A new type of adjuvant based on solid lipid nanoparticles: A study on cytotoxicity, compatibility and efficacy in chicken. *Altern. Lab. Anim.* **2002**, *30*, 443–458. [CrossRef]
39. Lv, Q.; Yu, A.; Xi, Y.; Li, H.; Song, Z.; Cui, J.; Cao, F.; Zhai, G. Development and evaluation of penciclovir-loaded solid lipid nanoparticles for topical delivery. *Int. J. Pharm.* **2009**, *372*, 191–198. [CrossRef]
40. Kang, J.-H.; Chon, J.; Kim, Y.-I.; Lee, H.-J.; Oh, D.-W.; Lee, H.-G.; Han, C.-S.; Kim, D.-W.; Park, C.-W. Preparation and evaluation of tacrolimus-loaded thermosensitive solid lipid nanoparticles for improved dermal distribution. *Int. J. Nanomed.* **2019**, *14*, 5381. [CrossRef]
41. Roussel, L.; Abdayem, R.; Gilbert, E.; Pirot, F.; Haftek, M. Influence of excipients on two elements of the stratum corneum barrier: Intercellular lipids and epidermal tight junctions. In *Percutaneous Penetration Enhancers Chemical Methods in Penetration Enhancement*; Springer: Berlin, Germany, 2015; pp. 69–90.
42. Lane, M.E. Skin penetration enhancers. *Int. J. Pharm.* **2013**, *447*, 12–21. [CrossRef]
43. Jain, S.; Addan, R.; Kushwah, V.; Harde, H.; Mahajan, R.R. Comparative assessment of efficacy and safety potential of multifarious lipid based Tacrolimus loaded nanoformulations. *Int. J. Pharm.* **2019**, *562*, 96–104. [CrossRef]
44. Jain, S.; Chourasia, M.; Masuriha, R.; Soni, V.; Jain, A.; Jain, N.K.; Gupta, Y. Solid lipid nanoparticles bearing flurbiprofen for transdermal delivery. *Drug Deliv.* **2005**, *12*, 207–215. [CrossRef] [PubMed]
45. Raza, K.; Singh, B.; Lohan, S.; Sharma, G.; Negi, P.; Yachha, Y.; Katare, O.P. Nano-lipoidal carriers of tretinoin with enhanced percutaneous absorption, photostability, biocompatibility and anti-psoriatic activity. *Int. J. Pharm.* **2013**, *456*, 65–72. [CrossRef] [PubMed]
46. Zeb, A.; Arif, S.T.; Malik, M.; Shah, F.A.; Din, F.U.; Qureshi, O.S.; Lee, E.-S.; Lee, G.-Y.; Kim, J.-K. Potential of nanoparticulate carriers for improved drug delivery via skin. *J. Pharm. Investig.* **2019**, *49*, 485–517. [CrossRef]
47. Trombino, S.; Serini, S.; Cassano, R.; Calviello, G. Xanthan gum-based materials for omega-3 PUFA delivery: Preparation, characterization and antineoplastic activity evaluation. *Carbohydr. Polym.* **2019**, *208*, 431–440. [CrossRef]
48. Bartosova, L.; Bajgar, J. Transdermal drug delivery in vitro using diffusion cells. *Curr. Med. Chem.* **2012**, *19*, 4671–4677. [CrossRef]
49. Zhuo, F.; Abourehab, M.A.; Hussain, Z. Hyaluronic acid decorated tacrolimus-loaded nanoparticles: Efficient approach to maximize dermal targeting and anti-dermatitis efficacy. *Carbohydr. Polym.* **2018**, *197*, 478–489. [CrossRef]
50. Liu, M.; Wen, J.; Sharma, M. Solid lipid nanoparticles for topical drug delivery: Mechanisms, dosage form perspectives, and translational status. *Curr. Pharm. Des.* **2020**, *26*, 3203–3217. [CrossRef]
51. Balamurugan, M. Chitosan: A perfect polymer used in fabricating gene delivery and novel drug delivery systems. *Int. J. Pharm. Pharm. Sci.* **2012**, *4*, 54–56.
52. Kelidari, H.; Saeedi, M.; Akbari, J.; Morteza-Semnani, K.; Gill, P.; Valizadeh, H.; Nokhodchi, A. Formulation optimization and in vitro skin penetration of spironolactone loaded solid lipid nanoparticles. *Colloids Surf. B Biointerfaces* **2015**, *128*, 473–479. [CrossRef]
53. Pooja, D.; Tunki, L.; Kulhari, H.; Reddy, B.B.; Sistla, R. Optimization of solid lipid nanoparticles prepared by a single emulsification-solvent evaporation method. *Data Brief* **2016**, *6*, 15–19. [CrossRef]

Article

Preparation of NLCs-Based Topical Erythromycin Gel: In Vitro Characterization and Antibacterial Assessment

Ameeduzzafar Zafar [1,*], Syed Sarim Imam [2,*], Mohd Yasir [3], Nabil K. Alruwaili [1], Omar Awad Alsaidan [1], Musarrat Husain Warsi [4], Shehla Nasar Mir Najib Ullah [5], Sultan Alshehri [2] and Mohammed M. Ghoneim [6]

[1] Department of Pharmaceutics, College of Pharmacy, Jouf University, Sakaka 72341, Al-Jouf, Saudi Arabia; nkalruwaili@ju.edu.sa (N.K.A.); osaidan@ju.edu.sa (O.A.A.)
[2] Department of Pharmaceutics, College of Pharmacy, King Saud University, Riyadh 11451, Riyadh, Saudi Arabia; salshehri1@ksu.edu.sa
[3] Department of Pharmacy, College of Health Science, Arsi University, Asella 396, Ethiopia; mohdyasir31@gmail.com
[4] Department of Pharmaceutics and Industrial Pharmacy, College of Pharmacy, Taif University, Taif 21944, Makkah, Saudi Arabia; mvarsi@tu.edu.sa
[5] Department of Pharmacognosy, College of Pharmacy, King Khalid University, Abha 61421, 'Asir, Saudi Arabia; shehlanasar2005@gmail.com
[6] Department of Pharmacy Practice, College of Pharmacy, Almaarefa University, Ad Diriyah 13713, Ar Riyad, Saudi Arabia; mghoneim@mcst.edu.sa
* Correspondence: azafar@ju.edu.sa (A.Z.); simam@ksu.edu.sa (S.S.I.)

Citation: Zafar, A.; Imam, S.S.; Yasir, M.; Alruwaili, N.K.; Alsaidan, O.A.; Warsi, M.H.; Mir Najib Ullah, S.N.; Alshehri, S.; Ghoneim, M.M. Preparation of NLCs-Based Topical Erythromycin Gel: In Vitro Characterization and Antibacterial Assessment. *Gels* 2022, 8, 116. https://doi.org/10.3390/gels8020116

Academic Editors: Maddalena Sguizzato and Filippo Rossi

Received: 30 December 2021
Accepted: 10 February 2022
Published: 13 February 2022

Publisher's Note: MDPI stays neutral with regard to jurisdictional claims in published maps and institutional affiliations.

Abstract: In the present study, erythromycin (EM)-loaded nanostructured lipid carriers (NLCs) were prepared by the emulsification and ultra-sonication method. EM-NLCs were optimized by central composite design using the lipid (A), pluronic F127 (B) and sonication time (C) as independent variables. Their effects were evaluated on particle size (Y_1) and entrapment efficiency (Y_2). The optimized formulation (EM-NLCs-opt) showed a particle size of 169.6 ± 4.8 nm and entrapment efficiency of 81.7 ± 1.4%. EM-NLCs-opt further transformed into an in-situ gel system by using the carbopol 940 and chitosan blend as a gelling agent. The optimized EM-NLCs in situ gel (EM-NLCs-opt-IG4) showed quick gelation and were found to be stable for more than 24 h. EM-NLCs-opt-IG4 showed prolonged drug release compared to EM in situ gel. It also revealed significant high permeation (56.72%) and flux (1.51-fold) than EM in situ gel. The irritation and hydration study results depicted no damage to the goat cornea. HET-CAM results also confirmed its non-irritant potential (zero score). EM-NLCs-opt-IG4 was found to be isotonic and also showed significantly ($p < 0.05$) higher antimicrobial activity than EM in situ gel. The findings of the study concluded that NLCs laden in situ gel is an alternative delivery of erythromycin for the treatment of bacterial conjunctivitis.

Keywords: erythromycin; ocular delivery; nanostructured lipid carrier; corneal permeation; antibacterial study

1. Introduction

The eye is the most sensitive organ of the body and problems associated with the eye directly affect vision. Most ocular illnesses, such as bacterial conjunctivitis, sympathetic ophthalmia, uveitis and retinal disease, are treated by the conventional ophthalmic eye drop. However, the eye drop has drawbacks, such as less ocular residence time and quick elimination from the ocular region. Only 5% of the administered dose is available for pharmacological activity due to the elimination of the administered dose in 2 min from the corneal surface by blinking, dilution with tear fluid, as well nasolacrimal drainage duct [1,2]. Therefore, efforts have been taken to enhance the ocular residence time, in order to achieve drug permeability. Viscous preparation, such as with conventional gel and ointment, enhances residence time on the eye surface, but has drawbacks such as stickiness, blurred vision, reflex blinking, as well as irritation [3,4]. The novel approach to developing a formulation with reduced administration frequency and greater residence time leads to an increase in patient compliance. Different novel formulations, including nanostructured lipid carriers (NLCs) [5], liposomes [6], niosomes [7], ocular inserts containing nanoparticles [8] and in situ gel systems [9], have been reported to increase bioavailability as well as therapeutic efficacy. These formulations have been reported to increase drug accumulation and retention at the target site [10].

Among these formulations, NLCs are widely reported ophthalmic formulations, due to their stability in the physiological environment. They consist of biocompatible solid and liquid lipids, as well as surfactants. They are also reported for high drug load without drug leakage [5], which enhances drug release and bioavailability [11]. Various studies published on ocular NLCs report a high corneal contact time in comparison to pure drug solution [5,12–14]. They report a significant enhancement in corneal permeation and a decrease in ocular clearance. The incorporation of NLCs into an in-situ gel system (stimuli sensitive agent) can further improve drug release, contact time, and minimize toxicity and stability. Research regarding the ciprofloxacin moxifloxacin and sertaconazole in situ gel system report prolonged release, residence time and increases in bioavailability and antimicrobial activity [15–17].

Erythromycin (EM) is a macrolide category antibiotic having broad-spectrum antimicrobial activity. It penetrates the bacterial cell membrane, binds to the ribosome (50s) reversibly, and stops protein synthesis. EM is also administered new-born babies to prevent ocular infection [18,19].

CP is polyacrylic acid in nature, biocompatible, non-toxic and a good bioadhesive polymer able to easily transform to gel from the sol system in a pH environment [20]. It also binds to the mucin by hydrogen bond and electrostatic force and improves ocular residence time [21]. Despite possessing outstanding, mucoadhesive properties, it may produce irritation and may damage the ocular tissue [21]. Therefore, a combination of CP with CS overcomes this limitation [22]. CS is a natural, biocompatible, biodegradable, bioadhesive polymer, and also exhibits antimicrobial activity [23,24]. There is no early research reporting on the EM-NLCs in situ gel system using the polymer blend CP and CS.

The present research is designed to develop EM-NLCs in situ gel using polymers such as carbopol (CP) and chitosan (CS) in different ratios. EM-NLCs were optimized by central composite experiment design using the solid lipid, liquid lipid and surfactant. The optimized formulation (EM-NLCs-opt) was further loaded into an in-situ gel (IG) system and then evaluated for gelling capacity, clarity, viscosity and pH. Finally, it was evaluated for in vitro drug release, mucoadhesive study, trans-corneal permeation, hydration study, ocular irritation, isotonicity and antimicrobial study.

2. Materials

Erythromycin (EM) was procured from Envee Drugs Pvt Ltd. (Gandhinagar, India) from India. Stearic acid, Compritol 888, Isopropyl myristate, Myristic acid, Lauric acid and Cetyl palmitate Pluronic F127, tween 20, Cremophor EL, Bile Salt, Span 20, chitosan and acetonitrile dialysis bag (MWCO 12kDa) were procured from Sigma Aldrich (St. Louis, MO, USA). Oleic acid, Peceol, sesame oil, Soya oil, olive oil, and coconut oil were procured from Loba Chemie (Mumbai, India). Carbopol 940 was obtained from Acros Organic (Newark, NJ, USA). Other chemicals used for study are analytical grade.

3. Experimental

3.1. Screening of Excipients

Selection of solid lipid, liquid lipid and surfactant was carried out based on the maximum solubility of EM. The excess amount of EM was dissolved in melted solid lipids (2 mL of stearic acid, compritol 888, isopropyl myristate, myristic acid, lauric acid and cetyl palmitate). The mixture was vortexed (2 min) and kept in a water bath shaker for 72 h to achieve supersaturation. The different liquid lipids—oleic acid, peceol, sesame oil, soya oil, olive oil, and coconut oil (1 mL) were taken in glass vials. The excess amount of

EM was added to each oil and vortexed for 5 min. Then, it was kept in an orbital shaker for 72 h at room temperature. Selection of surfactants was carried by adding the excess amount of EM to different surfactants, pluronic F127, tween 20, cremophor EL, bile salt, and span 20 (1 mL), vortexed and then kept in an orbital shaker for 72 h. All the tested samples were centrifuged at 6000 rpm for 30 min to separate the insoluble EM. The supernatant was collected and analysed for drug concentration by UV-Visible spectrophotometer (Genesys 10S UV-Vis, Thermo-scientific, Waltham, MA, USA) at 280 nm.

3.2. Selection Solid and Liquid Lipid Ratio by Miscibility

The selected solid and liquid lipids were mixed in different ratios (9:1, 8:2, 7:3, 6:4, 5:5) and miscibility was evaluated visually. The mixture was heated at 60 °C for 25 min and kept aside for cooling to room temperature. The mixture was assessed visually for phase separation and turbidity [25].

3.3. Optimization

The optimization of EM-NLCs was performed by using central composite design (CCD; Design-Expert® software, version 8.0.6, Stat-Ease, Minneapolis, MN, USA) to evaluate the influence of independent parameters (lipid, surfactant, sonication time) on dependent parameters particle size (nm, PS) and entrapment efficiency (%, EE). CCD is a type of factorial or fractional factorial design with centre points, amplified with a group of axial points (also called star points). CCD works on the principal of 8 factorial points, 6-star points, and 6 replicated centre points (for the statistical assessment), which were used for optimization [26]. As shown in Table 1, lipid (%, A), surfactant (%, B) and sonication time (min, C) were chosen as independent factors and their effect was observed on PS (nm, Y_1) and EE (%, Y_2). The design depicted the total number of runs of 2k + 2k + n, where k is the number of independent variables and n is the number of repetitions of experiments at the centre point [27]. For the prediction of the best formulation, the fitness of the model due to the analysis of variance was detected among the linear, two-factor (2F) interaction model and quadratic model [28]. For optimization, the focus was made to maximize the correlation coefficient value (R^2), i.e., multiple correlation coefficient, predicted correlation coefficient and adjusted correlation coefficient. The optimization was made by the point prediction method using the desirability function to obtain the minimum PS and maximum EE. The effect of selected independent factors on the dependent factors was assessed by the 3D images and polynomial equations obtained from the software [29]. The mathematical format of the polynomial equation was represented by equation (1):

$$Y = \beta_0 + \beta_1 A + \beta_2 B + \beta_3 C + \beta_{12} AB + \beta_{13} AC + \beta_{23} BC + \beta_{11} A^2 + \beta_{22} B^2 + \beta_{33} C^2 \quad (1)$$

Y, representing the theoretical value, β_0 is the intercept, β_1, β_2, β_3, represents the model coefficient. The statistical tests such as ANOVA and lack of fit test were carried out to determine the statistically significant model.

3.4. Formulation of Erythromycin Nanostructured Lipid Carrier (EM-NLCs)

EM-NLCs were prepared by the previously reported emulsification ultra-sonication method with slight modification [30]. The detailed composition of the formulations (EM-NLCs1—EM-NLCs20) is shown in Table 2. The solid lipid was melted at 80 °C (5 °C above the melting point). The liquid lipid and EM were added into melted solid lipid to form a homogenous mixture. Separately, a hot aqueous surfactant solution was prepared and heated to the same temperature (85 °C). The surfactant solution was added to the molten lipid mixture with continuous stirring to form the pre-emulsion. The prepared pre-emulsion homogenized (IKA T50 digital Homogenizer, Germany) at 7000 rpm for 2 min and then subjected to ultrasound using a probe sonicator (Qsonica-500 Sonicator, Newtown, CT, USA). The formed nanoemulsion was transferred to glass vials and kept at room temperature (25 °C) for recrystallization to form NLCs. The prepared EM-NLCs were stored in the vial for further characterization.

Table 1. Independent factors and their levels with the dependent factors used in the central composite design.

Independent Factors	Units	Levels			Star Points	
		Low (−1)	Medium (0)	High (+1)	−α (−1.68)	+α (+1.68)
Lipid (A)	(%)	2	4	6	0.64	7.36
Surfactant (B)	(%)	2	3.5	5	0.98	6.02
Sonication time (C)	(min)	2.5	5	7.5	0.8	9.2
Dependent factors		Goals				
Particle size	(nm)	Minimum				
Entrapment efficiency	(%)	Maximum				

Table 2. Composition of various batches and their predicted values obtained from central composite design along with their experimental values of EM-NLCs.

Formulation	Lipid (%)	Surfactant (%)	Sonication Time (min)	Particle Size (nm)	Entrapment Efficiency (%)
	A	B	C	Y_1	Y_2
EM-NLCs1	2	2	2.5	239.2 ± 5.2	73.5 ± 1.2
EM-NLCs2	6	2	2.5	350.6 ± 4.2	69.4 ± 1.3
EM-NLCs3	2	5	2.5	219.5 ± 6.2	71.5 ± 1.6
EM-NLCs4	6	5	2.5	261.5 ± 4.9	87.5 ± 2.2
EM-NLCs5	2	2	7.5	199.7 ± 3.7	78.5 ± 1.5
EM-NLCs6	6	2	7.5	319.4 ± 7.6	76.5 ± 1.9
EM-NLCs7	2	5	7.5	122.6 ± 6.1	70.5 ± 1.4
EM-NLCs8	6	5	7.5	201.3 ± 7.2	84.5 ± 1.6
EM-NLCs9	0.64	3.5	5	140.6 ± 3.4	70.5 ± 1.7
EM-NLCs10	7.36	3.5	5	285.3 ± 6.2	84.5 ± 1.9
EM-NLCs11	4	0.98	5	346.9 ± 7.9	73.5 ± 1.4
EM-NLCs12	4	6	5	177.3 ± 5.2	80.2 ± 1.6
EM-NLCs13	4	3.5	0.8	287.4 ± 6.3	76.5 ± 1.8
EM-NLCs14	4	3.5	9.2	177.5 ± 5.3	78.5 ± 1.4
* EM-NLCs15	4	3.5	5	195.5 ± 3.6	82.6 ± 1.2
* EM-NLC16	4	3.5	5	195.5 ± 3.6	82.6 ± 1.2
* EM-NLC17	4	3.5	5	195.5 ± 3.6	82.6 ± 1.2
* EM-NLC18	4	3.5	5	195.5 ± 3.6	82.6 ± 1.2
* EM-NLC19	4	3.5	5	195.5 ± 3.6	82.6 ± 1.2
* EM-NLC20	4	3.5	5	195.5 ± 3.6	82.6 ± 1.2

* Centre point having same composition.

3.5. Characterization

3.5.1. NLCs Evaluation

The particle size (PS), polydispersibility index (PDI), and zeta potential (ZP) were measured by the size analyzer (Malvern zeta sizer, Malvern, UK). The diluted samples (100-fold) were placed into the cuvette and analysed for PS and PDI. ZP was measured by taking the samples into a cuvette having an electrode. Each sample was analysed in triplicate and data were shown as mean ± SD.

3.5.2. Entrapment Efficiency (EE)

The entrapment efficiency of EM-NLCs was analysed by the indirect method. The prepared EM-NLCs were filled into a centrifugation tube and centrifuged at 10,000 rpm into a centrifuge (Hettich, Tuttlingen, Germany). The supernatant was separated, and the EM content was analysed by UV spectrophotometer (Genesys 10S UV-Vis, Thermo-scientific,

Waltham, MA, USA) at 280 nm. Then, the amount of EM entrapped into NLCs formulation was calculated by the below formula:

$$EE(\%) = \frac{\text{Total EM} - \text{Unentrapped EM}}{\text{Total EM}} \times 100 \tag{2}$$

3.5.3. Development of In Situ Gel of EM-NLCs (EM-NLCs-IG)

CCD optimized composition (EM-NLCs-IG) was converted to in situ gel (EM-NLCs-IG) by using the polymer blend chitosan and Carbopol 940. The fixed concentration of chitosan (0.4% w/v) was transferred to acetate buffer (pH 4.5) and kept overnight with stirring. The various concentrations of carbopol solution were prepared by dispersing in distilled water, as shown in Table 3. The chitosan solution (0.4% w/v) was added to carbopol solution and mixed to obtain a homogeneous dispersion. Finally, the lyophilized EM-NLCs (0.5%) were dispersed into carbopol-chitosan solution and stored for further study.

Table 3. Regression values of the selected responses during optimization.

Model	Y_1 (Particle Size, nm)				
	SD	**R^2**	**Adjusted R^2**	**Predicted R^2**	**p-Value**
Linear	31.8	7909	0.7517	0.6648	0.0073
2FI	29.8	8496	0.7802	0.5737	0.0198
Quadratic	11.6	0.9827	0.9671	0.8997	0.0001
	Y_2 (Entrapment efficiency, %)				
Linear	4.29	0.4887	0.3929	0.1700	<0.0001
2FI	2.32	0.7512	0.6364	0.2857	<0.0001
Quadratic	0.71	0.9913	0.9836	0.9252	<0.0011

3.5.4. Gelling Strength and Viscosity

The gelling strength of EM-NLCs-IG was evaluated using simulated tear fluid (STF, sodium chloride, sodium bicarbonate, and dehydrated calcium chloride). STF (2 mL) was taken into a glass vial and a drop of EM-NLCs-IG (sol form) was added. The temperature was maintained at $37 \pm 0.5\ ^\circ$C and visually evaluated for gelling time, as well as to convert again into sol form [31]. The viscosity of EM-NLCs-IG was analysed by the Brookfield viscometer (Fungilab, Barcelona, Spain) at 20 rpm.

3.5.5. Clarity, Optical Transmittance and pH Determination

The clarity of ophthalmic preparation was necessary due to the presence of visible particles that irritate the ocular tissue. The clarity was examined visually by keeping the sample behind the white backboard in sol as well as gel state. The transmitting of EM-NLCs-IG was examined by UV spectrophotometer at 480 nm using STF as blank. The pH of EM-NLCs-IGs was measured by pH meter (Digital pH meter, Hicon, New Delhi, India).

3.5.6. In Vitro Drug Release

The in vitro release of EM from EM-NLCs-opt, EM-NLCs-opt-IG4 and EM in situ gel (EM-IG) having the same dose of EM were analysed by the pre-treated dialysis bag method. The samples (1 mL; 0.5%) were transferred to a dialysis bag (molecular weight cut off, 12–14,000 kDa, Sigma Aldrich, St. Louis, MO, USA) and immersed into STF (100 mL), as release media. The release assembly was fixed on the thermostat magnetic stirrer with 100 rpm and a temperature of $37 \pm 0.5\ ^\circ$C. At a predetermined interval, 2 mL of release media was taken and replaced with fresh STF to maintain the concentration gradient. The released content was assessed on the UV-Vis spectrophotometer at 280 nm after appropriated dilution. The release value of the optimized EM-NLCs-opt-IG4 was fitted into various kinetic release models for finding the release pattern of EM.

3.5.7. Mucoadhesive Study

The mucoadhesive potential of EM-NLCs-opt-IG4 was performed by the physical balance method using goat cornea [32,33]. The freshly excised goat cornea was collected from the slaughterhouse and washed to remove adhered materials. Then, it was attached to the backside of the physical balance pan and fixed to the sample holder. The temperature was maintained at 37 ± 0.5 °C for the whole study. The sample was transferred to the sample holder and kept for a few minutes to attach to the cornea. The weight was added to the second pan of physical balance until the cornea was separated from the EM-NLCs-opt-IG4. The weight at which the sample separated from the balance is noted and bioadhesive potential was calculated by the given formula and denoted by dyne/cm^2.

3.5.8. Ex Vivo Goat Corneal Permeation

The freshly excised goat cornea was used for the permeation study and the result was compared to the amount of EM permeated from EM-NLCs-opt, EM-NLCs-opt-IG and EM-IG. The fresh eyeball was collected from a local slaughterhouse and stored in 0.9% NaCl solution at 4 °C. The cornea was separated with the sclera and washed with STF. The cornea was mounted between the donor and acceptor compartment of the diffusion cell (Franz diffusion, Logan FDC-6, Hudson, NJ, USA). The receptor media (STF) was filled to the acceptor compartment and the temperature was maintained at 37 ± 0.5 °C. The samples (equivalent to 0.5% EM) were filled into the donor compartment and at a definite time, the permeated sample (1 mL) was collected. The same volume of fresh blank STF was added to maintain the uniform volume. The sample was filleted through a filter (0.45 μm), diluted and then injected (20 μL) into the HPLC column for the analysis using the previously developed HPLC method [34]. HPLC instrument (HPLC, Auto sampler, Shimadzu LC10AD, Kyoto, Japan) with C18 column, mobile phase system 0.02 M potassium phosphate buffer and acetonitrile (60:40), a flow rate of 0.75 mL/min and detection at 254 nm was employed for the evaluation. The amount of drug permeated and flux was calculated.

3.5.9. Histopathological Examination

This study was carried out to observe any alteration in the internal corneal structure with the tested samples compared with the positive control. The tissue was treated similar to the ex vivo corneal permeation study for EM-NLCs-opt-IG4, negative control (0.9% NaCl) and positive control (1% w/v; SLS). The cornea was collected and stored in formalin solution (0.8%, v/v). The cornea was fixed into the solid block using paraffin after dehydration with alcohol. The microtome cutter was used to cut the cross-section and further stained with haematoxylin and eosin dye. The image was captured with a Motic digital optical microscope at $40\times$ magnification.

3.5.10. Corneal Hydration

The corneal hydration test was performed to determine the ocular tolerance of the cornea [35]. The goat cornea was taken after the permeation study and the initial weight was taken (W_1, wet weight). Then, it was placed into a hot air oven at 60 °C for 72 h and again weighted (W_2, dry weight). The corneal hydration (%) was calculated as per the given equation (Equation (3))

$$\text{Corneal hydration } (\%) = \frac{W1 - W2}{W2} \times 100 \tag{3}$$

3.5.11. HET-CAM Irritation Study

HET-CAM study is an in vitro method used to determine the irritation potential of the selected formulation (EM-NLCs-opt-IG4) and the result was compared with normal saline (negative control) and 0.1 N NaOH (positive control). The study was performed on chorioallantoic membrane (CAM) of a hen's egg, which is similar to the blood capillary of the eye [36]. The fertilized eggs were collected from a local poultry farm and placed in a box with the air chamber in an upward position. The eggs were incubated for nine days at 37 ± 1 °C and $50 \pm 1\%$ RH (Orbital incubator, Thermo Fisher Scientific, Waltham, MA, USA). On the 10th day, the eggs were removed from the incubator and shell was removed using forceps. The normal saline was added to remove white membrane without damaging the CAM. The samples of EM-NLCs-opt-IG4, negative control and positive control were added and the irritation scores were noted at a different time from the scale (0–0.9 non-irritant; 1–1.99 mild irritant; 2–2.99 moderate irritant; 3–<3 severe irritant) [37].

3.5.12. Sterility and Isotonicity Evaluation

Sterility was evaluated by using fluid thioglycollate and soybean casein digest medium. The fluid thioglycollate and soybean casein digest medium were prepared and sterilized at 121 °C for 15 min in an autoclave. The sample (EM-NLCs-opt-IG4) was inoculated in both mediums under aseptic conditions then incubated for 14 days and evaluated under aseptic conditions for turbidity and precipitation. Isotonicity of the EM-NLCs-opt-IG4 was evaluated on the goat blood sample and the result was compared with normal saline (0.9% NaCl). A drop of blood was taken on a glass slide and mixed with EM-NLCs-opt-IG4 and normal saline. The smear was prepared on a glass slide and further stained with Leishman's stain. The slide was kept aside to air-dry and the damage in red blood cells was observed under the microscope [38].

3.5.13. Antibacterial Activity

The antimicrobial susceptibility test of the developed formulation (EM-NLCs-opt, EM-NLCs-opt-IG4, EM-IG4) was evaluated by disk diffusion method on Gram-positive (*Staphylococcus aureus*) and Gram-negative (*Escherichia coli*) in nutrient agar medium. The bacterial strain was grown on the sterile Luria Bertani broth. The nutrient agar medium was prepared and sterilized at 121 °C. The bacterial culture of both microorganisms (100 µL, 10^6 CFU/mL) was mixed with the agar medium (10 mL), transferred into a Petri dish under aseptic conditions, and kept for solidification without any agitation. The samples (EM-NLCs-opt-IG4 and EM-IG) were soaked into a disk and placed in a Petri plate under aseptic conditions. The plate was placed into the incubator for 24 h in an inverted position and the zone of inhibition was measured for each sample by the graduated scale and the result was compared.

4. Result and Discussion

4.1. Screening of Solid and Liquid Lipid

The screening of solid lipid, liquid lipid and surfactant were performed for the selection of lipid and surfactant based on the maximum solubility, and results are expressed graphically in Figure 1. The solubility order of EM in solid lipids was found to be as stearic acid > compritol 888 > isopropyl myristate > myristic acid > lauric acid > cetyl palmitate (Figure 1A). The maximum solubility of EM was found to be in stearic acid (95.5 ± 9.5 mg/mL). EM showed solubility order in different liquid lipids as oleic acid > peceol > sesame oil > soya oil > olive oil > coconut oil (Figure 1B). The maximum solubility of EM was found to be in oleic acid (73.63 ± 3.65 mg/mL). Figure 1C shows the solubility of EM in various surfactants in order as pluronic F127 > tween 20 > cremophor EL > bile salt > span 20. The maximum solubility of EM was found in pluronic F127 as 97.87 ± 8.81 mg/mL. So, finally, stearic acid, oleic acid and pluronic F127 were selected as the final ingredients for the formulation of NLCs.

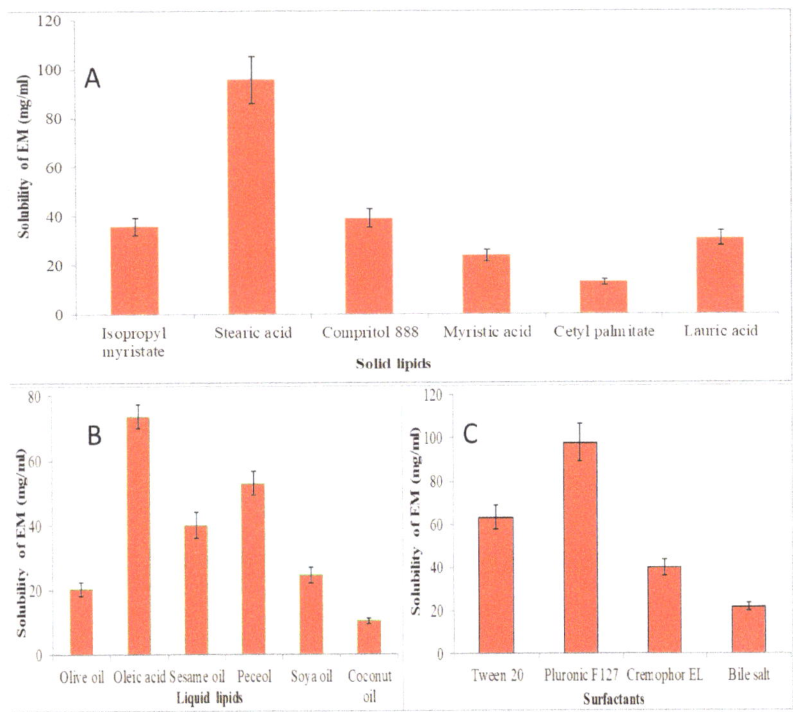

Figure 1. Solubility of EM in various solid lipids (**A**), liquid lipids (**B**), and surfactant (**C**). The study performed in triplicate and data shown as mean ± SD.

4.2. Selection Solid and Liquid Lipid Ratio by Miscibility

The solid (stearic acid) and liquid lipid (oleic acid) were mixed in various ratios (9:1, 8:2, 7:3, 6:4, 5:5) and visually observed for phase separation and turbidity. The final ratio of solid and liquid lipid at 6:4 exhibited no phase separation, turbidity as well as did not show any separated oil droplets on the filter paper.

4.3. Optimization

The formulation design is an important factor to consider the effect of factors, in addition to the interactions. In the present study, lipid (%, A), surfactant (%, B) and sonication time (min, C) were chosen as independent factors and their effect was observed on PS (nm, Y_1) and EE (%, Y_2) (Table 2).

4.4. Effect of Variables on Size (Y_1)

As shown in Table 2, the PS for different formulations were found in the range of 122.6 ± 6.1 nm (EM-NLCs7) to 350.6 ± 4.2 nm (EM-NLCs2). The lowest PS was found using lipid (6%), surfactant (2%) and sonication time (2.5 min). The highest PS shown by the formulation with lipid composition (2%), surfactant (5%) and sonication time (7.5 min). The result exhibited a significant variation in PS after varying the composition. The used three factors showed a considerable effect on the PS. The three-dimensional response surface plot (Figure 2A) and polynomial Equation (4) depicted the effect of independent variables on the PS (Y_1). The factors, i.e., lipid (A) showed a positive effect, surfactant (B) and sonication time (C) showed a negative effect on the PS. The increase in PS was observed with the increase in lipid content, as shown in EM-NLCs1 to EM-NLCs2 (Table 2). This might be due to the presence of excess lipids, which increase the viscosity of the dispersion [27]. The other reason may be incomplete emulsification, which may be responsible for the aggregation of the particle due to the lack of surfactant. The surfactant (B) demonstrated a negative effect on PS due to the reduction in interfacial tension between the aqueous and lipid phase, resulting in the reduction of PS (EM-NLCs1-EM-NLCs3). A similar type of finding was reported by Yang et al. in their research [39]. The third factor sonication time (C) also showed a negative effect. On increasing the duration of sonication, the PS of NLCs was decreased (EM-NLCs1- EM-NLCs5). This might be due to the increase in the energy to the system, which is responsible for the prevention of particle agglomeration [40]. The combined effect of the factors also depicted a significant effect on the PS. The second-order polynomial generated for PS is given below:

$$\text{Particle size } (Y_1) = 195.43 + 42.85A - 40.95B - 30.95C - 17.55AB + 6.88AC - 14.55BC + 6.63A^2 + 24.01B^2 + 13.53C^2 \tag{4}$$

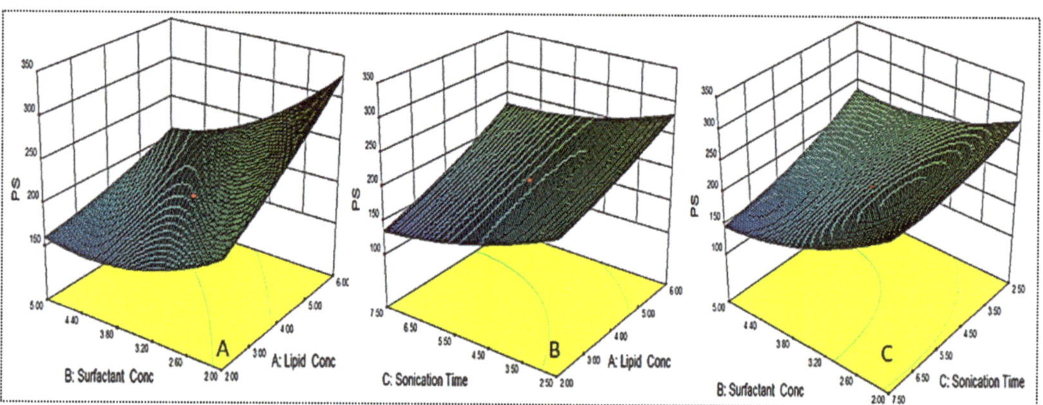

Figure 2. 3D response surface plot showing effect of factors lipid (**A**), surfactant (**B**) and sonication time (**C**) on the particle size (Y_1).

The data were fitted to various design models and the best fit model was found to be quadratic. The regression value ($R^2 = 0.9827$) was found to be maximum for this model. The lack of fit was found to be insignificant ($p = 0.6787$). The regression values of the selected responses during optimization were given in Table 3. The regression coefficient (R^2) of this equation was found to be 0.9827, indicating the good correlation between response and selected factors.

4.5. Effect of Variables on Encapsulation Efficiency

As shown in Table 2, the EE for different formulations was found in the range of 69.4 ± 1.3% (EM-NLCs2) to 87.5 ± 2.2% (EM-NLCs4). The minimum encapsulation is shown by the composition lipid (6%), surfactant (2%) and sonication time (2.5 min), and the maximum is shown by the composition lipid (6%), surfactant (5%) and sonication time (2.5 min). The result exhibited a significant variation in the encapsulation efficiency after varying the NLCs composition. The three-dimensional response surface plot (Figure 3) and the polynomial Equation (5) depicted the effect of independent variables on the EE (Y_2). The factors lipid (A), surfactant (B) and sonication time (C) showed a positive influence on the encapsulation efficiency. The increase in lipid concentration led to an increase in EE, as depicted in formulations EM-NLCs7 and EM-NLCs8, at a fixed level of surfactant (B) and sonication time (C). The possible reason for this effect is the availability of more space to accommodate the lipophilic drug within the NLCs. However, if the amount of lipid was increased beyond an optimum value (or insufficient amount of surfactant), the EE decreased, as found in EM-NLCs1 and EM-NLCs2 [40]. Similarly, in the presence of sufficient lipid with increased surfactant concentration, the EE also increased (EM-NLCs1 and EM-NLCs 14). This might be due to the higher solubility of the drug in the lipid and the presence of sufficient surfactant. On the other hand, at lower lipid content, the reverse results will cause a reduction in EE (EM-NLCs1 and EM-NLCs3). This effect may be due to the leakage of a drug into the external phase, reducing the EE. The third factor sonication time depicted the enhancement in EE at longer sonication duration (EM-NLCs1 and EM-NLCs5). This might be due to the high energy provided by sonication helping to prevent the drug from leaching from the NLCs system, thus increasing the EE [27]. The second-order polynomial generated for entrapment efficiency was found to be:

$$\text{Entrapment efficiency } (Y_2) = 82.61 + 4.00A + 1.93B + 0.92C + 3.61AB - 1.14AC - 2.14BC - 1.88A^2 - 2.11B^2 - 1.88C^2 \quad (5)$$

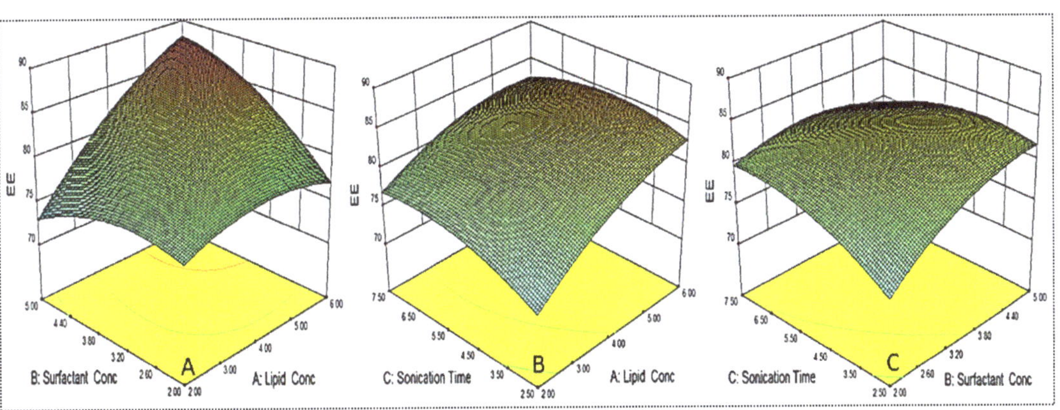

Figure 3. 3D response surface plot showing effect of factors lipid (**A**), surfactant (**B**) and sonication time (**C**) on the encapsulation efficiency (Y_2).

The independent variables exhibited a positive effect on entrapment efficiency (Y_2) with the quadratic ($R^2 = 0.9913$) model as the best fit model. The lack of fit was found to be insignificant ($p = 0.6171$) with F-value of 2.67. The regression values of the all model for each response are shown in Table 3. The regression coefficient (R^2) of the quadratic model was found to be 0.9913, indicating the good correlation between response and selected factors. A close agreement in the regression value was observed for the actual and predicted quadratic model.

4.6. Point Prediction

The optimized formulation (EM-NLCs-opt) was selected from the point prediction method after slight modification in the independent variables and their effect was observed on PS (Y_1) and EE (Y_2), as shown in Table 4. The optimum composition of EM-NLCs-opt formulation is found as lipid (3.5%), surfactant (4%) and sonication time (5.5 min). The actual value of PS and EE of EM-NLCs-opt formulation was found to be 169.6 ± 4.8 nm (Figure 4A), and 81.7 ± 1.4%, respectively. The predicted PS and EE were found to be 167.4 nm and 83.8%, respectively. The close agreement between the practical and predicted value confirms the validation of the model. PDI (0.14) and ZP (−23 mV) values of the optimized formulation were found under the standard limit. The practical value of PDI (<0.5) and ZP (±30 mV) indicates that prepared NLCs have shown homogeneous distribution and high stability. The value lower and greater than the standard gives the unstable composition. The morphology of EM-NLCs-opt was evaluated by TEM and the image showed spherical shape particles with uniform distribution (Figure 4B). There was no aggregation observed, which further supports the PDI evaluation.

Table 4. Point prediction optimization by central composite design.

Formulations	Lipid: Surfactant: Sonication Time	Actual Value		Predicted Value	
		Y_1 (nm)	Y_2 (%)	Y_1 (nm)	Y_2 (%)
EM-NLC-opt	3.5:4:5.5	169.6 ± 4.8	81.7 ± 1.4	167.4	83.8

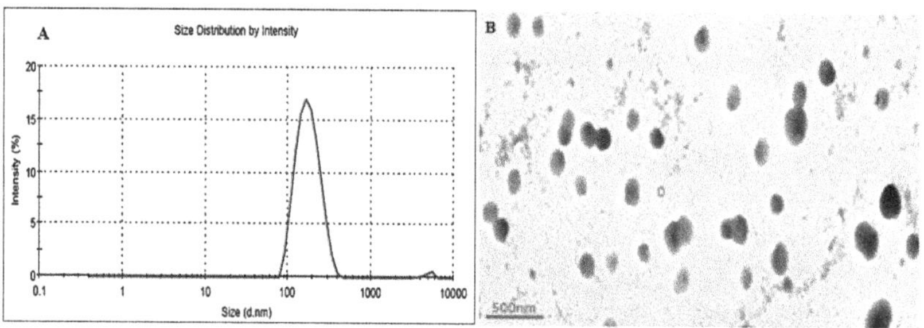

Figure 4. (**A**). Particle size and (**B**). TEM image of EM-NLCs-opt.

4.7. Development of EM-NLCs In Situ Gel

The optimized formulation (EM-NLCs-opt) was converted to in situ gel (EM-NLCs-opt-IG) using different concentrations of carbopol and a fixed concentration of chitosan. The prepared gels were evaluated for different parameters to select the optimum gelling agent concentration (Table 5).

Table 5. Physicochemical parameter evaluation of EM-NLCs in situ gel system. The study performed in triplicate and data shown as mean ± SD.

Code	Carbopol (%)	Chitosan (%)	pH	Optical Transmittance (%)	Drug Content (%)	Gelling Strength	Viscosity (cP)	
							Solution	With STF
EM-NLCs-opt2-IG1	0.1	0.4	6.54 ± 0.4	94.8 ± 1.1	98.7 ± 3.1	−	6.2 ± 1.1	10.5 ± 2.1
EM-NLCs-opt2-IG2	0.2	0.4	6.43 ± 0.6	95.8 ± 0.8	98.8 ± 2.1	+	38.5 ± 3.7	141.3 ± 2.4
EM-NLCs-opt2-IG3	0.3	0.4	6.24 ± 0.2	95.7 ± 1.1	98.2 ± 2.7	++	50.4 ± 1.7	176.4 ± 2.5
EM-NLCs-opt2-IG4	0.4	0.4	6.16 ± 0.2	98.5 ± 0.4	99.6 ± 3.7	+++	87.4 ± 2.3	216.3 ± 1.3
EM-NLCs-opt2-IG5	0.5	0.4	6.02 ± 0.7	96.6 ± 0.9	99.6 ± 1.4	++++	98.5 ± 2.1	254.8 ± 1.1

(−) no gel formation, (+) gel formation in 3 min and disappeared <10 min, (++) gel formation in <5 s and disappeared after 2 h, (+++) gel formation in <5 s and stable for >24 h, (++++) gel formation in <5 s, hard gel and stable for >24 h.

4.8. Characterization of EM-NLCs In Situ Gel

4.8.1. Gelling Strength

Gelling strength is an important parameter for the in situ gel system as it directly influences the residence time in the ocular region. It is indicated by gel formation on contact with STF at physiological pH and remains stable for a long time. The gelling capacity of different EM-NLCs-IG formulations is shown in Table 5. The different concentrations of gelling agents (carbopol and chitosan) were used to prepare EM-NLCs-opt-IG1 to EM-NLCs-opt-IG5 and the gelling strength was represented by the negative and positive signs. EM-NLCs-opt-IG1 showed no gelation, EM-NLCs-opt-IG2, EM-NLCs-opt-IG3 indicates slight gelation but have different gelation times. EM-NLCs-opt-IG2 showed gelation time of 3 min but disappeared after a few minutes (<10 min) and EM-NLCs-opt-IG3 showed gelation quickly (5 s) and stable for 2 h. However, EM-NLCs-opt-IG4 exhibited quick gelation (5 s) and were stable for more than 24 h and EM-NLCs-opt-IG5 depicted the gelation time (<5 s), and it forms the thick gel and was found stable for more than 24 h. From the result, it was observed that with the increase in the concentration of carbopol the gelling power increases. Gel formation took place by electrostatic repulsion between an ionic functional group of polymers (pH-sensitive) on increasing the pH when in contact with tear fluid [41,42].

4.8.2. Viscosity Measurement

Due to the high shear rate during the blinking ($0.03 \, S^{-1}$) of the eyelid, most of the administered formulation dose was eliminated from the cul-de-sac [43]. So, the optimum viscosity of the in situ gel system needed was not affected by the shear rate and tear fluid flow. Viscosity also influenced the gelling strength and residence time, in addition to prolonged releases. The result of the viscosity of EM-NLCs-opt-IGs formulations in normal and physiological conditions (STF) is shown in Table 5. The result of the study showed that the viscosity of formulation increases with the increase in carbopol concentration at a fixed chitosan concentration in both conditions (normal and STF). The viscosity of EM-NLCs-opt-IG4 prepared with carbopol (0.4%) and chitosan (0.4%) increased after adding it into STF, due to the crosslinking of the polymer with the ion present in STF [44]. The EM-NLCs-opt-IG4 exhibited the viscosity of 216.32 ± 1.35 cp with STF, and the gel remained stable for more than 24 h, hence it was selected as the optimized formulation.

4.8.3. In Vitro Drug Release Study

The comparative drug release study of EM-NLCs-opt, EM-NLCs-opt-IG4 and EM-IG was performed, and the result is shown graphically in Figure 5. The release order of EM was found as EM-NLCs-opt ($65.3 \pm 3.9\%$) > EM-NLCs-opt-IG4 ($56.6 \pm 4.4\%$) > EM-in situ gel ($27.32 \pm 4.1\%$) in 6 h. However, EM-NLCs-opt, EM-NLCs-opt-IG4, EM-IG exhibited sustained release, i.e., $91.3 \pm 3.8\%$, $76.3 \pm 4.5\%$ and $41.2 \pm 5.1\%$ in 24 h. The formulation EM-NLCs-opt showed a higher EM release than EM-NLCs-opt-IG4. This may be due to surface deposition of EM on the surface of NLCs, as well as the direct contact of EM-NLCs with release media. However, EM-NLCs-opt-IG4 showed slightly lower EM release than EM-NLCs-opt due to the presence of an extra layer of gelling agent on the NLCs. The drug (EM) needs to cross the gel matrix, so the lesser EM releases. The slower release of EM is ideal for the ophthalmic topical formulation, which helps to achieve better therapeutic activity. EM-NLCs-opt and EM-NLCs-opt-IG4 showed significantly higher release of EM due to the increase in the solubility of EM in the presence of the lipid and surfactant. The drug release from EM-IG was found to be significantly low due to the poor solubility of EM, high viscosity, and formation of insoluble gel matrix hindering the release of EM in STF media [45]. The release data of the EM-NLCs-opt-IG4 were fitted in different kinetic models to determine the best fit model. The regression coefficient (R^2) of kinetic model, i.e., zero-order ($R^2 = 0.7039$), first-order ($R^2 = 0.8435$), Higuchi ($R^2 = 0.8887$), Korsmeyer—Peppas model ($R^2 = 0.9269$) and Hixon—Crowell model ($R^2 = 0.8185$). The maximum R^2 value was found to be for the Korsmeyer—Peppas model and considered as the best fit model. The $n = 0.48$ ($0.43 < n < 0.85$) indicated non-Fickian diffusion (anomalous) mechanism of drug release [46].

4.8.4. Mucoadhesive Study

The mucoadhesive force of EM-NLCs-opt-IG4 was determined by the physical balance method and the value was found to be 1087.38 dyne/cm^2. The value was found to be significantly higher than the tear film shear force (150 dyne/cm^2). The high mucoadhesiveness of EM-NLCs-opt-IG4 is due to the combination of gelling polymers (carbopol and chitosan). It indicates that EM-NLCs-opt-IG4 have a high residence time and are not eliminated by tear fluid flow and blinking.

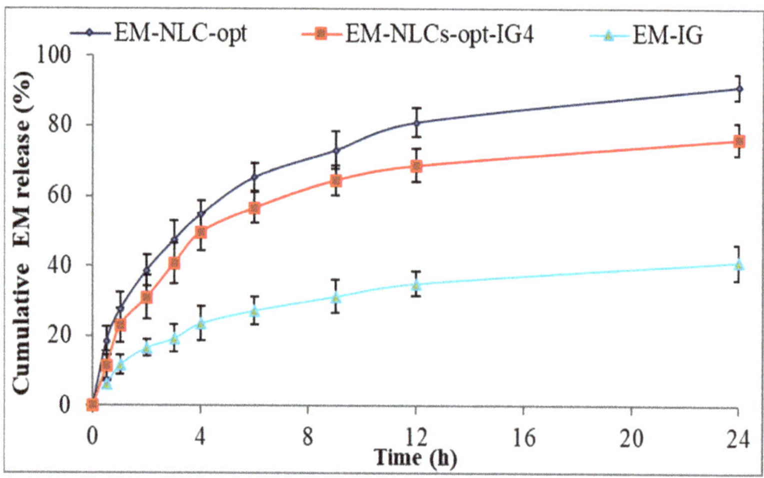

Figure 5. In vitro drug release profile of EM-NLCs-opt, EM-NLCs-opt-IG4, and EM-in situ gel (EM-IG). The study performed in triplicate and data shown as mean ± SD.

4.8.5. Ex Vivo Corneal Permeation Study

Figure 6 shows the ex-vivo permeation result of the EM-NLCs-opt, EM-NLCs-opt-IG4 and EM-IG on goat cornea. EM corneal permeation (%) was found to be 68.2 ± 5.4% from EM-NLCs-opt, 56.7 ± 4.5% from EM-NLCs-opt-IG4 and 37.6 ± 4.5% from EM-IG. The highly significant ($p < 0.001$) permeation was achieved from EM-NLCs-opt and EM-NLCs-opt-IG4 than EM-IG. A significant ($p < 0.05$) difference was observed between EM-NLCs-opt and EM-NLCs-opt-IG4. The permeation data was also used to calculate the flux. The result showed the flux value of 172.15 µg cm^{-2}/h from EM-NLCs-opt, 143.24 µg cm^{-2}/h from EM-NLCs-opt-IG4 and 94.91 µg cm^{-2}/h from EM-IG. EM-NLCs-opt-IG4 exhibited lesser permeation and flux than EM-NLCs-opt due to the formation of a gel network matrix in STF. However, EM-NLCs-opt-IG4 exhibited significant ($p < 0.05$) high permeation and flux than EM-IG. The high permeation and flux of EM-NLCs-opt-IG4 than EM-IG is due to nano-metric size, presence of surfactant (helps to solubilize), presence chitosan (bioadhesive polymer) and carbopol (helps to open the tight junction of lipid membrane). The corneal permeation enhances due to the increase in residence time and prevents the loss of drugs by tear fluid. It can also develop a film over corneal epithelium and release the drug slowly [47]. Nano-size range of particles can be internalized in the carnal cell by a receptor-mediated endocytosis uptake mechanism that can increase the permeation [16].

4.8.6. Histopathological Examination

Histopathology study on goat cornea showed the cornea treated with EM-NLCs-opt-IG4 and negative control (0.9% NaCl) had typical corneal structures (Figure 7A,B), and cells retained their normal morphology. It was concluded that, upon treatment, the epidermal layer remains unchanged. The cornea treated with the positive control (1% *w/v*; SLS) showed damage to the destruction of corneal epithelial cells (Figure 7C). Finally, based on histopathology study, it was established that the EM-NLCs-opt-IG4 was found safe for ocular delivery.

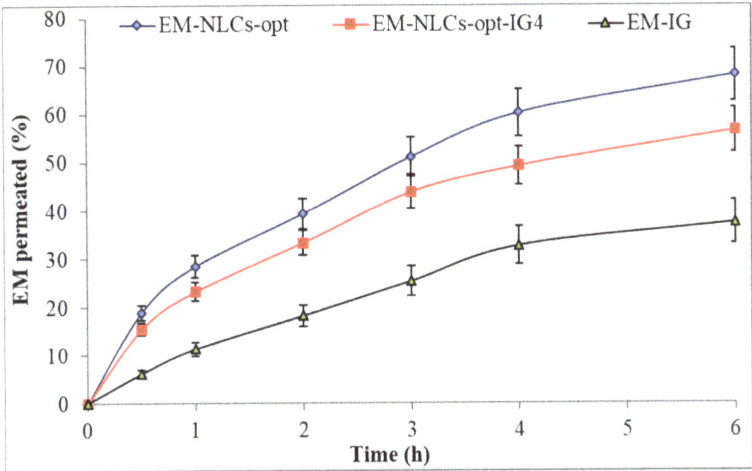

Figure 6. Ex vivo corneal permeation of EM-NLCs-opt, EM-NLCs-opt-IG4 and EM-In situ gel (EM-IG). The study performed in triplicate and data shown as mean ± SD.

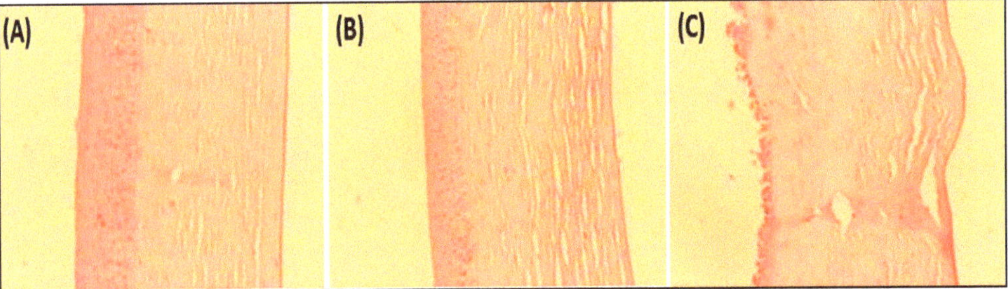

Figure 7. Histopathological image of (**A**). EM-NLCs-opt-IG4, (**B**). Negative control (0.9% NaCl) and (**C**). Positive control as SLS (1% w/v).

4.8.7. Corneal Hydration

This study determined the damage in goat cornea after the application of the formulation. EM-NLCs-opt-IG4 treated cornea showed a hydration value of 79.34% after the study. The value was found to be within the standard limit of 75–80% [48]. So, from the study, we can say that the developed formulation was found to be safe for ocular administration.

4.8.8. HET CAM Irritation Study

HET-CAM (irritation) study was performed for EM-NLCs-opt-IG4, normal saline (0.9% NaCl, negative control) and positive control 0.1N sodium hydroxide) using a fertilized hen egg's (Figure 8). The developed CAM after incubation is similar to the artery and vein of an eye. EM-NLCs-opt-IG4 and normal saline-treated CAM showed no irritation. There was no haemorrhage and bleeding from the artery and vein of CAM (no breaking) observed in 5 min. and the score was found to be closer to zero. In addition, the positive control (0.1N NaOH) treated CAM depicted a score of 6 and indicated haemorrhage of artery and vein of CAM in 5 min. So, the positive control treated CAM is considered a severe irritant. The study revealed that the developed EM-NLCs-opt-IG4 formulation was found to be non-irritant and safe for ophthalmic administration.

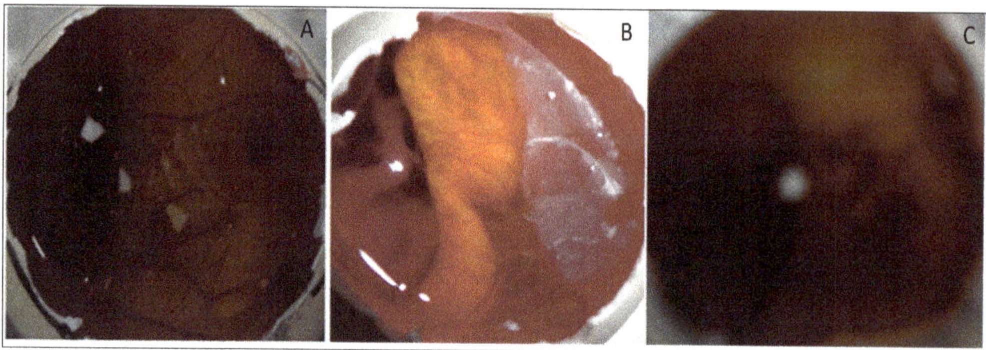

Figure 8. HET-CAM image of (**A**). EM-NLCs-opt-IG4, (**B**). Negative control (0.9% NaCl) and (**C**). Positive control (0.1 N NaOH).

4.8.9. Sterility and Isotonicity Study

The sterility test was performed to evaluate the microbial growth in EM-NLCs-opt-IG4. The sterility study result showed no microbial growth in fluid thioglycolate and soybean casing digest media after placing the formulation for 14 days. Figure 9 depicted the isotonicity test of EM-NLCs-opt-IG4 and the result compared with the normal saline (0.9% NaCl). No damage in red blood cells was observed (shrinking and swelling) after treatment with EM-NLCs-opt-IG4. The result was found to be the same for normal saline-treated cells. From the study, it can be revealed that EM-NLCs-opt-IG4 formulation was found to be isotonic.

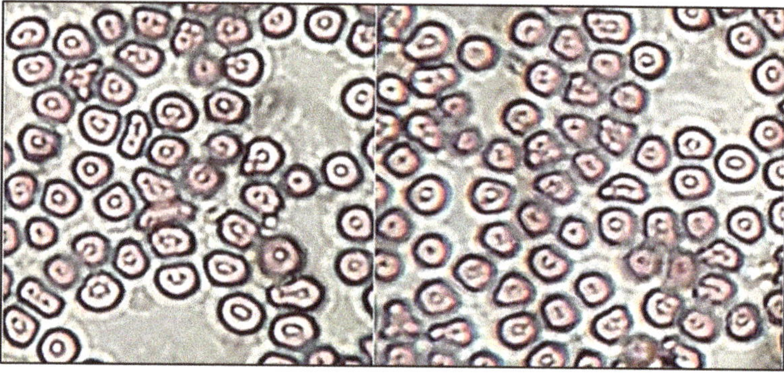

Figure 9. Isotonicity image of EM-NLCs-opt-IG4 and normal saline treated blood cells.

4.8.10. Antimicrobial Activity

Antibacterial evaluation study EM-NLCs-opt-IG4 and normal EM-IG were assessed on Gram-positive bacteria (*S. aureus*) and Gram-negative (*E. coli*) bacterial the using cup plate method and data was depicted in Figure 10. EM-NLCs-opt-IG4 exhibited ZOI of 16 ± 1 mm (12 h) and 19 ± 1 mm (24 h) against *S. aureus*. EM-IG showed ZOI of 10 ± 1 mm (12 h) and 13 ± 2 mm (24 h) against *S. aureus*. However, EM-NLCs-opt-IG4 showed ZOI against *E. coli* is of 17 ± 1 mm and 21 ± 2 mm in 12 h and 24 h, respectively. Furthermore, EM-IG showed ZOI against *E. coli* is 11 ± 1 mm and 14 ± 2 mm in 12 h and 24 h, respectively. The results revealed that EM-NLCs-opt-IG4 exhibited significantly ($p < 0.001$) higher activity than EM-IG due to more solubility of EM in the NLCs gel system due to the presence of the lipid and surfactant. This may enhance the permeation to cell wall of bacteria and inhibited protein synthesis by arresting ribosome. It was also observed that EM is more effective in Gram-negative bacteria than Gram-positive bacteria.

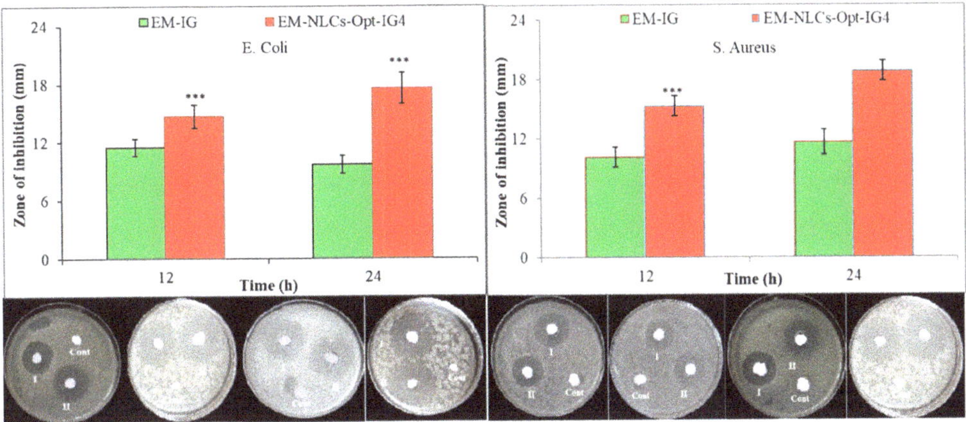

Figure 10. Comparative antimicrobial results of EM-in situ gel (EM-IG) and EM-NLCs-opt-IG4 against *S. aureus* and *E. coli* at 12 h and 24 h study. The study performed in triplicate and data shown as mean $\pm$ SD. *** $p < 0.001$ considered as highly significant than EM in situ gel.

5. Conclusions

Erythromycin loaded NLCs were successfully prepared by the emulsification ultrasonication method and further optimized by central composite design using lipid (stearic acid+ oleic acid), surfactant (Pluronic F127) and sonication time. EM-NLCs-opt showed nanometric size with high entrapment efficiency. EM-NLCs-opt formulation was successfully incorporated in situ gel using pH-sensitive carbopol 940 and chitosan. EM-NLCs-opt-IG4 showed frequent gelation, stable for more than 24 h and also depicted significantly enhanced bioadhesion. EM-NLCs-opt-IG4 showed a sustained release profile with high ex vivo goat permeation than EM in situ gel. It also exhibited good tolerability and irritation established by the hydration study, histopathology and HET-CAM test. EM-NLCs-opt-IG4 showed better antimicrobial activity due to the nano-size of the NLCs and high release of EM. It was concluded that NLCs incorporated in situ gel are a better alternative carrier for the improvement of precorneal residence time and therapeutic efficacy.

Author Contributions: Conceptualization and methodology, A.Z.; software and data curation, S.S.I.; validation and formal analysis, M.Y.; M.H.W.; investigation and resources, N.K.A.; O.A.A.; writing—original draft preparation, A.Z.; writing—review and editing, S.N.M.N.U. and M.M.G.; visualization and supervision, O.A.A.; project administration, S.A.; funding acquisition, A.Z. All authors have read and agreed to the published version of the manuscript.

Funding: Deanship of Scientific Research at Jouf University for funding this work through research grant DSR-2021-01-0331.

Institutional Review Board Statement: Not applicable.

Informed Consent Statement: Not applicable.

Data Availability Statement: Not applicable.

Acknowledgments: The authors extend their appreciation to the Deanship of Scientific Research at Jouf University for funding this work through research grant DSR-2021-01-0331.

Conflicts of Interest: The authors declare no conflict of interest.

References

1. Gholizadeh, S.; Wang, Z.; Chen, X.; Dana, R.; Annabi, N. Advanced nanodelivery platforms for topical ophthalmic drug delivery. *Drug Discov. Today* **2021**, *26*, 1437–1449. [CrossRef] [PubMed]
2. Baig, M.S.; Ahad, A.; Aslam, M.; Imam, S.S.; Aqil, M.; Ali, A. Application of Box-Behnken design for preparation of levofloxacin-loaded stearic acid solid lipid nanoparticles for ocular delivery: Optimization, in vitro release, ocular tolerance, and antibacterial activity. *Int. J. Biol. Macromol.* **2016**, *85*, 258–270. [CrossRef] [PubMed]
3. Agrawal, A.; Das, M.; Jain, S. In situ gel systems as 'smart' carriers for sustained ocular drug delivery. *Expert Opin. Drug Deliv.* **2012**, *9*, 383–402. [CrossRef]
4. Maulvi, F.A.; Shetty, K.H.; Desai, D.T.; Shah, D.O.; Willcox, M.D.P. Recent advances in ophthalmic preparations: Ocular barriers, dosage forms and routes of administration. *Int. J. Pharm.* **2021**, *608*, 121105. [CrossRef] [PubMed]
5. Lakhani, P.; Patil, A.; Wu, K.W.; Sweeney, C.; Tripathi, S.; Avula, B.; Taskar, P.; Khan, S.; Majumdar, S. Optimization, stabilization, and characterization of amphotericin B loaded nanostructured lipid carriers for ocular drug delivery. *Int. J. Pharm.* **2019**, *572*, 118771. [CrossRef]
6. Ferreira, K.S.A.; Santos, B.M.A.D.; Lucena, N.P.; Ferraz, M.S.; Carvalho, R.S.F.; Duarte Júnior, A.P.; Magalhães, N.S.S.; Lira, R.P.C. Ocular delivery of moxifloxacin-loaded liposomes. *Arq. Bras. Oftalmol.* **2018**, *81*, 510–513. [CrossRef]
7. Ameeduzzafar; Alruwaili, N.K.; Imam, S.S.; Alotaibi, N.H.; Alhakamy, N.A.; Alharbi, K.S.; Alshehri, S.; Afzal, M.; Alenezi, S.K.; Bukhari, S.N.A. Formulation of Chitosan Polymeric Vesicles of Ciprofloxacin for Ocular Delivery: Box-Behnken Optimization, In Vitro Characterization, HET-CAM Irritation, and Antimicrobial Assessment. *AAPS PharmSciTech* **2020**, *21*, 167. [CrossRef]
8. Taghe, S.; Mirzaeei, S.; Alany, R.G.; Nokhodchi, A. Polymeric Inserts Containing Eudragit® L100 Nanoparticle for Improved Ocular Delivery of Azithromycin. *Biomedicines* **2020**, *8*, 466. [CrossRef]
9. Shelley, H.; Rodriguez-Galarza, R.M.; Duran, S.H.; Abarca, E.M.; Babu, R.J. In Situ Gel Formulation for Enhanced Ocular Delivery of Nepafenac. *J. Pharm. Sci.* **2018**, *107*, 3089–3097. [CrossRef]
10. Janagam, D.; Wu, L.; Lowe, T. Nanoparticles for drug delivery to the anterior segment of the eye. *Adv. Drug Deliv. Rev.* **2017**, *122*, 31–64. [CrossRef]
11. Zhang, W.; Li, X.; Ye, T.; Chen, F.; Sun, X.; Kong, J.; Yang, X.; Pan, W.; Li, S. Design, characterization, and in vitro cellular inhibition and uptake of optimized genistein-loaded NLC for the prevention of posterior capsular opacification using response surface methodology. *Int. J. Pharm.* **2013**, *454*, 354–366. [CrossRef]
12. Liu, D.; Li, J.; Pan, H.; He, F.; Liu, Z.; Wu, Q.; Bai, C.; Yu, S.; Yang, X. Potential advantages of a novel chitosan-N-acetylcysteine surface modified nanostructured lipid carrier on the performance of ophthalmic delivery of curcumin. *Sci. Rep.* **2016**, *6*, 28796. [CrossRef] [PubMed]
13. El-Salamouni, N.S.; Farid, R.M.; El-Kamel, A.H.; El-Gamal, S.S. Nanostructured lipid carriers for intraocular brimonidine localisation: Development, in-vitro and in-vivo evaluation. *J. Microencapsul.* **2018**, *35*, 102–113. [CrossRef] [PubMed]
14. Abd-Elhakeem, E.; El-Nabarawi, M.; Shamma, R. Lipid-based nano-formulation platform for eplerenone oral delivery as a potential treatment of chronic central serous chorioretinopathy: In-vitro optimization and ex-vivo assessment. *Drug Deliv.* **2021**, *28*, 642–654. [CrossRef]
15. Tavakoli, N.; Taymouri, S.; Saeidi, A.; Akbari, V. Thermosensitive hydrogel containing sertaconazole loaded nanostructured lipid carriers for potential treatment of fungal keratitis. *Pharm. Dev. Technol.* **2019**, *24*, 891–901. [CrossRef]
16. Youssef, A.; Dudhipala, N.; Majumdar, S. Ciprofloxacin Loaded Nanostructured Lipid Carriers Incorporated into In-Situ Gels to Improve Management of Bacterial Endophthalmitis. *Pharmaceutics* **2020**, *12*, 572. [CrossRef]
17. Gade, S.; Patel, K.K.; Gupta, C.; Anjum, M.M.; Deepika, D.; Agrawal, A.K.; Singh, S. An Ex Vivo Evaluation of Moxifloxacin Nanostructured Lipid Carrier Enriched In Situ Gel for Transcorneal Permeation on Goat Cornea. *J. Pharm. Sci.* **2019**, *108*, 2905–2916. [CrossRef] [PubMed]
18. Freitas, P.R.; de Araújo, A.C.J.; Barbosa, C.R.; Muniz, D.F.; Tintino, S.R.; Ribeiro-Filho, J.; Siqueira Júnior, J.P.; Filho, J.M.B.; de Sousa, G.R.; Coutinho, H.D.M. Inhibition of Efflux Pumps by Monoterpene (α-pinene) and Impact on *Staphylococcus aureus* Resistance to Tetracycline and Erythromycin. *Curr. Drug Metab.* **2021**, *22*, 123–126. [CrossRef]
19. Olajuyigbe, O.O.; Animashaun, T. Synergistic activities of amoxicillin and erythromycin against bacteria of medical importance. *Pharmacologia* **2012**, *3*, 450–455. [CrossRef]
20. Gupta, S.; Vyas, S.P. Carbopol/chitosan based pH triggered in situ gelling system for ocular delivery of timolol maleate. *Sci. Pharm.* **2010**, *78*, 959–976. [CrossRef]
21. Wu, Y.; Liu, Y.; Li, X.; Kebebe, D.; Zhang, B.; Ren, J.; Lu, J.; Li, J.; Du, S.; Liu, Z. Research progress of in-situ gelling ophthalmic drug delivery system. *Asian J. Pharm. Sci.* **2019**, *14*, 1–15. [CrossRef] [PubMed]
22. Sheshala, R.; Kok, Y.Y.; Ng, J.M.; Thakur, R.R.; Dua, K. In situ gelling ophthalmic drug delivery system: An overview and its applications. *Recent Pat. Drug Deliv.* **2015**, *9*, 237–248. [CrossRef] [PubMed]
23. Kong, M.; Chen, X.-G.; Xing, K.; Park, H.-J. Antimicrobial properties of chitosan and mode of action: A state of the art review. *Int. J. Food Microb.* **2010**, *144*, 51–63. [CrossRef] [PubMed]
24. Modi, D.; Mohammad Warsi, M.H.; Garg, V.; Bhatia, M.; Kesharwani, P.; Jain, G.K. Formulation development, optimization, and in vitro assessment of thermoresponsive ophthalmic pluronic F127-chitosan *in situ* tacrolimus gel. *J. Biomater. Sci. Polym. Ed.* **2021**, *32*, 1678–1702. [CrossRef]

25. Cirri, M.; Maestrini, L.; Maestrelli, F.; Mennini, N.; Mura, P.; Ghelardini, C.; Di Cesare Mannelli, L. Design, characterization and in vivo evaluation of nanostructured lipid carriers (NLC) as a new drug delivery system for hydrochlorothiazide oral administration in pediatric therapy. *Drug Deliv.* **2018**, *25*, 1910–1921. [CrossRef]

26. Hao, J.; Wang, F.; Wang, X.; Zhang, D.; Bi, Y.; Gao, Y.; Zhao, X.; Zhang, Q. Development and optimization of baicalin-loaded solid lipid nanoparticles prepared by coacervation method using central composite design. *Eur. J. Pharm. Sci.* **2012**, *47*, 497–505. [CrossRef]

27. Kollipara, S.; Bende, G.; Movva, S.; Saha, R. Application of rotatable central composite design in the preparation and optimization of poly(lactic-co-glycolic acid) nanoparticles for controlled delivery of paclitaxel. *Drug Dev. Ind. Pharm.* **2010**, *36*, 1377–1387. [CrossRef]

28. Ye, Q.; Li, J.; Li, T.; Ruan, J.; Wang, H.; Wang, F.; Zhang, X. Development and evaluation of puerarin-loaded controlled release nanostructured lipid carries by central composite design. *Drug Dev. Ind. Pharm.* **2021**, *47*, 113–125. [CrossRef]

29. Velmurugan, R.; Selvamuthukumar, S. Development and optimization of ifosfamide nanostructured lipid carriers for oral delivery using response surface methodology. *Appl. Nanosci.* **2016**, *6*, 159–173. [CrossRef]

30. Makoni, P.A.; Khamanga, S.M.; Walker, R.B. Muco-adhesive clarithromycin-loaded nanostructured lipid carriers for ocular delivery: Formulation, characterization, cytotoxicity and stability. *J. Drug Deliv. Sci. Technol.* **2021**, *61*, 102171. [CrossRef]

31. Tavakoli, M.; Mahboobian, M.M.; Nouri, F.; Mohammadi, M. Studying the ophthalmic toxicity potential of developed ketoconazole loaded nanoemulsion *in situ gel* formulation for ophthalmic administration. *Toxicol. Mech. Methods* **2021**, *31*, 572–580. [CrossRef] [PubMed]

32. Katiyar, S.; Pandit, J.; Mondal, R.S.; Mishra, A.K.; Chuttani, K.; Aqil, M.; Ali, A.; Sultana, Y. In situ gelling dorzolamide loaded chitosan nanoparticles for the treatment of glaucoma. *Carbohydr. Polym.* **2014**, *102*, 117–124. [CrossRef] [PubMed]

33. Upadhayay, P.; Kumar, M.; Pathak, K. Norfloxacin Loaded pH Triggered Nanoparticulate in-situ Gel for Extraocular Bacterial Infections: Optimization, Ocular Irritancy and Corneal Toxicity. *Iran. J. Pharm. Res.* **2016**, *15*, 3–22. [PubMed]

34. Wardrop, J.; Ficker, D.; Franklin, S.; Gorski, R.J. Determination of erythromycin and related substances in enteric-coated tablet formulations by reversed-phase liquid chromatography. *J. Pharm. Sci.* **2000**, *89*, 1097–1105. [CrossRef]

35. Nagarwal, R.C.; Kumar, R.; Pandit, J.K. Chitosan coated sodium alginate-chitosan nanoparticles loaded with 5-FU for ocular delivery: In vitro characterization and in vivo study in rabbit eye. *Eur. J. Pharm. Sci.* **2012**, *47*, 678–685. [CrossRef] [PubMed]

36. Bagley, D.M.; Waters, D.; Kong, B.M. Development of a 10-day chorioallantoic membrane vascular assay as an alternative to the Draize rabbit eye irritation test. *Food Chem. Toxicol.* **1994**, *32*, 1155–1160. [CrossRef]

37. ICCVAM-Recommended Test Method Protocol: Hen's Egg Test—Chorioallantoic Membrane (HET-CAM) Test Method. *ICCVAM Test Method Eval. Rep.* **2010**, *13*, B30–B38.

38. Hiremath, S.S.P.; Dasankoppa, F.S.; Nadaf, A.; Jamakandi, V.G.; Mulla, J.S.; Sholapur, H.N.; Aezaz, A. Formulation and Evaluation of a Novel *In Situ* Gum Based Ophthalmic Drug Delivery System of Linezolid. *Sci. Pharm.* **2008**, *76*, 515–532. [CrossRef]

39. Yang, G.; Wu, F.; Chen, M.; Jin, J.; Wang, R.; Yuan, Y. Formulation design, characterization, and in vitro and in vivo evaluation of nanostructured lipid carriers containing a bile salt for oral delivery of gypenosides. *Int. J. Nanomed.* **2019**, *14*, 2267–2280. [CrossRef]

40. Behbahani, E.S.; Ghaedi, M.; Abbaspour, M.; Rostamizadeh, K. Optimization and characterization of ultrasound assisted preparation of curcumin-loaded solid lipid nanoparticles: Application of central composite design, thermal analysis and X-ray diffraction techniques. *Ultrason. Sonochem.* **2017**, *38*, 271–280. [CrossRef]

41. Tinu, T.S.; Thomas, L.; Kumar, A. Polymers used in ophthalmic in situ gelling system. *Int. J. Pharm. Sci. Rev. Res.* **2013**, *30*, 176–183.

42. Hamman, J.H. Chitosan based polyelectrolyte complexes as potential carrier materials in drug delivery systems. *Mar. Drugs.* **2010**, *8*, 1305–1322. [CrossRef] [PubMed]

43. Abraham, S.; Furtado, S.; Bharath, S.; Basavaraj, B.V.; Deveswaran, R.; Madhavan, V. Sustained ophthalmic delivery of ofloxacin from an ion-activated in situ gelling system. *Pak. J. Pharm. Sci.* **2009**, *22*, 175–179. [PubMed]

44. Shukr, M.H.; Ismail, S.; El-Hossary, G.G.; El-Shazly, A.H. Design and evaluation of mucoadhesive in situ liposomal gel for sustained ocular delivery of travoprost using two steps factorial design. *J. Drug Deliv. Sci. Technol.* **2021**, *61*, 102333. [CrossRef]

45. Fahmy, U.A.; Ahmed, O.A.A.; Badr-Eldin, S.M.; Aldawsari, H.M.; Okbazghi, S.Z.; Awan, Z.A.; Bakhrebah, M.A.; Alomary, M.N.; Abdulaal, W.H.; Medina, C.; et al. Optimized Nanostructured Lipid Carriers Integrated into In Situ Nasal Gel for Enhancing Brain Delivery of Flibanserin. *Int. J. Nanomed.* **2020**, *15*, 5253–5264. [CrossRef]

46. Yu, S.; Li, Q.; Li, Y.; Wang, H.; Liu, D.; Yang, X.; Pan, W. A novel hydrogel with dual temperature and pH responsiveness based on a nanostructured lipid carrier as an ophthalmic delivery system: Enhanced trans-corneal permeability and bioavailability of nepafenac. *New J. Chem.* **2017**, *41*, 3920–3929. [CrossRef]

47. Ritger, P.L.; Peppas, N.A. A simple equation for description of solute release I. Fickian and non-fickian release from non-swellable devices in the form of slabs, spheres, cylinders or discs. *J. Control. Release* **1987**, *5*, 23–36. [CrossRef]

48. Fouda, N.H.; Abdelrehim, R.T.; Hegazy, D.A.; Habib, B.A. Sustained ocular delivery of Dorzolamide-HCl via proniosomal gel formulation: In-vitro characterization, statistical optimization, and in-vivo pharmacodynamic evaluation in rabbits. *Drug Deliv.* **2018**, *25*, 1340–1349. [CrossRef]

Article

Development and Evaluation of Clove and Cinnamon Supercritical Fluid Extracts-Loaded Emulgel for Antifungal Activity in Denture Stomatitis

Meenakshi Srinivas Iyer [1], Anil Kumar Gujjari [1], Sathishbabu Paranthaman [2], Amr Selim Abu Lila [3,4], Khaled Almansour [4], Farhan Alshammari [4], El-Sayed Khafagy [5,6], Hany H. Arab [7] and Devegowda Vishakante Gowda [2,*]

[1] Department of Prosthodontics, JSS Dental College and Hospital, Mysuru 570015, India; dr.meenakshis@jssuni.edu.in (M.S.I.); dr.anilkumarg@jssuni.edu.in (A.K.G.)
[2] Department of Pharmaceutics, JSS College of Pharmacy, Mysuru 570015, India; sathishbabu.p94@gmail.com
[3] Department of Pharmaceutics and Industrial Pharmacy, Faculty of Pharmacy, Zagazig University, Zagazig 44519, Egypt; a.abulila@uoh.edu.sa
[4] Department of Pharmaceutics, College of Pharmacy, University of Hail, Hail 81442, Saudi Arabia; kh.almansour@uoh.edu.sa (K.A.); frh.alshammari@uoh.edu.sa (F.A.)
[5] Department of Pharmaceutics, College of Pharmacy, Prince Sattam Bin Abdulaziz University, Al-kharj 11942, Saudi Arabia; e.khafagy@psau.edu.sa
[6] Department of Pharmaceutics and Industrial Pharmacy, Faculty of Pharmacy, Suez Canal University, Ismailia 41552, Egypt
[7] Department of Pharmacology and Toxicology, College of Pharmacy, Taif University, P.O. Box 11099, Taif 21944, Saudi Arabia; h.arab@tu.edu.sa
* Correspondence: dvgowda@jssuni.edu.in; Tel.: +91-966-316-2455

Citation: Iyer, M.S.; Gujjari, A.K.; Paranthaman, S.; Abu Lila, A.S.; Almansour, K.; Alshammari, F.; Khafagy, E.-S.; Arab, H.H.; Gowda, D.V. Development and Evaluation of Clove and Cinnamon Supercritical Fluid Extracts-Loaded Emulgel for Antifungal Activity in Denture Stomatitis. *Gels* **2022**, *8*, 33. https://doi.org/10.3390/gels8010033

Academic Editors: Maddalena Sguizzato, Rita Cortesi and Rachel Yoon Chang

Received: 15 December 2021
Accepted: 1 January 2022
Published: 4 January 2022

Publisher's Note: MDPI stays neutral with regard to jurisdictional claims in published maps and institutional affiliations.

Abstract: Denture stomatitis (DS), usually caused by Candida infection, is one of the common denture-related complications in patients wearing dentures. Clove and cinnamon oils have been acknowledged for their anti-inflammatory, antimicrobial activity, and antifungal effects in the oral cavity. The aim of this study, therefore, was to prepare clove/cinnamon oils-loaded emulgel and to assess its efficacy in treating *Candida albicans*-associated denture stomatitis. Central composite design was adopted to formulate and optimize clove/cinnamon extracts-loaded emulgel. The formulated preparations were assessed for their physical appearance, particle size, viscosity, spreadability, and in-vitro drug release. In addition, in-vivo therapeutic experiments were conducted on 42 patients with denture stomatitis. The prepared emulgel formulations showed good physical characteristics with efficient drug release within 3 h. In addition, in-vivo antifungal studies revealed that the optimized formula significantly ($p < 0.001$) reduced Candida colony counts from the denture surface, compared to commercially available gel (240.38 ± 27.20 vs. 398.19 ± 66.73 CFU/mL, respectively). Furthermore, the optimized formula and succeeded in alleviating denture stomatitis-related inflammation with a better clinical cure rate compared to commercially available gel Collectively, herbal extracts-loaded emulgel might be considered an evolution of polyherbal formulations and might represent a promising alternative to the existing allopathic drugs for the treatment of denture stomatitis, with better taste acceptability and no side effects.

Keywords: *Candida albicans*; clove extracts; cinnamon extracts; denture stomatitis; super critical fluid extraction

1. Introduction

Denture stomatitis is a frequent condition affecting denture wearers that causes inflammation and redness of the oral mucosal tissues covered by the denture [1]. Despite its prevalence, the etiology of denture stomatitis is not completely understood. Nevertheless, denture stomatitis is often associated with Candidal colonization, especially *Candida albicans*, a normal commensal flora of the oral cavity [2]. Along with denture surface colonization, during swallowing and flushing action of saliva, patients can aspire the microorganisms from denture plaque, which could expose patients to unexpected systemic infection especially in immunocompromised patients. Currently, antifungal medications that can combat *Candida albicans* are commonly used to alleviate stomatitis symptoms [2,3]. Nevertheless, despite the availability of an increasing array of over-the-counter anti-fungal drugs, there is an alarming decrease in the success rate of treatment. This might be attributed to the development of anti-fungal resistance, the diverse resistance profile of Candida species, or the lack of patient compliance [4,5].

Historically, herbal medicines have a long history and were utilized in ancient Chinese, Egyptian, Greek, and Indian medicine for a variety of therapeutic purposes [6]. According to the World Health Organization, 80 percent of the world's population still relies mostly on traditional medicines for health treatment. Furthermore, numerous researches have recently demonstrated a paradigm shift from conventional synthetic drugs to natural substances, such as phytochemicals, for the treatment of many diseases, including fungal infections [7,8]. The desirable properties exhibited by phytomedicines are accredited to their phyto-constituents, such as saponins, tannins, alkaloids, flavonoids, terpenoids, and sesquiterpenes, which work in synergy to produce the desired effect.

For many decades, essential oils and herbal extracts have been employed for various purposes, such as pharmaceuticals, alternative medicine, and natural therapies. Furthermore, polyherbal preparations, containing active substances from one or more herbs, has been repeatedly investigated for their synergistic therapeutic effect [9,10]. When the active phytochemical constituents of individual plants are inadequate to provide the desired therapeutic effects, combining multiple herbs in a certain ratio could significantly enhance the therapeutic effect while minimizing toxicity. Consequently, formulations with the pharmacodynamic synergism of active constituents exhibiting similar therapeutic activity is an added advantage over a single herbal formulation.

Gels represents a distinct class of semisolid dosage forms that have a wide spectrum of applications for the treatment of various diseases. Gels possess excellent advantages, such as compatibility with various excipients; spreadability; ease of withdrawing; and emollient, non-staining, greaseless, and thixotropic nature [11]. In addition, compared to ointments or creams, gels usually have a higher aqueous component that allows greater dissolution of drugs and permits easy migration of the drug through the vehicle [12]. Nevertheless, despite the many benefits of gels, one significant constraint is the delivery of hydrophobic drugs. To circumvent this limitation, emulsion hydrogels, synonymously called emulgels, were introduced, allowing even hydrophobic drug moieties to benefit from the special features of gels [13,14]. Polymer plays crucial roles in the formulation of emulgels. Because of their gelling capacity, polymers ensure the formulations to be stable by altering the surface and interfacial tension and enhancing the viscosity of the water phase. The gelling agent in the water phase converts typical emulsions into emulgels. Badam gum, a plant exudate obtained from *Terminalia catappa*, is a natural polymer that has been frequently used for the formulation of muco-adhesive drug-delivery systems [15]. Being natural polymer, it exhibits certain advantages over synthetic polymers, namely that it is nontoxic, readily available, economic, and potentially biodegradable.

The aim of this study, therefore, was to formulate clove/cinnamon extracts-loaded emulgels. The prepared emulgel formulations were optimized using central composite design and were analyzed for physical and chemical properties. In addition, the anti-fungal activity of optimized emulgel formulation were assessed clinically against Candida-associated denture stomatitis in geriatric denture wearers.

2. Results and Discussion

2.1. Percentage Yield of Clove and Cinnamon Extracts

Supercritical fluid extraction (SCFE) technique is the process of extracting one component from a mixture [16]. Several supercritical fluids have been adopted for the extraction of lipid including ethane, ethylene, ethanol, methanol, benzene, and CO_2 [17]. In this study, CO_2 was utilized as the supercritical extracting solvent to prepare clove and cinnamon extracts from clove buds and cinnamon barks, respectively. Supercritical carbon dioxide extraction is considered an eco-friendly method where it excludes the use of typical organic solvents [18]. Our results indicated that the percentage yield of clove and cinnamon extracts were 18.9 and 1.9%, respectively, at critical temperature of 37 °C, critical pressure of 300 bar, and an extraction time of 120 min.

2.2. Minimum Inhibitory Concentration (MIC) of Clove and Cinnamon Extracts

Many plant products, including plant extract and essential oils, have been acknowledged for their antifungal activities [19,20]. In this study, to assess the antifungal efficacies of clove and cinnamon extracts, the MICs of clove and cinnamon extracts alone and in combination against *C. albicans* isolates were estimated. Both extracts showed antifungal activity against *C. albicans* with MICs of 512 µg/mL for clove extract and 64 µg/mL for cinnamon extract. Similar findings were reported by Stevic et al., who investigated the antifungal potential of 16 selected essential oils against 21 fungi [21]. In addition, the in-vitro interaction of clove and cinnamon extracts against *C. albicans* was examined at concentrations ranged from 32–512 µg/mL for clove extract and 4–64 µg/mL for cinnamon extract (Table S1). The combinations exhibited different interactions against fungal isolates with FICIs ranging from 0.13–2. The combination of clove and cinnamon extracts, at concentrations of 256 and 32 µg/mL, respectively, exhibited additive effect with 99.45% growth inhibition detected at FICI of 1. Most importantly, no regrowth of *C. albicans* was observed upon exposure to clove and cinnamon extracts over 24 h. Of note, combined treatment with clove and cinnamon extracts significantly decreased the MICs values against *C. albicans*, for clove extract from 512 to 256 µg/mL and for cinnamon extract from 64 to 32 µg/mL. Accordingly, combination therapy could be used for expanding the antimicrobial spectrum, reducing toxicity, and decreasing antimicrobial resistance during treatment.

2.3. Solubility and Emulsification Studies

First, solubility studies of clove and cinnamon extracts were conducted in different surfactants and co-surfactants to select the appropriate surfactant/co-surfactant (S_{mix}) mixture to be used for emulsification process. As shown in Figure 1A, maximum solubility of clove and cinnamon extracts was observed in Tween 80 and PEG 400 compared to other systems containing Tween 20, Span 20, or Span 80. Accordingly, a surfactant/co-surfactant (S_{mix}) mixture composing of Tween 80 and PEG 400 was selected for further emulsification processes.

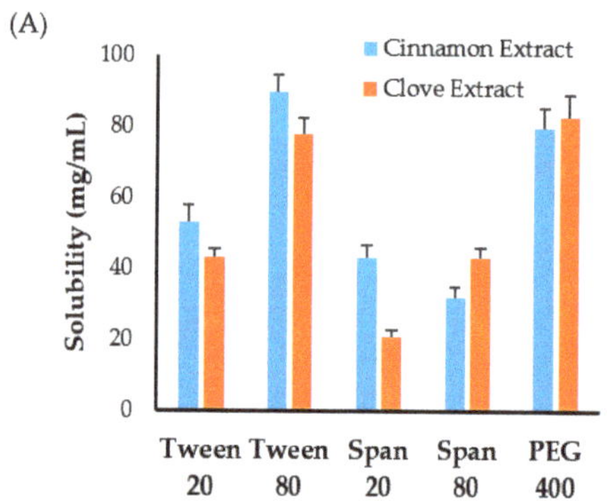

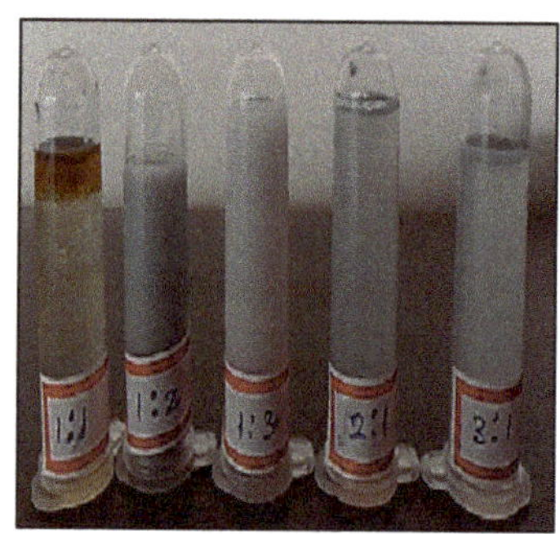

Figure 1. (**A**) Solubility of clove and cinnamon extracts in different surfactants and co-surfactants. (**B**) Emulsion formation from various S_{mix} ratio.

Next, to study the effect of surfactant mixture (S_{mix}) ratio on the emulsification process, different ratios of Tween 80 and PEG 400 were mixed together. It was clear that increasing co-surfactant (PEG 400) concentration from 1:1 to 1:2 or 1:3 resulted in an increase in the turbidity of the system (Figure 1B). On the other hand, increasing surfactant (Tween 80) concentration in surfactant/co-surfactant (S_{mix}) mixture from 1:1 to 3:1 while keeping co-surfactant ratio constant resulted in a good and stable emulsions (Figure 1B). Accordingly, the formulation consisting of a minimal S_{mix} ratio (2:1) producing good and stable emulsion was selected for the preparation of emulgel.

2.4. Formulation Screening

Quality by design (QbD) is a systematic approach to pharmaceutical development that is commonly used to develop a distinctive product based on pre-determined parameters and studying their effect on certain responses in order to get an optimized formula [22]. In the current study, central composite design (CCD) was adopted as a tool to formulate and to optimize clove/cinnamon extracts-loaded emulgels. A total of 9 runs were prepared, and the effect of two independent variables, namely polymer concentration (% *w/w*) and S_{mix} ratio (% *v/v*), on the response parameters, namely particle size (nm) and drug content (%), was investigated. In the present design, the optimum formulation was derived by keeping responses of drug content at a maximum level and particle size at a minimum level. The observed responses in CCD of prepared emulgel are summarized in Table 1.

Table 1. Observed responses in central composite design of Emulgel.

Formulation Code	X1 Polymer Conc. (% *w/w*)	X2 S_{mix} Ratio (% *v/v*)	R1 Globule Size (nm)	R2 Drug Content Clove (%)	R3 Drug Content Cinnamon (%)
F1	1	10	198.0 ± 14.7	91.6 ± 0.23	92.1 ± 0.12
F2	3	11	224.2 ± 12.3	97.9 ± 0.09	96.4 ± 0.24
F3	5	10	624.1 ± 12.5	92.5 ± 0.17	93.7 ± 0.12
F4	3	4	347.7 ± 14.1	89.6 ± 0.22	90.1 ± 0.14
F5	1	5	280.4 ± 14.5	86.4 ± 0.15	86.3 ± 0.28
F6	3	7.5	387.9 ± 12.7	91.8 ± 0.31	89.2 ± 0.23
F7	6	7.5	631.2 ± 16.4	90.4 ± 0.32	91.2 ± 0.16
F8	1	7.5	258.8 ± 8.4	89.8 ± 0.12	90.1 ± 0.15
F9	5	5	424.1 ± 10.3	89.1 ± 0.20	86.4 ± 0.34

All data represent the mean ± SD of three independent experiments.

2.4.1. Response Analysis for Optimization of Emulgel
Effect of Independent Variables on Globule Size

Globule size is an important parameter for evaluating the formulated emulgels. The smaller the particle size, the larger the interfacial surface area for drug absorption. Globule size of the developed emulgels was determined, and the results are summarized in Table 1. The globule size of all the formulations was in the range of 198.7 ± 14.7 nm to 631.2 ± 16.4 nm (Table 1). As evident from literature, formulations having particle size <1000 nm are found stable [23]. Hence, the globule size of all the formulations was within the limit. F7 showed the highest particle size, whereas F1 showed the least particle size among the formulations. It was apparent that there is a direct relation between the globule size of the developed emulgels and polymer concentration and S_{mix} ratio. Increment in polymer concentration while keeping S_{mix} ratio constant resulted in a remarkable increase in globule size. On the other hand, increasing S_{mix} ratio while keeping polymer concentration constant resulted in a significant reduction in globule size. Similar results were reported by Chuacharoen et al., who reported that the average particle sizes of curcumin emulsions was decreased by increasing surfactant concentration from 10 to 30% [24]. The impact of polymer concentration and S_{mix} ratio on globule size is depicted by 3D-response surface plot shown in Figure 2A.

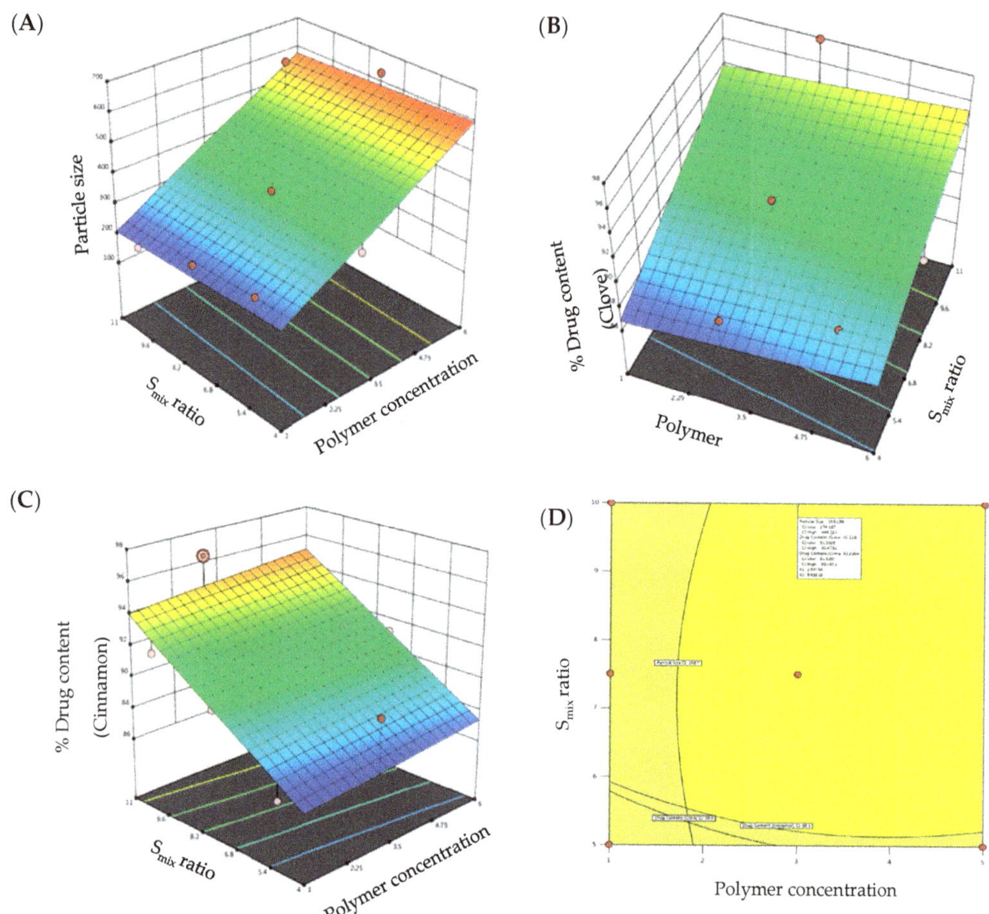

Figure 2. 3D-surface plots of (**A**) particle size; (**B**) drug content of clove; (**C**) drug content of cinnamon; and (**D**) overlay plot of optimized emulgel formulation.

The regression equation verified the effect of formulation variables on globule size. It was obvious that polymer concentration exerted a positive effect on globule size, while S_{mix} ratio exhibited a negative effect on globule size (Equation (1)):

$$\text{Particle Size (R1)} = 165.564 + 74.04 \text{ polymer concentration} - 2.76 \, S_{mix} \text{ ratio} \quad (1)$$

The obtained F-value of this model was 9.81, and *p* value < 0.0500 implies that the model is significant.

Effect of Independent Variables on Drug Content

Despite the many benefits of gels, one significant constraint is the delivery of hydrophobic drug. To address this constraint, emulgels were created. Drug content of clove and cinnamon in the prepared emulgel formulations was found to range from 86.4 to

97.9% for clove and 86.3 to 96.4% for cinnamon, indicating efficient drug loading within the emulgel. The model F-values of clove drug content (R2) and cinnamon drug content (R3) were 6.28 and 8.28, respectively, implying that the model is significant. The estimated *p*-values were less than 0.05, indicating that the both model terms are significant. Equations (2) and (3), in terms of actual factors, can be used to make predictions about the response for given levels of each factor. As shown in Figure 2B,C, both polymer concentration and S_{mix} ratio displayed a monotonous impact on drug content.

$$\% \text{ Drug Content Clove (R2)} = 82.51 + 0.269 \text{ polymer concentration} + 1.02\, S_{mix} \text{ ratio} \qquad (2)$$

$$\% \text{ Drug Content Cinnamon (R3)} = 81.64 + 0.21 \text{ polymer concentration} + 1.107\, S_{mix} \text{ ratio} \qquad (3)$$

2.4.2. Optimized Formulation

The CCD design was useful to analyze the obtained polynomial model, which was used to optimize the process of formulating emulgels with fewer formulations. Further, desirability and overlay plots were used to optimize the designed formulations. Optimized formulation was obtained by setting the constraints such that the obtained formulation should have minimal particle size and maximum drug content. Concentrations recommended by DoE to achieve the above-mentioned goals were depicted in the form of an overlay plot (Figure 2D) with desirability near to 1. The optimized formulation, depicted in the overlay plot as a form of a flag, was obtained at a polymer concentration of 2.6% and S_{mix} ratio of 9%. The observed globule size, clove drug content, and cinnamon drug content were 321.2 nm, 96.65%, and 95.94%, respectively, which were close to the predicted values (359.2 nm, 93.52%, and 93.33%) for the optimized formula with a desirability of 1.

2.5. Evaluation of Emulgel Formulations

2.5.1. The Physical Appearance of Emulgel Formulations

Emulgel formulations were white, viscous preparations with homogenous, smooth texture, free from grittiness. The physical appearance, homogeneity, grittiness, and smoothness of all formulations are summarized in Table S2.

2.5.2. Determination of pH

The pH of all the formulations was found to be in the range of 5.91 ± 0.09 to 6.72 ± 0.16 (Table 2). There was no substantial change in the pH among all formulations. Accordingly, all the prepared formulations could be considered convenient for oral application.

Table 2. Physicochemical parameters of clove/cinnamon extracts-loaded emulgel formulations.

Formulation Code	pH	Viscosity (cPs)	Spreadability (g cm/s)	Extrudability (%)	Muco-adhesion (dyne/cm²)
F1	6.23 ± 0.12	2820.1 ± 0.86	90.16 ± 0.09	84.26 ± 0.07	5.8 ± 0.14
F2	6.26 ± 0.06	4090.2 ± 0.21	89.03 ± 0.19	79.50 ± 0.15	8.5 ± 0.04
F3	6.72 ± 0.16	6224.8 ± 0.13	81.16 ± 0.09	89.50 ± 0.57	13.9 ± 0.07
F4	6.23 ± 0.06	4012.3 ± 0.62	87.67 ± 0.08	78.53 ± 0.15	8.0 ± 0.11
F5	6.50 ± 0.10	4040.0 ± 0.53	86.83 ± 0.37	77.00 ± 0.06	10.2 ± 0.15
F6	6.43 ± 0.15	4204.9 ± 0.36	63.5 ± 0.20	82.33 ± 0.57	10.4 ± 0.04
F7	5.91 ± 0.09	7204.9 ± 0.36	76.33 ± 0.07	66.43 ± 0.15	15.4 ± 0.74
F8	6.33 ± 0.05	4010.0 ± 0.33	84.33 ± 0.17	74.50 ± 0.57	9.7 ± 0.42
F9	5.91 ± 0.09	6426.9 ± 0.39	76.33 ± 0.07	86.50 ± 0.57	13.9 ± 0.07

Data represent mean ± SD of three independent experiments.

2.5.3. Measurement of Viscosity

The viscosity values of the prepared gel formulations ranged from 2820.1 ± 0.86 to 7204.9 ± 0.36 (Table 2). Increasing polymer concentration from 1 to 6% significantly increased the viscosity of the formulated emulgel. Similar results are reported by Kotwal et al., who demonstrated that increasing the concentration of poloxamer 407 resulted in a significant increase in the gelling property of the gel [25]. Most importantly, by increasing spindle speed (shear), the viscosity of the gels was decreased. These results stated that all the developed formulations had a non-Newtonian, pseudoplastic, and/or dilatant behavior at a temperature of 25 °C. This type of rheological behavior indicates an appropriate spreadability.

2.5.4. Determination of Spreadability

The spreading coefficient depends on the polymer concentration in the formulation, and it dictates how easy the emulgel will spread upon applying a small amount of shear. Generally, relatively low value of spreading coefficient enhances emulgel application. All the tested formulae showed good spreadability with a spreadability coefficient ranging from 63.5 ± 0.20 to 90.16 ± 0.09 gm cm s^{-1} (Table 2).

2.5.5. Extrudability

Extrudability of the gel from the tube is a very important parameter during application and it affects patient acceptance. Gels with high consistency may not extrude from the tube, whereas low-viscous gels may flow quickly. Consequently, suitable consistency is required to extrude the gel from the tube. The extrudability percentage of different formulations, summarized in Table 3, ranged from 66.43 ± 0.15 to 89.50 ± 0.57, which indicates good to fair extrudability.

Table 3. Demographic details of study participants.

Variable			Group		Total	Test Statistics	*p*-Value
			Test	Control			
Mean age			61.57 ± 7.8	62.86 ± 6.5	-	$t = 0.58$	0.564
Age groups (years)	<50	Frequency	3	1	4		
		Percent	14.3	4.8	9.5		
	51–60	Frequency	5	7	12		
		Percent	23.8	33.3	28.6	$X^2 = 1.333$	0.721
	61–70	Frequency	10	10	20		
		Percent	47.6	47.6	47.6		
	70+	Frequency	3	3	6		
		Percent	14.3	14.3	14.3		
Gender	Male	Frequency	17	18	35		
		Percent	81.0	85.7	83.3	$X^2 = 0.155$	0.694
	Female	Frequency	4	3	7		
		Percent	19	14.3	16.7		

Table 3. *Cont.*

Variable			Group		Total	Test Statistics	*p*-Value
			Test	Control			
Use of denture (years)	<5	Frequency	5	4	9	$X^2 = 10.48$	0.005
		Percent	23.8	19.0	21.4		
	>5	Frequency	16	17	33		
		Percent	76.2	21	78.6		

		Test Group (A)				Control Group (B)				*p*-Value	
Inflammation stages	Denture Age	I	II	III	Total	I	II	III	Total	A	B
	<5	50.0%	50.0%	0.0%	100.0%	33.3%	50.0%	16.7%	100.0%	0.334	0.156
	>5	20.0%	66.7%	13.3%	100.0%	40.0%	53.3%	6.7%	100.0%		

Variable				Test Group (A)			Control Group (B)			*p* value	
				YES	NO	Total	YES	NO	Total	A	B
Nocturnal denture wear	Type of DS	I	Count	6	0	6	8	0	8	0.590	0.645
				(46.2%)	(0.0%)	(28.6%)	(57.1%)	(0.0%)	(38.1%)		
		II	Count	7	6	13	6	5	11		
				(53.8%)	(75.0%)	(61.9%)	(42.9%)	(71.4%)	(52.4%)		
		III	Count	0	2	2	0	2	2		
				(0.0%)	(25.0%)	(9.5%)	(0.0%)	(28.6%)	(9.5%)		

2.5.6. In-Vitro Muco-Adhesion

A significant characteristic of the oral gel is the adhesion to the mucosa. Generally, enhanced gel adhesion results in increased contact time with the mucosa and prolongation of drug contact and clinical efficacy. In-vitro muco-adhesion test revealed that increasing the concentration of polymer-forming gel (badam gum) in F3, F7, and F9 could significantly increase the muco-adhesion of the formulation. Badam gum has a higher molecular weight and could form a large adhesive surface with the mucin and consequently give good muco-adhesiveness [15].

2.5.7. FTIR Spectrum

FTIR spectra of pure clove and cinnamon extract, physical mixture, and optimized emulgel formulation were recorded. The peaks corresponding to the functional groups of pure clove and cinnamon extract were in correlation with those observed in the FTIR spectra of physical mixture and emulgel formulation (Figure 3). The broad absorption peaks at around 3290 to 3680.30 cm^{-1} were assigned to the phenolic OH stretching vibration. The absorption peaks positioned at 1612 cm^{-1}, 1516 cm^{-1}, 1429 cm^{-1}, and 1240 cm^{-1} are assigned to the C-C, C=O, C=C (aromatic), and C-O (aryl ether), respectively. From the results, it was evident that no chemical interactions were observed between pure clove and cinnamon extract and its physical mixture and emulgel formulation. Therefore, it can be inferred that the selected pure clove and cinnamon extract was compatible with the selected S_{mix} and polymer.

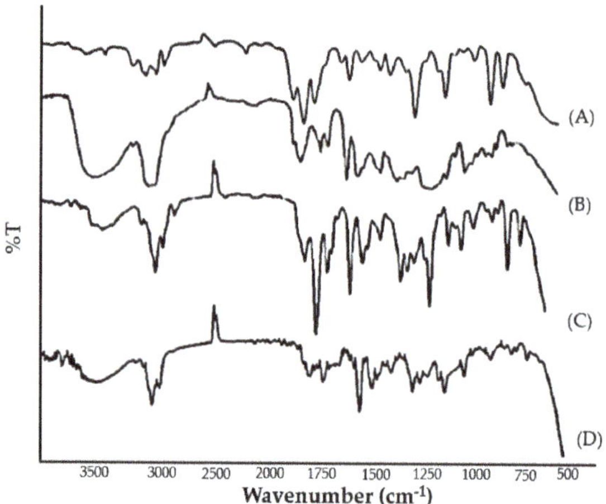

Figure 3. FTIR spectrum of (**A**) pure cinnamon extract, (**B**) pure clove extract, (**C**) physical mixture, and (**D**) clove/cinnamon extracts—loaded emulgel.

2.6. In-Vitro Drug Release Studies

The in-vitro release pattern was carried out for the optimized formulation and was compared with that of a marketed preparation using Franz diffusion cell. As shown in Figure 4, the amount of drugs (clove and cinnamon) released from the optimized formula was comparable to the amount of clotrimazole released from the marketed gel formulation, with about 50% of the loaded drug released within 50 min. In addition, nearly 100% of loaded drugs were released from both formulae after 180 min, suggesting the efficient release of clove and cinnamon extracts from the optimized formula and nullifying the possible detrimental effect of gel-forming polymer on drug release from emulgel formulation.

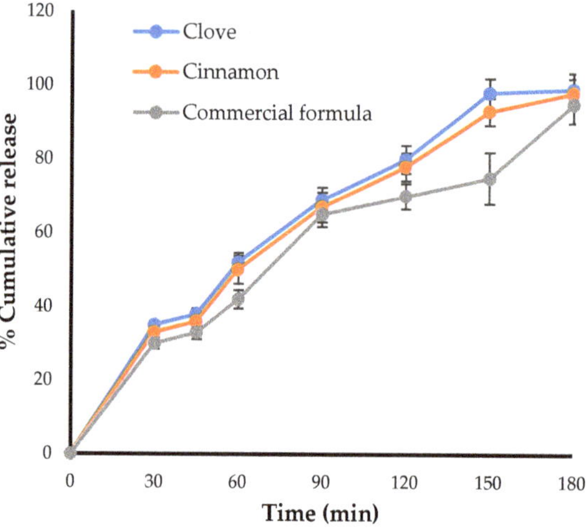

Figure 4. Drug release profile for optimized emulgel formula and marketed (Candid®) gel.

2.7. Stability Studies

Stability testing was conducted for the optimized emulgel formulation at different storage conditions, and the results are summarized in Table S3. It was obvious that non-significant changes were detected in the physical appearance, pH, and drug content of the stored emulgel formulation over one and three months compared to fresh preparation ($p > 0.05$). The results confirmed the stability of the formulated clove/cinnamon extracts-loaded emulgel.

2.8. Randomized Clinical Trial
2.8.1. Study Design

A randomized double-blind controlled study was adopted to assess and compare the antifungal potential of clove/cinnamon extracts-loaded emulgel with a commercially available antifungal gel (Candid®) for the treatment of denture stomatitis. A total of 42 volunteers were randomly divided into two groups: one receiving the optimized emulgel formulation and the second receiving Candid® gel. The mean ages of test and control groups were 61.57 ± 7.8 and 62.86 ± 6.5, respectively. The CONSORT Flowchart is depicted in Figure S1. Both groups presented male predominance, and the chi-square test revealed no significant association between age or gender among both the groups (Table 3). In addition, although denture age was not related to the degree of inflammation in both the test ($p = 0.334$) and control ($p = 0.156$) groups, it was observed that 66.7% of patients in the test group and 53.3% patients in the control group had Newton's Type II inflammation. Complete dentures with full palatal coverage could encourage Candidal colonization due to an increase in an acrylic volume covering the palatal surface, which could provide a large area for adhesion of microorganisms and decreased cleansing action by saliva and tongue. Accordingly, there was an association between the duration of denture wear and the occurrence and/or progression of denture stomatitis inflammation stage. Furthermore, 53.8% of patients in the test group and 42.9% patients in the control group who practiced nocturnal denture wear showed Newton's Type II denture stomatitis.

2.8.2. Antifungal Efficacy of Clove/Cinnamon Extracts-Loaded Emulgel

To assess the antifungal efficacy of optimized formula of clove/cinnamon extracts-loaded emulgel, patients were treated with either the optimized formula (Group A) or a commercially available marketed gel (Candid®; Group B) on days 7, 14, and 21 after the first visit. Swab samples were collected from palatal and denture surface at baseline and after the intervention at 7, 14, and 21 days. Comparisons of the microbial growth, pre-, and post-intervention in terms of colony-forming units (CFU per mg of plaque) were assessed for each group. As depicted in Figure 5, treatment with either the test formula or the commercially available gel (Candid®) significantly reduced microbial growth on both palatal and denture surfaces compared to control (untreated) group. In addition, such reduction of microbial growth in both groups was significantly correlated with the frequency of dosing; microbial growth inhibition was much more observed in both treated groups on day 21 post treatment following three treatments as compared to microbial growth on days 7 or 14 post intervention (Table 4). The microbial growth on palatal and denture surface post-test formula application were 91.76 ± 61 CFU/mL and 240.38 ± 27 CFU/mL, respectively, on day 21 post intervention as compared to 432.8 ± 236 CFU/mL and 815.8 ± 145 CFU/mL on day 7 post intervention, respectively.

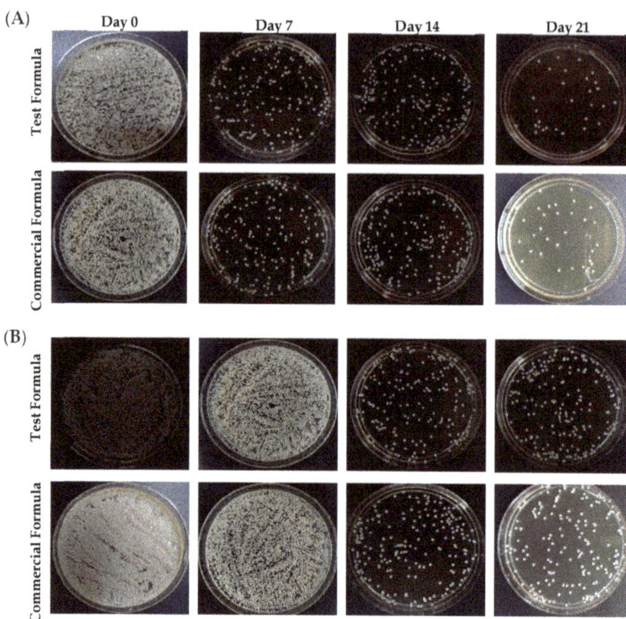

Figure 5. Mean fungal growth reduction (CFU/ mL) from the (**A**) palatal surface and (**B**) denture surface upon treatment with either test formula or a commercially available gel on days 0, 7, 14, and 21 post treatment.

Table 4. Pre-intervention and post-intervention microbial growth with Group A and Group B from palatal and denture surface ranges.

Test Group	Group A		Group B	
	Palatal Surface	**Denture Surface**	**Palatal Surface**	**Denture Surface**
Baseline (Day 0)	649.71 ± 282.82	1092.04 ± 150.12	677.38 ± 303.21	1160.76 ± 194.72
Post Treatment				
Day 7	432.80 ± 236.42	815.80 ± 145.04	449.80 ± 211.81	815.23 ± 95.4328
Day 14	228.85 ± 136.85	572.42 ± 151.42	231.61 ± 115.77	568.66 ± 136.85
Day 21	91.76 ± 61.07	240.38 ± 27.20	66.38 ± 39.11	398.19 ± 66.73

Next, intergroup comparison were conducted to compare the antifungal efficiency of test formula (Group A) with that of a commercially available gel (Candid®; Group B). The percentage reduction of colony-forming units (CFU) of test formulation and commercially available marketed gel groups was calculated and was used as an assessment parameter for comparison. As summarized in Table 5, both test and commercial formulations efficiently suppressed microbial growth from palatal and denture surfaces in a dose frequency-dependent manner. The percentage reduction of microbial growth on both palatal and denture surfaces for test formula was much higher at day 21 post treatment (three sequential treatments) compared to those observed on either day 7 (one treatment) or day 14 (two sequential treatments). Most importantly, the percentage reduction of CFU, on day 21 post treatment, on denture surface of test formulation was superior to that of the commercially available marketed gel. Collectively, these results emphasized the efficient anti-fungal potential of test formula against Candida species.

Table 5. Percentage reduction of CFU in Group A and B.

Day	Reduction of CFU in Percentage			
	Palatal Surface		Denture Surface	
	Group A	Group B	Group A	Group B
Baseline vs. 7 days	33.38%	33.59%	25.2%	27.06%
Baseline vs. 14 days	64.78%	64.80%	47.58%	49.36%
Baseline vs. 21 days	85.9%	90.20%	77.98%	71.68%
7 days vs. 14 days	47.2%	48.51%	29.80%	30.03%
14 days vs. 21 days	59.9%	71.34%	70.05%	44.03%
7 days vs. 21 days	78.7%	85.2%	58.07%	30.80%

2.8.3. Clinical Response to Treatment with Clove/Cinnamon Extracts-Loaded Emulgel

The response to treatment with either clove/cinnamon extracts-loaded emulgel or the commercially available gel (Candid®) in patients suffering from denture stomatitis was evaluated clinically. As depicted in Figure 6, treatment with optimized emulgel formula significantly enhanced the clinical outcomes in patients suffering from denture stomatitis as manifested by a remarkable improvement in erythema of palatal mucosa. In addition, as summarized in Table 6, treatment with optimized test formula significantly alleviated denture stomatitis-related inflammation, compared to Candid®-treated group, in a treatment duration-dependent manner. The complete cure rate percentage in patients treated with the optimized formula increased from 19% on day 7 post treatment to 47.6% on day 21 post treatment. Most importantly, there was a remarkable shift from Newton's stages II/III denture stomatitis to Newton's stage I denture stomatitis, and no patients suffering from either Newton's stage II/III denture stomatitis were observed on day 21 post treatment. On the other hand, more than 20% of Candid®-treated group still suffered from Newton's stage II denture stomatitis even following three weeks' treatment. These results emphasized that the optimized emulgel formula could provide superior clinical remission of Newton's classification and reduction of fungal burden compared to the commercially tested gel (Candid®). Chi-square test was conducted for test and control groups to evaluate the influence of applied treatment on clinical improvement and found statistically significant improvement in test group when compared to the control group (Table S4).

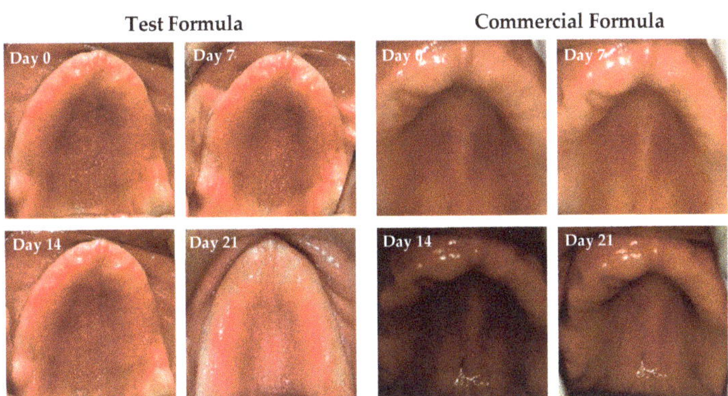

Figure 6. Pictorial representation of clinical response to treatment of Group A and B on Stage III and Stage II conditions on days 0, 7, 14, and 21 post treatment.

Table 6. Clinical response to treatment between test and control groups.

Group	Duration	Newton's Classification			
		I	**II**	**III**	**Cure**
Group A (Test formula)	0 (baseline)	6 (18.8%)	13 (44.8%)	2 (100.0%)	0 (0.0%)
	7 days	6 (18.8%)	11 (37.9%)	0 (0.0%)	4 (19.0%)
	14 days	9 (28.1%)	5 (17.2%)	0 (0.0%)	7 (33.3%)
	21 days	11 (34.4%)	0 (0.0%)	0 (0.0%)	10 (47.6%)
Group B (Candid®)	0 (baseline)	9 (22.5%)	10 (47.6%)	2 (100.0%)	0 (0.0%)
	7 days	13 (32.5%)	5 (23.8%)	0 (0.0%)	3 (14.3%)
	14 days	11 (27.5%)	1 (4.8%)	0 (0.0%)	9 (42.9%)
	21 days	7 (17.5%)	5 (23.8%)	0 (0.0%)	9 (42.9%)

2.8.4. Trial Adherence

To address treatment adherence, dosing compliance was calculated by weighing the tubes at baseline and one week, two weeks, and three weeks of the usage of gels. Patients who consumed more than 5 g of gel during treatment were considered adherents [26]. Greater adherence was shown for the test group, where the mean usage of gel was 7.83 ± 0.85 g at 7 days post usage, 7.63 ± 0.54 g at 14 days post usage, and 5.73 ± 0.738 g at 21 days post usage. On the other hand, commercial gel usage was 5.78 ± 0.79 g, 5.82 ± 0.93 g, and 5.04 ± 0.64 g at days 7, 14, and 21 post usage, respectively. Based on these results, we expect a high degree of compliance to the optimized test formula.

2.8.5. Taste Acceptability

Taste acceptability represents one of the crucial parameters that might affect patience compliance for the treatment. In the study, based on the Hedonic scale [27], 81% of investigated patients in the test group showed greater likeliness for the formulated gel. By contrary, 76.2% of patients treated with marketed gel disliked the gel. This result indicates the greater taste acceptability of optimized emulgel formula compared to marketed gel. This might explain with the improved treatment adherence in patients treated with test formula compared to marketed product.

Although the key factor in the treatment of chronic infection such as denture stomatitis is efficient denture cleaning, the denture cleansing action and mechanism of biofilm regrowth seems unclear [28,29]. In this study, most of the patients used to clean their dentures with soap and brush, but denture stomatitis was still detected, which concluded that only mechanical brushing does not cure denture stomatitis [30]. Many commercially available cleansers are ineffective at inhibiting Candida biofilms [29]. With the global threat of antimicrobial resistance, a paradigm shift to research into plant extracts and/or plants essential oils has been revisited. The essential oil targets multiple microbes, thus exhibiting broad-spectrum antimicrobial activity with little or no occurrence of antimicrobial resistance [31,32]. In this study, we revealed for the first time the efficacy of clove oil and cinnamon bark oil incorporated polyherbal emulgel treatment for *Candida albicans*-associated denture stomatitis.

3. Conclusions

In this study, we succeeded to develop and optimize clove and cinnamon supercritical fluid extracts-loaded emulgel and compared its antifungal potential in denture stomatitis with a commercially available marketed product. The inter-group comparison between the two formulations inferred a comparable reduction in CFU percentage from the palatal surface between both groups. In addition, both formulations efficiently alleviated denture stomatitis-related inflammation in a treatment duration-dependent manner with comparable clinical cure rates. Accordingly, herbal extracts-loaded emulgel might be considered an evolution of indigenous polyherbal formulations and might represent a promising alternative to the existing allopathic drugs for the treatment of denture stomatitis, with better taste acceptability and no side effects.

4. Materials and Methods

4.1. Materials

Clove bud, cinnamon bark, and almond gum were purchased from the Govindaraja Shetty and Sons (Mysuru, India). Clotrimazole was procured from Sigma Aldrich (St. Louis, MO, USA). The freeze-dried form of the microorganism *Candida Albicans* (ATCC 10231) and human gingival fibroblast (HGF) (ATCC®PCS-201-018TM) were obtained from American Type Culture Collection. Fetal Bovine Serum and Dulbecco's modified eagle medium (DMEM) were purchased from Gibco (Fort Worth, TX, USA). All other solvents and reagents were of analytical grade.

4.2. Preparation of Supercritical Fluid Plant Extract

All the procured and authenticated clove buds and cinnamon barks were dried in shade, cleaned by hand, sorted, and powdered. A weight of 10 kg of powered individual herbs were then loaded in the supercritical fluid (SCF) extractor unit. SCF extraction was done using CO_2 gas without any co-solvents. Supercritical CO_2 was fed to the extractor through a high-pressure pump (300 bar) at 37 ± 0.5 °C, which was above its critical temperature and pressure. The extract-laden CO_2 was then sent to a separator (60–120 bar) via a pressure-reduction valve. The temperature and pressure were reduced so that the extract precipitated into the separator, and gaseous CO_2 was released into the atmosphere [33].

4.3. Quantification of Herbal Extracts Constituents by Gas Chromatography

Gas chromatography (GC; Shimadzu GC- 2014, Tokyo, Japan) with a flame ionization detector was used for the detection of standard active compounds (Eugenol, Cinnamaldehyde, and benzaldehyde) in the herbal extracts. Gas chromatograph was equipped with an Rtx®-5 fused-silica column (30-m 0.25-mm id, film thickness 0.25 μm). The oven temperature of GC was programmed to an increasing temperature from 80 to 230 °C at a time rate of 6 °C/min and at 230 °C for 2 min. Detector and injector temperatures were set at 230 °C. The helium and nitrogen were adjusted at a linear velocity of 24 mL/min. A total of 1 μL of the samples (clove and cinnamon extracts dissolved in HPLC-grade chloroform) were injected into the GC using split mode with a split ratio of 10 to 1. Stock solutions were prepared by precisely weighing 10 mg of eugenol, cinnamaldehyde, or benzaldehyde and dissolving them in 10 mL of HPLC grade chloroform. A series of aliquots (2 to 10 μg/mL of eugenol, cinnamaldehyde, or benzaldehyde) were prepared by serially diluting stock solution and was quantified with GC to construct standard calibration curves [34,35].

4.4. Antifungal Activity of the Herbal Extracts

The minimum inhibitory concentrations (MICs) of either clove or cinnamon extracts alone or in combination against *C. albicans* isolates were determined with a broth microdilution method as described by the CLSI guidelines [36]. Briefly, in a 96-well plates, 100 μL of serial dilutions of clove extract, cinnamon extract alone or clove/cinnamon extracts, having the concentrations of 512, 256, 128, 64, 32, 16, 8, 4, 2, and 1 µg/mL, were added to each well. A drug-free well served as a negative control. Then, 100 µL aliquots of inoculum of the test strain, *C. Albicans*, adjusted to 1.5×10^6 CFU/mL equal to 0.5 McFarland, were taken aseptically and then added to each well and kept for incubation at $35 \pm 2\,^{\circ}\mathrm{C}$ under anaerobic conditions for 24 h. After the specified incubation time, the growth inhibition was determined both by visual inspection.

The in-vitro interaction of clove and cinnamon extracts against *C. Albicans* was interpreted in terms of the fractional inhibitory concentration index (FICI) as follows [37]:

$$\mathrm{FICI} = \mathrm{FIC}_{clove} + \mathrm{FIC}_{cinnamon} = (\text{MIC of clove in combination/MIC of clove alone}) + (\text{MIC of cinnamon in combination/MIC of cinnamon alone}).$$

The interpretation of the FICI was defined as FICI of ≤ 0.5 for synergy, $0.5 < \mathrm{FICI} \leq 1.0$ for additive effect, and $\mathrm{FICI} > 1.0$ for antagonism.

4.5. Solubility and Emulsification Studies

The solubility of clove and cinnamon extracts was determined in various surfactants (Tween 20, Tween 80, Span 20, and Span 80) and co-surfactant (PEG 400) by adding an excess amount of extract in 1 mL of the selected vehicle. In a typical procedure, an excess of the drug (clove and cinnamon extracts) was mixed in the respective systems using a vortex mixer. The mixtures obtained were set aside for 72 h and centrifuged for 10 min at 5000 rpm. Then, 0.5 mL of supernatant was drawn out, diluted, and analyzed for both clove and cinnamon solubility using GC as aforementioned.

To select the best suitable ratio of surfactant and co-surfactant (S_{mix} ratio) to obtain the desired emulgel, emulsification studies were performed. Tween 80 and PEG 400 were selected as surfactant and co-surfactant, respectively. Five different S_{mix} ratios were evaluated (1:1, 1:2, 1:3, 2:1, and 3:1) for the emulsifying ability. The obtained solutions were mixed using a vortex mixer to form homogenous blends.

4.6. Preparation of Emulgel

4.6.1. Experimental Design

To demonstrate the response surface model by attaining different combination of values, design expert software (Version 12, Stat-Ease Inc. and Minneapolis, MN, USA) was employed. A 2-factor, 2-level central composite design was used to design the optimized procedure to formulate various emulgel formulations and to investigate the impact of two formulation variables, namely polymer concentration and surfactant/co-surfactant (S_{mix}) ratio on formulations parameters, namely particle size AN drug content. The experimental design matrix of the central composite design is summarized in Table 7. A total 9 runs were performed to obtain the optimized formula and to achieve the desired responses.

Table 7. Variables in center-composite design for preparation and optimization of emulgel.

Factors	Levels	
Independent Variable	**Low**	**High**
X_1 = Polymer (% w/w)	1	6
X_2 = S_{mix} (% v/v)	4	11
Dependent variable	**Goals**	
R_1 = Globule size (nm)	Decrease	
R_2 = Drug Content clove (%)	Increase	
R_2 = Drug Content cinnamon (%)	Increase	

4.6.2. Formulation of Clove/Cinnamon Extracts-Loaded Emulgel

Different formulations were prepared using varying amounts of gel-forming polymer (badam gum; 1–6% w/w) and different ratios of surfactant/co-surfactant mixture (S_{mix};

5–10% v/v). Briefly, specified weighs of clove and cinnamon extracts at a ratio of 1:8, corresponding to their MIC values, were mixed with different ratios of Tween 80 and PEG 400 (S_{mix}) mixture, forming an oily phase. Varying amounts of gel-forming polymer (badam gum) were soaked in water and served as the aqueous phase. The oily phase was added to the aqueous phase portion wise and was homogenized at 15,000 rpm for 15 min to form different emulgel formulations [11].

4.7. Evaluation of Clove/Cinnamon Extracts-Loaded Emulgel

4.7.1. Physical Examination

Appearance, color, homogeneity, grittiness, and smoothness of the prepared emulgel formulations were tested both visually and by touch.

4.7.2. Determination of pH

The pH of different emulgel formulation was measured by standardized pH meter (MW802, Milwaukee Instruments, Szeged, Hungary). The pH measurements were performed in triplicates, and the mean value was calculated.

4.7.3. Viscosity Measurement

The viscosity (in cPs) of the formulated emulgels was determined by Brookfield viscometer using spindle number T-F at 25 °C. The viscosity measurements were performed in triplicates, and the mean value was tabulated.

4.7.4. Determination of Spreadability

As per the International standards for Harmonization (ICH) guidelines, 1 g of emulgel sample was sandwiched in between two glass plates (20 cm × 20 cm). A definite load (500 g) was applied on the upper glass plate for 1 min. Later, the radius of the spread gel was measured. The spreadability (g cm/s) was measured by using the following equation:

$$S = (m \times l)/t$$

where S is the spreadability (g cm/s), m is the mass of the weight applied (gm), l is the radius of the spread gel (cm), and t is the time taken (s).

4.7.5. Extrudability

The emulgel formulations were placed in standard capped collapsible aluminum tubes crimped at the end, and the weights of each tube was recorded. The tubes were placed between two glass slides and were clamped. A definite weight of 500 g was placed over the slides, and the amount of the emulgel extruded from tubes was weighed [38]. The percent of the extruded gel was calculated, and the extrudability is recognized as excellent if extrudability > 90%, good if extrudability > 80%, and fair if extrudability > 70%.

4.7.6. Determination of Globule Size of Emulgel Formulations

The globule size of all emulgel formulations was measured by the internal light scattering technique using a Malvern instrument (Malvern Nano ZS90, Malvern Instrument Ltd., Worcestershire, UK). To avoid multiple scattering, concentrated emulsions were diluted in (1:1000) with deionized water before analysis. The droplet size distribution of each gel was measured trice, and the mean droplet diameter was reported as the average.

4.7.7. Drug Content Study

The drug content of all the formulations was measured by dissolving an accurately weighing 100 mg of gel formulations in 10 mL of HPLC-grade chloroform. The solution was filtered through 0.45-mm membrane filter, and further serial dilutions were made and subjected for drug content uniformity by using gas chromatography.

4.7.8. In-Vitro Muco-adhesion Measurement

In-vitro muco-adhesion for emulgel formulations was evaluated by a modified tensiometer technique adopted from Fisher's tensiometer [39]. Briefly, 200 mg of emulgel were taken on a mica disk and were placed on the tensiometer, allowing the gel to come in contact with 1% (w/v) sodium alginate solution for 5 min. Later, the gel was removed from the solution of sodium alginate at 0.2 inches/min rate. The adhesion force between the mica disk and the sodium alginate solution was employed as the blank. The detachment force was calculated in terms of dyne/cm^2, and the study was performed in triplicates.

4.7.9. Fourier-Transform Infrared (FTIR) Analysis

Fourier-transform infrared (FTIR) spectra of both pure extract, physical mixtures of clove and cinnamon, and emulgel were recorded using Shimadzu 8400S spectrometer (Tokyo, Japan) to evaluate the compatibility between the selected extracts and S$_{mix}$. A physical mixture was prepared by employing the sodium chloride (NaCl) plate method, where (NaCl) was used to hold samples and scanned at a resolution of 4 cm^{-1} in the range of 4000–400 cm^{-1}.

4.8. In-Vitro Drug Release Studies

The in-vitro drug release studies were conducted using a Franz diffusion cell. Briefly, cellophane membrane (12,000–14,000 MW) previously soaked in a phosphate buffer of pH 7.4, was fixed in between donor and receiver cells. Then, 100 mg of the optimized emulgel formulation or a commercially available gel (Candid®) were placed into donor compartment. Phosphate buffer pH 7.4 was used as a dissolution media. The temperature of the cell was kept at 37 ± 0.5 °C by circulating water jacket, and the buffer was stirred continuously at 100 rpm. At definite time points (0, 15, 30, 45, 60, 90, 120, 150, and 180 min), aliquot samples were collected and were replaced with equal volumes of fresh medium to maintain the sink condition. The collected samples containing clove and cinnamon were extracted by chloroform, filtered through a 0.45-mm membrane filter, further diluted, and finally analyzed by gas chromatography as aforementioned. The amount of clotrimazole released from Candid® gel was determined by HPLC as previously described [40]. Briefly, HPLC (Shimadzu, Kyoto, Japan) equipped with C18 (150 mm × 4.6 mm, 5 µm) column was used. The mobile phase consists of acetonitrile and water (70:30% v/v). The flow rate was 1.0 mL/min, and the UV detection was set at 210 nm. Clotrimazole was eluted at 5.6 min. Clotrimazole concentration was estimated from a pre-constructed calibration curve of clotrimazole at various concentrations.

4.9. Stability Studies

The stability study was carried out in accordance with ICH guidelines. Briefly, the optimized emulgel formulation was packed in collapsible tubes and kept for three months at varied temperature and humidity settings, namely 25 ± 2 °C and 40 ± 2°C and 65% RH, for 3 months and was then tested for various parameters, such as appearance, pH, and drug content.

4.10. Randomized Clinical Trial

4.10.1. Ethical clearance

The study proposal was submitted for approval and clearance was obtained from the ethical committee (No. ECR/1170/inst/KA/2019), JSS Dental College and Hospital under JSS AHER, Mysuru, India.

4.10.2. Study Design and Study Setting

A randomized, double-blind, controlled clinical trial, conducted at department of Prosthodontics, JSS Dental College and Hospital, was adopted to evaluate the antifungal potential of optimized clove/cinnamon extracts-loaded emulgel against Candida species and to compare it with a commercially available gel in patients with denture stomatitis.

Microbial swabs from selected patients with denture stomatitis were collected in the department of Prosthodontics, JSS Dental College and Hospital, and the microbial testing was carried out at department of Microbiology, JSS College of Life Sciences, Mysuru, India. All data regarding estimation of sample size, selection criteria, and randomization are provided as supplementary materials.

4.10.3. Biological Sample Collection and Analysis

Participants with more than 50 colonies were considered positive for fungal infection and were subjected to biological sample collection and analysis. For biological sample collection, a swab stick (ETO, sterile) was used to collect biofilms from the oral palatal and denture surfaces. Swabs were mixed separately with saline diluted according to the defined protocol. The swabs were immediately transferred to the Microbiology Department, College of Life Sciences, for plating and culture analysis. Briefly, 50 μL of the inoculum (1/100 dilution sample) were pipetted out and were plated onto Sabouraud dextrose agar (SDA) plates supplemented with chloramphenicol (0.1 g/mL medium) and incubated at 37 °C for 48 h. The procedure was done inside a laminar airflow chamber to ensure an even distribution of the sample.

4.10.4. Interventions and Blinding

A total of 42 patients, divided into two groups, were subjected to this study. The tubes with the medicaments were placed in a container and randomly divided and numbered into two groups: Group A (Optimized emulgel) and Group B (Commercially available gel (Candid® gel)). The tubes were labeled using a PVC stick-on label with coding and were segregated according to grouping. The tubes were then distributed as per participant's number and group. Verbal instructions and a written protocol were given to the volunteers in each study group for ensuring the appropriate application of gels under test.

To evaluate the fungal burden, on days 7, 14, and 21 after the first visit, oral palatal and denture swab samples were obtained and were assessed for the number of colony-forming units per milliliter (CFU/mL) using colony count software (Image J) (Figure S2). Briefly, digital photographs of the cultures were captured and analyzed for fungal colonies per swab (CFU/swab) using Open CFU software. The number of fungal CFU/mL was calculated by multiplying the number of colonies in each sample by the dilution factor [41]. The clinical and mycological examination were conducted before and after treatment, and the clinical effectiveness at each stage of treatment was correlated to clinical alterations in the severity and reduction in the colony counts.

4.10.5. Treatment Adherence

The sequentially numbered tubes with the medicaments were weighed before the study and on the 7th, 14th, and 21st days to estimate the patient adherence to treatment. Participants using more than 5 g gel throughout the study period were recognized as adherents. The appointments of the participants were arranged to avoid communication between participants to minimize research outcome bias.

4.10.6. Taste Acceptability

Subjective evaluation of the participants was conducted to analyze taste acceptability of products under investigation, using a Hedonic scale adapted by Gacula et al. [27], which ranged from liked extremely to dislike very much.

4.11. Statistical Analysis

The paired *t*-test was used to evaluate the reduction of fungal burden in both Group A and Group B. The CFU/mL reduction was analyzed by repeated measure ANOVA. Differences between treatment groups in clinical cure and improvement rates were tested for significance by the chi-square test. The acceptability of the products was compared using the

Mann–Whitney test and Crammers for the association. All the analysis was performed using SPSS version 18.0 software. A value of $p < 0.05$ was considered statistically significant.

Supplementary Materials: The following supporting information can be downloaded at: https://www.mdpi.com/article/10.3390/gels8010033/s1, Data S1: Estimation of sample size, selectin criteria and randomization; Figure S1: CONSORT Chart, Figure S2: Colony counting on SDA by using Colony Count Software, Table S1: Drug interactions of clove and cinnamon extracts against *C. albicans* in vitro, Table S2: Physical appearance of Emulgel; Table S3: Stability data of clove/cinnamon extracts-loaded emulgel at $25 \pm 2\ ^\circ\mathrm{C}$ and $40 \pm 2\ ^\circ\mathrm{C}$, Table S3: Chi-Square tests of Clinical response to treatment between both the groups.

Author Contributions: Conceptualization, A.S.A.L. and D.V.G.; methodology, M.S.I., A.K.G., S.P., K.A., and F.A.; software, E.-S.K. and H.H.A.; validation, E.-S.K. and H.H.A.; formal analysis, K.A., F.A., M.S.I., and A.K.G.; investigation, M.S.I., A.K.G., S.P., and K.A.; and F.A.; resources, E.-S.K. and H.H.A.; data curation, K.A. and F.A.; writing—original draft preparation, M.S.I., A.K.G., and S.P.; writing—review and editing, A.S.A.L. and D.V.G.; supervision, D.V.G.; project administration, M.S.I. and D.V.G. All authors have read and agreed to the published version of the manuscript.

Funding: The current work was supported in part by Department of Biotechnology, Government of India and JSS Academy of Higher Education and Research, Mysore, Karnataka, India, project BT/PR20803/TRM/120/118/2016.

Data Availability Statement: The data presented in this study are available on request from the Corresponding author.

Acknowledgments: This work was supported by Taif University Researchers Supporting Project number (TURSP-2020/29), Taif University, Taif, Saudi Arabia.

Conflicts of Interest: The authors declare no conflict of interest.

References

1. Gendreau, L.; Loewy, Z.G. Epidemiology and Etiology of Denture Stomatitis. *J. Prosthodont.* **2011**, *20*, 251–260. [CrossRef] [PubMed]
2. Pereira-Cenci, T.; Cury, A.D.B.; Crielaard, W.; Cate, J.M.T. Development of Candida-associated denture stomatitis: New insights. *J. Appl. Oral Sci.* **2008**, *16*, 86–94. [CrossRef] [PubMed]
3. Lalla, R.V.; Dongari-Bagtzoglou, A. Antifungal medications or disinfectants for denture stomatitis. *Evidence-Based Dent.* **2014**, *15*, 61–62. [CrossRef] [PubMed]
4. Deorukhkar, S.C.; Saini, S. Echinocandin Susceptibility Profile of Fluconazole Resistant Candida Species Isolated from Blood Stream Infections. *Infect. Disord. Drug Targets* **2016**, *16*, 63–68. [CrossRef]
5. Bolla, P.K.; Meraz, C.A.; Rodriguez, V.A.; Deaguero, I.; Singh, M.; Yellepeddi, V.K.; Renukuntla, J. Clotrimazole Loaded Ufosomes for Topical Delivery: Formulation Development and In-Vitro Studies. *Molecules* **2019**, *24*, 3139. [CrossRef]
6. Pan, S.-Y.; Litscher, G.; Gao, S.-H.; Zhou, S.-F.; Yu, Z.-L.; Chen, H.-Q.; Zhang, S.-F.; Tang, M.-K.; Sun, J.-N.; Ko, K.-M. Historical Perspective of Traditional Indigenous Medical Practices: The Current Renaissance and Conservation of Herbal Resources. *Evidence-Based Complement. Altern. Med.* **2014**, *2014*, 1–20. [CrossRef]
7. Cordell, G.A.; Colvard, M.D. Natural Products and Traditional Medicine: Turning on a Paradigm. *J. Nat. Prod.* **2012**, *75*, 514–525. [CrossRef]
8. Karimi, A.; Majlesi, M.; Rafieian-Kopaei, M. Herbal versus synthetic drugs; beliefs and facts. *J. Nephropharmacology* **2015**, *4*, 27–30.
9. Palla, A.H.; Amin, F.; Fatima, B.; Shafiq, A.; Rehman, N.U.; Haq, I.U.; Gilani, A.-U. Systematic Review of Polyherbal Combinations Used in Metabolic Syndrome. *Front. Pharmacol.* **2021**, *12*, 752926. [CrossRef]
10. Parasuraman, S.; Thing, G.; Dhanaraj, S. Polyherbal formulation: Concept of ayurveda. *Pharmacogn. Rev.* **2014**, *8*, 73–80. [CrossRef]
11. Khullar, R.; Kumar, D.; Seth, N.; Saini, S. Formulation and evaluation of mefenamic acid emulgel for topical delivery. *Saudi Pharm. J.* **2011**, *20*, 63–67. [CrossRef]
12. Hoare, T.R.; Kohane, D.S. Hydrogels in drug delivery: Progress and challenges. *Polymer* **2008**, *49*, 1993–2007. [CrossRef]
13. Eid, A.M.; Hawash, M. Biological evaluation of Safrole oil and Safrole oil Nanoemulgel as antioxidant, antidiabetic, antibacterial, antifungal and anticancer. *BMC Complement. Med. Ther.* **2021**, *21*, 159. [CrossRef] [PubMed]
14. Talat, M.; Zaman, M.; Khan, R.; Jamshaid, M.; Akhtar, M.; Mirza, A.Z. Emulgel: An effective drug delivery system. *Drug Dev. Ind. Pharm.* **2021**, *47*, 1–7. [CrossRef] [PubMed]
15. Mylangam, C.K.; Beeravelli, S.; Medikonda, J.; Pidaparthi, J.S.; Kolapalli, V.R.M. Badam gum: A natural polymer in mucoadhesive drug delivery. Design, optimization, and biopharmaceutical evaluation of badam gum-based metoprolol succinate buccoadhesive tablets. *Drug Deliv.* **2014**, *23*, 195–206. [CrossRef] [PubMed]

16. Khaw, K.-Y.; Parat, M.-O.; Shaw, P.N.; Falconer, J.R. Solvent Supercritical Fluid Technologies to Extract Bioactive Compounds from Natural Sources: A Review. *Molecules* **2017**, *22*, 1186. [CrossRef] [PubMed]

17. Capuzzo, A.; Maffei, M.E.; Occhipinti, A. Supercritical Fluid Extraction of Plant Flavors and Fragrances. *Molecules* **2013**, *18*, 7194–7238. [CrossRef]

18. Wu, T.; Han, B. Supercritical Carbon Dioxide (CO_2) as Green Solvent. In *Green Chemistry and Chemical Engineering*; Han, B., Wu, T., Eds.; Springer New York: New York, NY, USA, 2019; pp. 173–197.

19. Sales, M.D.C.; Costa, H.B.; Fernandes, P.M.B.; Ventura, J.A.; Meira, D.D. Antifungal activity of plant extracts with potential to control plant pathogens in pineapple. *Asian Pac. J. Trop. Biomed.* **2016**, *6*, 26–31. [CrossRef]

20. Soliman, S.; Alnajdy, D.; El-Keblawy, A.; Mosa, K.A.; Khoder, G.; Noreddin, A.M. Plants' natural products as alternative promising anti-Candida drugs. *Pharmacogn. Rev.* **2017**, *11*, 104–122. [CrossRef]

21. Stević, T.; Berić, T.; Šavikin, K.; Sokovic, M.; Godevac, D.; Dimkić, I.; Stanković, S. Antifungal activity of selected essential oils against fungi isolated from medicinal plant. *Ind. Crop. Prod.* **2014**, *55*, 116–122. [CrossRef]

22. Yu, L.X.; Amidon, G.; Khan, M.A.; Hoag, S.W.; Polli, J.; Raju, G.K.; Woodcock, J. Understanding Pharmaceutical Quality by Design. *AAPS J.* **2014**, *16*, 771–783. [CrossRef]

23. Singla, S.; Aggarwal, G. Proniosomes for effective topical delivery of clotrimazole: Development, characterization and performance evaluation. *Asian J. Pharm. Sci.* **2012**, *7*, 257–262.

24. Chuacharoen, T.; Prasongsuk, S.; Sabliov, C.M. Effect of Surfactant Concentrations on Physicochemical Properties and Functionality of Curcumin Nanoemulsions under Conditions Relevant to Commercial Utilization. *Molecules* **2019**, *24*, 2744. [CrossRef]

25. Kotwal, V.; Bhise, K.; Thube, R. Enhancement of iontophoretic transport of diphenhydramine hydrochloride thermosensitive gel by optimization of pH, polymer concentration, electrode design, and pulse rate. *AAPS PharmSciTech* **2007**, *8*, 320–325. [CrossRef]

26. Weatherell, J.A.; Robinson, C.; Nattress, B. Site-specific variations in the concentrations of substances in the mouth. *Br. Dent. J.* **1989**, *167*, 289–292. [CrossRef] [PubMed]

27. Gacula, M.; Rutenbeck, S.; Pollack, L.; Resurreccion, A.V.; Moskowitz, H.R. The just-about-right intensity scale: Functional analyses and relation to hedonics. *J. Sens. Stud.* **2007**, *22*, 194–211. [CrossRef]

28. Faot, F.; Cavalcanti, Y.W.; Bertolini, M.D.M.E.; Pinto, L.; Da Silva, W.J.; Cury, A.A.D.B. Efficacy of citric acid denture cleanser on the Candida albicans biofilm formed on poly(methyl methacrylate): Effects on residual biofilm and recolonization process. *BMC Oral Health* **2014**, *14*, 77. [CrossRef]

29. Freitas-Fernandes, F.S.; Cavalcanti, Y.W.; Filho, A.P.R.; Silva, W.J.; Cury, A.A.D.B.; Bertolini, M.M. Effect of daily use of an enzymatic denture cleanser on Candida albicans biofilms formed on polyamide and poly(methyl methacrylate) resins: An in vitro study. *J. Prosthet. Dent.* **2014**, *112*, 1349–1355. [CrossRef] [PubMed]

30. Walker, L.A.; Gow, N.; Munro, C.A. Fungal echinocandin resistance. *Fungal Genet. Biol.* **2010**, *47*, 117–126. [CrossRef] [PubMed]

31. Oliveira, J.D.A.J.d.A.; Da Silva, I.C.G.; Trindade, L.A.; Lima, E.O.; Carlo, H.; Cavalcanti, A.L.; De Castro, R.D. Safety and Tolerability of Essential Oil fromCinnamomum zeylanicumBlume Leaves with Action on Oral Candidosis and Its Effect on the Physical Properties of the Acrylic Resin. *Evidence-Based Complement. Altern. Med.* **2014**, *2014*, 1–10. [CrossRef]

32. Vairappan, C.S.; Nagappan, T.; Kulip, J. The Essential Oil Profiles and Antibacterial Activity of Six Wild Cinnamomum species. *Nat. Prod. Commun.* **2014**, *9*, 1387–1389. [CrossRef]

33. García-González, C.A.; Concheiro, A.; Alvarez-Lorenzo, C. Processing of Materials for Regenerative Medicine Using Supercritical Fluid Technology. *Bioconjugate Chem.* **2015**, *26*, 1159–1171. [CrossRef]

34. Guan, W.; Li, S.; Yan, R.; Tang, S.; Quan, C. Comparison of essential oils of clove buds extracted with supercritical carbon dioxide and other three traditional extraction methods. *Food Chem.* **2007**, *101*, 1558–1564. [CrossRef]

35. Li, Y.-Q.; Kong, D.-X.; Wu, H. Analysis and evaluation of essential oil components of cinnamon barks using GC–MS and FTIR spectroscopy. *Ind. Crop. Prod.* **2013**, *41*, 269–278. [CrossRef]

36. Gong, Y.; Liu, W.; Huang, X.; Hao, L.; Li, Y.; Sun, S. Antifungal Activity and Potential Mechanism of N-Butylphthalide Alone and in Combination with Fluconazole against Candida albicans. *Front. Microbiol.* **2019**, *10*, 1461. [CrossRef]

37. Odds, F.C. Synergy, antagonism, and what the chequerboard puts between them. *J. Antimicrob. Chemother.* **2003**, *52*, 1. [CrossRef] [PubMed]

38. Abdallah, M.H.; Lila, A.S.A.; Anwer, M.K.; Khafagy, E.-S.; Mohammad, M.S.; Soliman, M. Formulation, Development and Evaluation of Ibuprofen Loaded Nano-transferosomal Gel for the Treatment of Psoriasis. *J. Pharm. Res.* **2019**, *31*, 1–8. [CrossRef]

39. Aslani, A.; Ghannadi, A.; Najafi, H. Design, formulation and evaluation of a mucoadhesive gel from Quercus brantii L. and coriandrum sativum L. as periodontal drug delivery. *Adv. Biomed. Res.* **2013**, *2*, 21. [CrossRef] [PubMed]

40. Ferreira, P.G.; Lima, C.G.D.S.; Noronha, L.L.; De Moraes, M.C.; Silva, F.D.C.D.; Viçosa, A.L.; Futuro, D.O.; Ferreira, V.F. Development of a Method for the Quantification of Clotrimazole and Itraconazole and Study of Their Stability in a New Microemulsion for the Treatment of Sporotrichosis. *Molecules* **2019**, *24*, 2333. [CrossRef] [PubMed]

41. Belazi, M.; Velegraki, A.; Fleva, A.; Gidarakou, I.; Papanaum, L.; Baka, D.; Daniilidou, N.; Karamitsos, D. Candidal overgrowth in diabetic patients: Potential predisposing factors. *Mycoses* **2005**, *48*, 192–196. [CrossRef]

Article

Quality by Design for Optimizing a Novel Liposomal Jojoba Oil-Based Emulgel to Ameliorate the Anti-Inflammatory Effect of Brucine

Marwa H. Abdallah [1,2,]*, Heba S. Elsewedy [3], Amr S. AbuLila [1,2], Khaled Almansour [1], Rahamat Unissa [1], Hanaa A. Elghamry [2] and Mahmoud S. Soliman [1,4]

[1] Department of Pharmaceutics, College of Pharmacy, University of Ha'il, Ha'il 81442, Saudi Arabia; a.abulila@uoh.edu.sa (A.S.A.); kh.almansour@uoh.edu.sa (K.A.); ru.syed@liveuohedu.onmicrosoft.com (R.U.); m.soliman@uoh.edu.sa (M.S.S.)
[2] Department of Pharmaceutics and Industrial Pharmacy, Faculty of Pharmacy, Zagazig University, Zagazig 44519, Egypt; hanaaelghamry@yahoo.com
[3] Department of Pharmaceutical Sciences, College of Clinical Pharmacy, King Faisal University, Alhofuf 31982, Saudi Arabia; helsewedy@kfu.edu.sa
[4] Department of Pharmaceutics and Industrial Pharmacy, Faculty of Pharmacy, Al-Azhar University, Cairo 11651, Egypt
* Correspondence: mh.abdallah@uoh.edu.sa

Citation: Abdallah, M.H.; Elsewedy, H.S.; AbuLila, A.S.; Almansour, K.; Unissa, R.; Elghamry, H.A.; Soliman, M.S. Quality by Design for Optimizing a Novel Liposomal Jojoba Oil-Based Emulgel to Ameliorate the Anti-Inflammatory Effect of Brucine. *Gels* **2021**, *7*, 219. https://doi.org/10.3390/gels7040219

Academic Editors: Maddalena Sguizzato, Rita Cortesi and Rachel Yoon Chang

Received: 30 October 2021
Accepted: 16 November 2021
Published: 18 November 2021

Publisher's Note: MDPI stays neutral with regard to jurisdictional claims in published maps and institutional affiliations.

Abstract: One of the recent advancements in research is the application of natural products in developing newly effective formulations that have few drawbacks and that boost therapeutic effects. The goal of the current exploration is to investigate the effect of jojoba oil in augmenting the anti-inflammatory effect of Brucine natural alkaloid. This is first development of a formulation that applies Brucine and jojoba oil int a PEGylated liposomal emulgel proposed for topical application. Initially, various PEGylated Brucine liposomal formulations were fabricated using a thin-film hydration method. (2^2) Factorial design was assembled using two factors (egg Phosphatidylcholine and cholesterol concentrations) and three responses (particle size, encapsulation efficiency and *in vitro* release). The optimized formula was incorporated within jojoba oil emulgel. The PEGylated liposomal emulgel was inspected for its characteristics, *in vitro*, *ex vivo* and anti-inflammatory behaviors. Liposomal emulgel showed a pH of 6.63, a spreadability of 48.8 mm and a viscosity of 9310 cP. As much as 40.57% of Brucine was released after 6 h, and drug permeability exhibited a flux of 0.47 μg/cm²·h. Lastly, % of inflammation was lowered to 47.7, which was significant effect compared to other formulations. In conclusion, the anti-inflammatory influence of jojoba oil and Brucine was confirmed, supporting their integration into liposomal emulgel as a potential nanocarrier.

Keywords: PEGylated liposomes; emulgel; anti-inflammatory; Brucine; jojoba oil; transdermal drug delivery

1. Introduction

Nanotechnology is a technique that is concerned with manipulating materials into new matters within a nano-range [1]. Predominantly, nanotechnology is applied in many fields, especially the medical field through drug delivery systems. Currently, most the active pharmaceutical ingredients suffer from low solubility and bioavailability, which could be avoided via applying nanotechnology in developing drug delivery systems. A number of nanocarriers have been successfully developed, including nanoparticles, nanoemulsion, ethosome, niosome and liposome [2].

Liposome is a nanocarrier that gained a great concern since it offers a variety of advantages over conventional systems and free drugs [3]. It represents a biodegradable phospholipid bilayer system with a spherical shape characterized by its high efficiency, ability to improve drug solubility and capability to incorporate both hydrophilic and hydrophobic drugs [4]. Currently, liposomes are applied in many aspects such as in vaccination, cancer therapy, anti-inflammatory [5] and bacterial and fungal infection [6]. It could be delivered via different routes of administration including oral, parenteral transdermal and topical delivery [7].

One obstacle that researchers might face after developing liposomes is related to their stability issues [8]. In order to avoid such hurdles, it is better to shield liposomes with certain hydrophilic polymers to form steric hindrance barrier that could improve their stability [9]. Poly-ethylene glycol (PEG) is a widely used polymer that is adopted for coating pharmaceutical formulations in order to enhance their stability and/or protect the encapsulated drug [10]. Interestingly, PEGylation was extensively applied for intravenous drug targeting; nevertheless, recent studies have revealed the impact of PEGylated liposome in topical applications [11,12].

Developing liposomes for topical drug delivery system could face certain difficulties related to their low viscosities and improper application to the skin [13]. Integrating liposomes into emulgel formulation could possibly increase the viscosity of the formulation and facilitate its application over the affected area [14]. Emulgel is a combination of emulsion and gel such that it gains the benefit of both systems. Topical application of emulgel has shown several advantages such as being thixotropic, straightforward spreadability and good patient compliance [15]. Emulgel has been investigated for incorporating different drugs to be applied for various purposes such as mefenamic acid for analgesic and anti-inflammatory activity [16], flaxseed extracts for wound healing [17] and Metronidazole for antifungal and antibacterial activity [15].

Right now, the use of many natural products for treating several diseases has been established since they are safe and exhibit substantial pharmacological behavior [18]. Brucine is a natural substance that is obtained from seeds of *Nux vomica* tree and being one of its essential alkaloids [19]. It has been reported to influence the cardiovascular system and has antitumor, antibacterial, analgesic and anti-inflammatory effects [20]. However, its low solubility constitutes a great hurdle to its efficacy, which necessitates the finding of new formulations that could overcome the drawbacks and boost its activity.

Another category of natural product is plant oils, which are regarded as a nourishing source of good health and characterized by their availability and cheapness [21]. One of these oils is jojoba oil, obtained from *Simmondsia chinensis* seeds [22]. It is yellow in color and comprises numerous components, mainly flavonoids, polyphenols and alkaloids. Recently, jojoba oil has been extensively used in skin care products [23]. In addition, it exhibited great effectiveness in handling skin disorders such as dermatitis, eczema, acne and psoriasis, besides its potential for promoting wound healing [24]. Moreover, it can be used as a therapy in treating cancer, kidney and liver dysfunction, and acute lung injury [25], and it has exhibited certain anti-inflammatory activities [26]. Jojoba oil can be used in combination with several nanocarriers to form emulsion, microemulsion [27], nanoemulsion and emulgel [25]. Assimilation of jojoba-oil-based emulgel with liposomal formulation of Brucine could support the topical delivery and effectiveness of the drug.

Quality by design (QbD) is a systematic and organized technique for developing a unique product depending on pre-established factors and studying their influence on certain responses in order to attain the most optimized formula. It includes several tools, such as three-level full factorial, central composite design (CCD) and Box–Behnken design [28]. The most commonly used of them is Central Composite Design (CCD), which helps in selecting the optimized formula and in predicting the model relying on statistical analysis of variance (ANOVA) and definite equations [29].

In these contexts, our target was to develop PEGylated liposomal formulation containing Brucine that was subjected to optimization by implementing a 2^2 full factorial design and discovering the impact of definite independent variables on the examined dependent variables. The optimized PEGylated Brucine-loaded liposome was combined with the developed jojoba oil based emulgel. Then, the PEGylated liposomal emulgel loaded with Brucine was analyzed for physical and chemical properties. Ultimately, the developed liposomal emulgel was examined for its skin permeation and its anti-inflammatory influence through an exploration of the initiative of incorporating jojoba oil with Brucine.

2. Results and Discussion

2.1. Experimental Design

2.1.1. Fitting the Model

Operating CCD software gave rise to generating a matrix of 11 examinations distributed as four factorial points, four axial and three central points. Table 1 displays the impact of every independent variable on the experiential dependent variables of diverse PEGylated liposome formulations.

Table 1. The independent variables used for optimizing different PEGylated liposomal formulations and the detected dependent variables.

Formulation	Independent Variables		Dependent Variables		
	A (mg)	B (mg)	R_1 (nm)	R_2 (%)	R_3 (%)
F1	75	17.07	248 ± 2.0	51.4 ± 3.8	61.2 ± 4.9
F2	100	15	271 ± 5.3	65.8 ± 3.9	59.2 ± 4.1
F3	50	15	207 ± 1.9	45.8 ± 1.6	72.5 ± 3.7
F4	110.35	10	287 ± 4.5	68.3 ± 3.5	56.3 ± 3.9
F5	75	10	229 ± 3.2	54.2 ± 1.9	64.3 ± 5.1
F6	50	5	202 ± 3.5	49.3 ± 3.0	75.4 ± 2.6
F7	75	10	227 ± 2.9	55.0 ± 1.3	65.0 ± 4.6
F8	75	2.92	210 ± 2.7	60.0 ± 2.5	71.2 ± 4.9
F9	39.64	10	191 ± 2.9	41.6 ± 2.1	79.2 ± 5.5
F10	75	10	223 ± 3.2	56.2 ± 2.3	67.1 ± 4.5
F11	100	5	262 ± 3.6	71.6 ± 1.6	62.8 ± 4.4

A: EPC concentration; B: cholesterol concentration; R_1: particle size; R_2: EE%; R_3: % drug *in vitro* release.

2.1.2. Analysis of the Design

In order to predict and recognize the model, statistical analysis of data should be performed. It was understood that the model's *p*-values should be significant, and this is achieved if the value being less than 0.05. Additionally, lower F-values of the responses could result in more error in the model, and thus greater F-values are more desirable. Another vital parameter is the lack of fit, which is preferred to be non-significant in order to fit the data with the model [30].

It is clear in Table 2 that *p*-values of all the dependent variables R_1, R_2, and R_3 remained smaller than 0.0001, which confirmed that the effect of the independent variables on the observed dependent ones was significant [31]. Moreover, the model's F-values in R_1, R_2, and R_3 seemed to be significant. Further, lack of fit in the three cases, R_1, R_2, and R_3, exhibited non-significant values ($p > 0.05$) of 8.53, 4.39 and 1.75 and consistent *p*-values of 0.1086, 0.1971 and 0.4069 for R_1, R_2 and R_3, respectively.

Table 2. Results of statistical analysis of all dependent variables R_1, R_2 and R_3.

Source	R_1		R_2		R_3	
	F-Value	*p*-Value	F-Value	*p*-Value	F-Value	*p*-Value
Model	72.58	<0.0001 *	119.61	<0.0001 *	71.93	<0.0001 *
A-EPC	135.92	<0.0001 *	223.17	<0.0001 *	127.83	<0.0001 *
B-Cholesterol	9.24	<0.0161 *	16.04	0.0039 *	16.03	0.0039
Lack of Fit	8.53	0.1086	4.39	0.1971	1.75	0.4069
R^2 analysis						
R^2	0.9478		0.9676		0.9899	
Adjusted R^2	0.9347		0.9595		0.9797	
Predicted R^2	0.8948		0.9329		0.9341	
Adequate Precision	22.3246		28.6059		29.1829	

A, EPC concentration (mg); B, cholesterol concentration (mg); R_1, particle size (nm); R_2, EE (%); R_3, *in vitro* release (%); * significant.

2.2. Characterization

2.2.1. Influence of the Independent Variables on Particle Size (R_1)

Particle size of the developed PEGylated Brucine-loaded liposomes is considered one of the essential independent variables that were estimated as apparent in Table 1. It was settled that particle size of all formulations ranged between 191 ± 2.9 nm and 287 ± 4.5 nm for F9 and F4, respectively. In the data, it was obvious that increasing both EPC and cholesterol concentration resulted in a remarkable increase the particle size of all the formulations. This could be attributed to the increased thickness of the liposomal layers as more cholesterol will be dispersed within the lipid bilayers [32]. This was in accordance with Wu et al., who noticed that papain liposomes showed a decrease in the particle size upon decreasing phospholipid concentration [33]. The following equation emphasized the previous findings, as it explains the positive influence of A and B independent variables on the detected R_1 response, where the positive sign point to a synergistic action, while negative one defines antagonistic effect [34]:

$$R_1 = 118.108 + 1.29882A + 1.6935B$$

Furthermore, Figure 1a,b shows 2D Contour and 3D-response surface plots that help in elucidating the influence of A and B independent variables on the R_1 response. Figure 1c and facts in Table 2 clarified the linearity of the data through the interrelation between the adjusted R^2 value (0.9347) and the predicted one (0.8948).

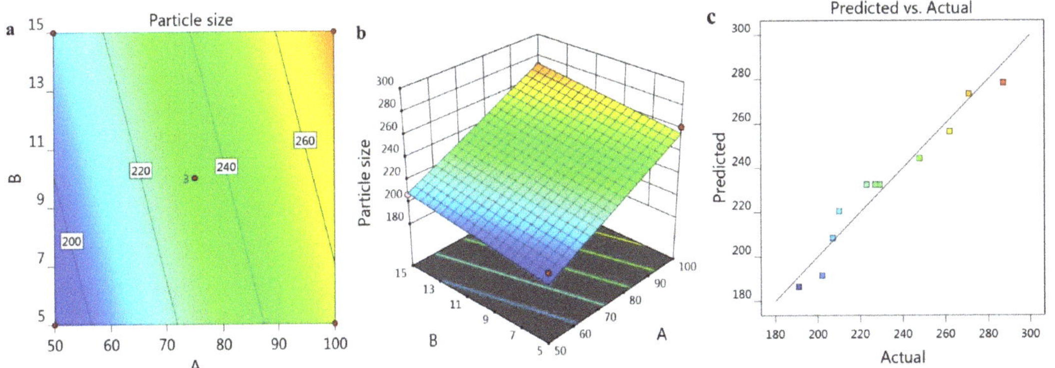

Figure 1. Representing (**a**) 2D-Contour plot, (**b**) 3D-Response Surface Plot, and (**c**) linear correlation plot between predicted against actual values.

2.2.2. Influence of the Independent Variables on EE (R_2)

Encapsulation efficiency for all developed PEGylated liposomal preparations were implemented, and as apparent in Table 1, it ranged from 41.6 ± 2.1 to 71.6 ± 1.6% for F9 and F11, respectively. Referring to the data, it was noticed that percentage of encapsulation efficiency increased upon increasing EPC concentration, which proved a positive influence of variable A. This is probably ascribed to the larger particle size of the formed liposome upon increasing EPC concentration, which gives more space for the drug to be encapsulated. Our findings are in agreement with those of Tefas et al., who established that encapsulation of curcumin and doxorubicin significantly increased upon increasing phospholipid concentration [35]. On the other hand, cholesterol concentration exerted a negative influence on the % of EE, where decreasing cholesterol concentration with constant EPC concentration resulted in a remarkable increase in percentage EE. Interestingly, cholesterol and hydrophobic drugs favor staying in the hydrophobic area of the liposomal membrane, and thus competition between them could occur and lower encapsulation accordingly attained [36]. Wu et al. confirmed our results since the study revealed that using more cholesterol leads to lowering the stability and rigidity of the liposome [37]. Remarkably, it was revealed from the results that there is a correlation between liposomal particle size, and its capability to encapsulate drug since increasing particle size resulted in considerable increment in % of EE as declared in a previous study [38]. The mathematical equation obtained from the design proved our prospects as it clarifies the synergistic influence of A and antagonistic action of B independent variables on the comeback of R_2.

$$R_2 = 31.6342 + 0.400298A - 0.536556B$$

The impact of the independent variables A and B on R_2 is graphically represented in Figure 2a,b exhibiting 2D Contour and 3D-response surface plot. Furthermore, Figure 2c supports the linearity of the data where the adjusted R^2 value (0.9595) was in correlation with the predicted one (0.9329).

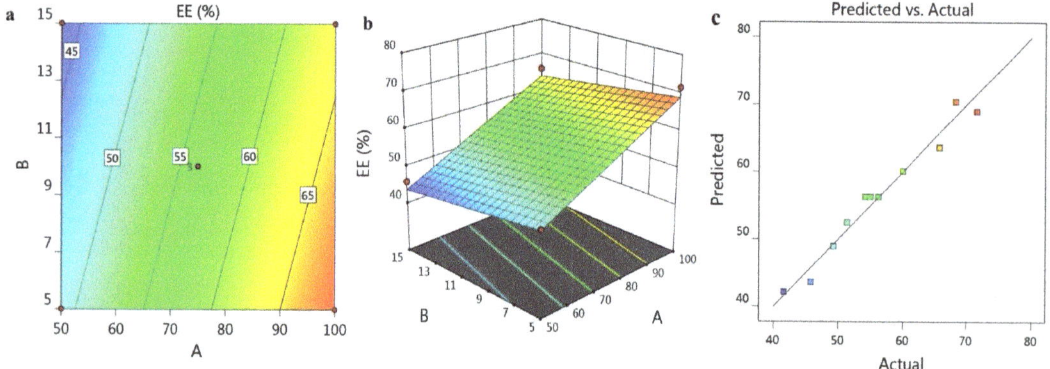

Figure 2. Representing (**a**) 2D-Contour plot, (**b**) 3D-Response Surface Plot and (**c**) linear correlation plot between predicted against actual values.

2.2.3. Influence of the Independent Variables on *In Vitro* Release (R_3)

The *in vitro* release characteristic of Brucine from all liposomal formulations under investigation was efficiently evaluated. The results were depicted in Figure 3 and Table 1. The study lasted for 6 h and the % of drug released ranged between 56.3 ± 3.9 and 79.2 ± 5.5%. It was obvious that increasing both EPC and cholesterol concentration exerted a negative antagonistic effect on the dependent variable R_3. This could be due to the formerly revealed fact that large liposomes were obtained upon increasing concentration of A and B independent variables. Larger nanocarriers usually provide larger surface area and accordingly lower the percentage of Brucine released [39]. Another explanation could be accredited to the presence of DSPE-PEG in the liposomal formulation that offers more stability for the formulation in addition to the rigidization formed at the surface of the liposome by DSPE-PEG that led to a decrease in the % of drug *in vitro* release [40].

The former data obtained from *in vitro* release investigation are demonstrated in Figure 4a,b screening 2D-Contour and 3D-Response Surface Plots. Further, the linearity of the data was corroborated as presented in Figure 4c displaying the linear correlation plot where, the observed adjusted R^2 value was 0.9797 and the predicted one was 0.9341. Both values of R^2 are in reasonable agreement with each other as the variation between them was less than 0.2. In addition, the mathematical equation provided below would confirm the interrelation among the independent variables and their examined dependent one:

$$R_3 = 93.763 - 0.291427A - 0.516053B$$

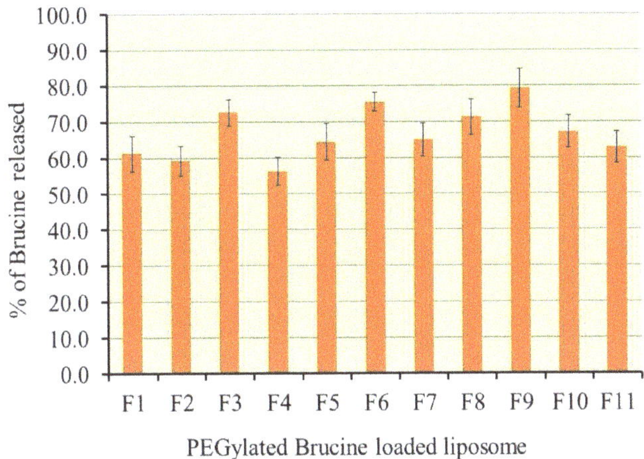

Figure 3. *In vitro* release of Brucine from different PEGylated liposome formulations in phosphate buffer pH 7.4 at 37 °C. Results are expressed as mean ± SD of three experiments.

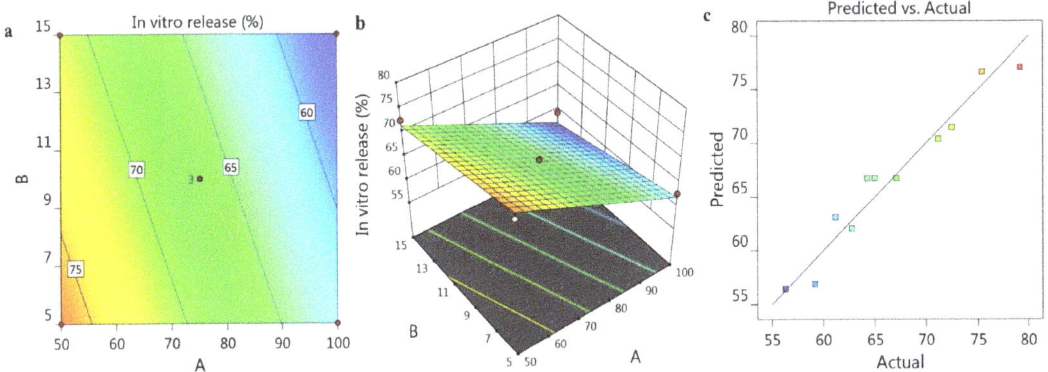

Figure 4. Representing (**a**) 2D-Contour plot, (**b**) 3D-Response Surface Plot and (**c**) linear correlation plot between predicted against actual values.

2.3. Optimizing the Developed PEGylated Liposomal Formulations Using CCD

Following constructing the design using CCD, it is very important to specify the optimum formulation via point prediction method and relying on the greater desirability obtained to provide the best formulation with appropriate features [41]. In the numerical optimization, the independent and dependent variables were guided toward requisite goals that help in selecting ideal formula. In the current investigation, A and B were adjusted to be within range in addition to R_1, while the criteria shifted toward maximizing both R_2 and R_3. As a consequence, A and B values were proposed to be 75.64 and 5 mg, respectively, in addition to the expected values for the optimized formulation that seemed to fulfill the optimization process as illustrated in Table 3 and the desirability value (0.574) as in Figure 5. The data of the optimal point suggested from the software were used to develop a new optimized formulation, and upon comparing its observed result, it was found to be very close to the expected one. The distribution curve of the optimized PEGylated Brucine liposomal formulation is exhibited in Figure 6, demonstrating the particle size (221.8 ± 2.04) and allied PDI (0.289 ± 0.62).

Table 3. Predicted and observed results of the optimized PEGylated Brucine liposomal formulation.

Independent Variables	Symbol	Goal
EPC	A	In range
Cholesterol	B	In range
Dependent variables	Predicted results	Observed results
R_1 (nm)	224.8 ± 7.8	221.8 ± 3.01
R_2 (%)	59.22 ± 1.89	57.66 ± 3.06
R_3 (%)	69.14 ± 1.82	67.96 ± 2.65

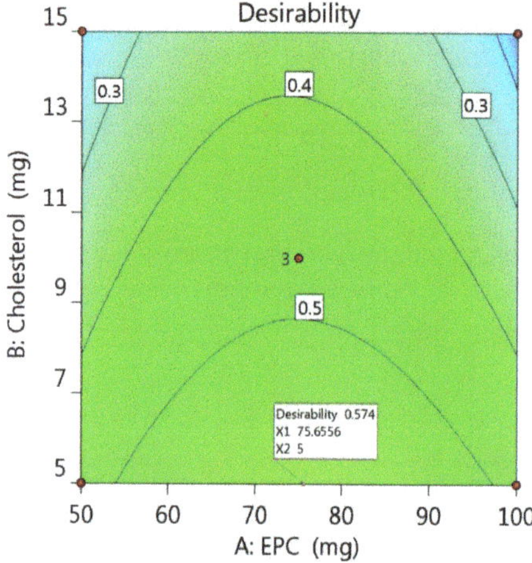

Figure 5. Desirability graph presenting the influence of independent variables A and B on the overall dependent variables R_1, R_2 and R_3.

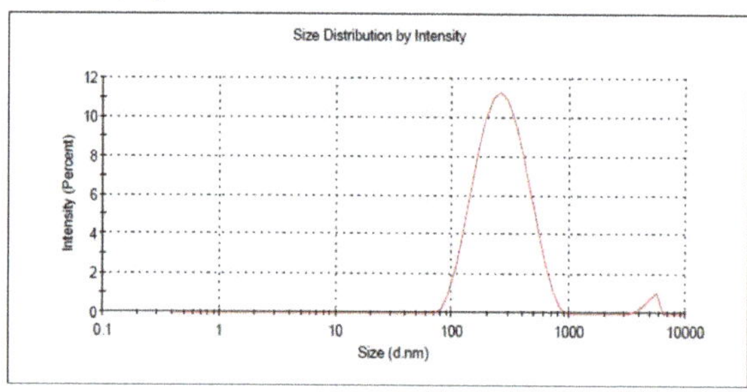

Figure 6. Particle size and PDI of optimized PEGylated Brucine liposomal formulation.

2.4. Stability Study of the Optimized PEGylated Brucine Liposomal Formulation

Estimating stability of the optimized liposomal formulation was accomplished over 1 and 3 months following storage at 4 ± 1 °C and at 25 ± 1 °C, and results are discussed in Figure 7. The results revealed that non-significant differences ($p < 0.05$) were observed upon comparing fresh preparation with that following storage, which considered confirmed evidence for the formulation stability and warrants the efficiency of liposome as a nanocarrier.

This stability could be attributed to the presence of DSPE-PEG since it provides a steric hindrance for the liposomal membrane [11].

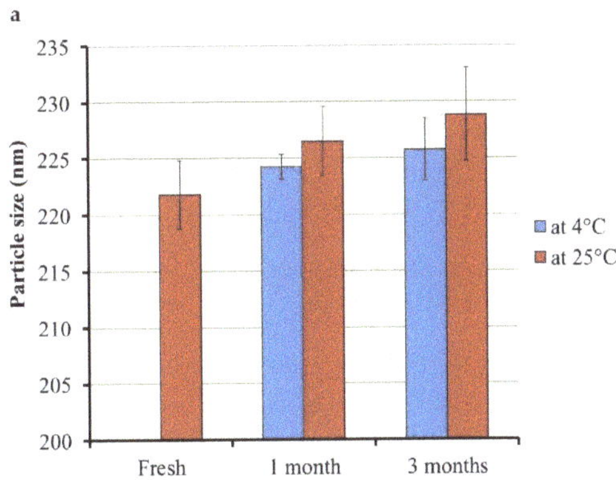

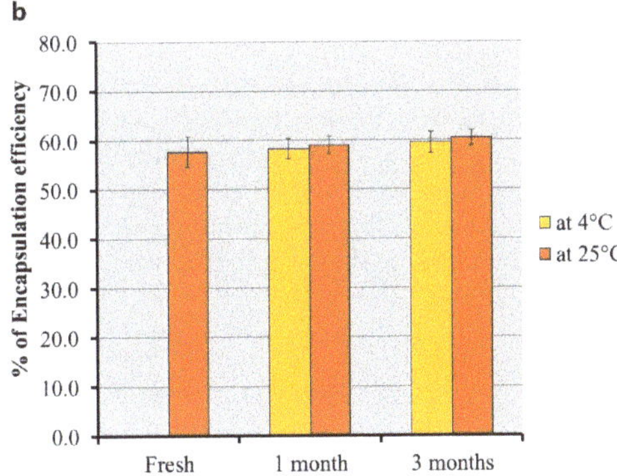

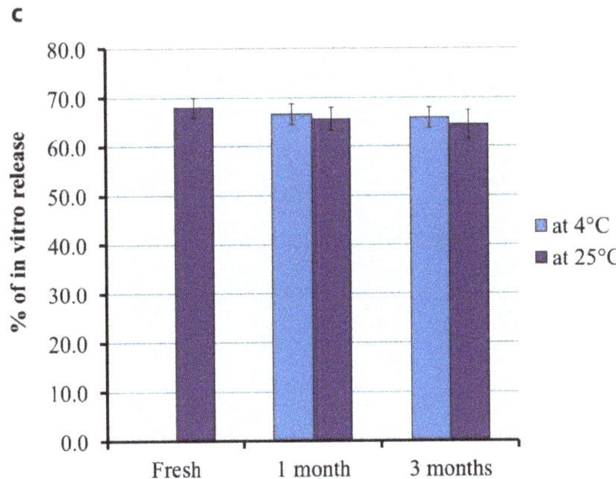

Figure 7. Outline of stability study for optimized PEGylated Brucine liposomal formulation for 1 and 3 months at 4 °C and 25 °C in terms of (**a**) particle size; (**b**) EE and (**c**) *in vitro* drug release, compared to freshly prepared formulation. Results are expressed as mean ± SD of three experiments.

As stated by the preceding attained results, the optimized Brucine liposomal formulation was assimilated with the pre-formulated jojoba oil-based emulgel via gentle stirring in order to develop a novel liposomal jojoba oil-based emulgel encapsulating Brucine, which was used in certain evaluations.

2.5. Estimating the Features of Developed PEGylated Liposomal Emulgel Encapsulating Brucine

The formulation characterization was demonstrated in terms of several aspects. Physical inspection revealed that the preparation appeared as a smooth, homogenous emulgel whose screening for physical appearance is convenient. With reference to pH value validation, it was recorded as 6.63 ± 0.25, which was compatible with pH of the skin and appropriate to preclude any skin irritation. Spreadability of the formulation was 48.8 ± 2.7 mm in addition to the viscosity, which was 9310 ± 336 cP, indicating adequate result for topical preparation to be easily applied over the skin.

2.6. In Vitro Drug Release from Liposomal Emulgel

The behavior of drug release from Brucine suspension, liposomal formulation and optimized liposomal emulgel fabricated with jojoba oil was successfully prolonged for 6 h in phosphate buffer pH 7.4, and the release profile is displayed in Figure 8. It is obvious that almost 97.6 ± 3.8% of Brucine was released from free drug suspension within almost 3 h. The *in vitro* release of Brucine from PEGylated liposomal formulation and optimized liposomal emulgel was 57.53 ± 5.85 and 40.57 ± 4.82%, respectively, which displayed significant lower profile compared to Brucine released from suspension ($p < 0.05$). Alternatively, the *in vitro* release of Brucine from PEGylated liposomal formulation is significantly higher than that released from optimized PEGylated liposomal emulgel ($p < 0.05$). This could be attributed to the presence of jojoba oil and gelling agent in the emulgel formulation, which provides greater viscosity of the preparation and consequently lowers the rate of encapsulated drug diffusion [42].

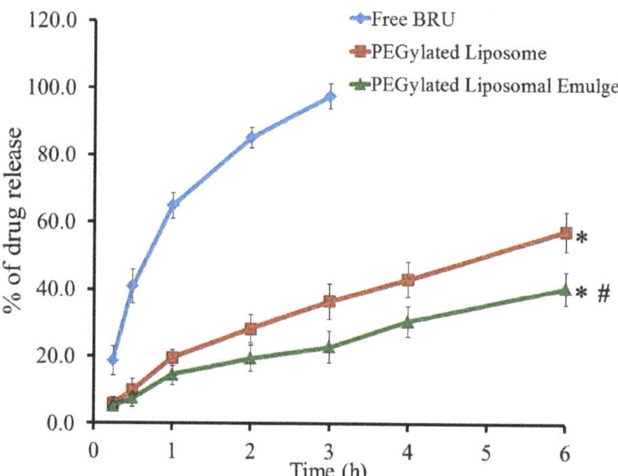

Figure 8. *In vitro* release outline of Brucine from free Brucine suspension, PEGylated liposome and PEGylated liposomal emulgel in phosphate buffer pH 7.4 at 37 °C. Results are identified with respect to the mean + SD of three experiments. * $p < 0.05$ compared to free drug; # $p < 0.05$ compared to PEGylated liposome formulation.

2.7. Permeation Studies

Permeation studies were executed for Brucine formulations over 6 h across rat skin membranes, and the results, expressed as the amount permeated, are presented in Figure 9.

In addition, the permeability parameters represented by SSTF and ER values are displayed in Table 4. A statistically significantly lower amount of Brucine was permeated from free Brucine suspension (0.202 ± 0.015 µg/cm^2·h) ($p < 0.05$), which is comparable with other formulations under examination. On the other hand, the SSTF of Brucine from PEGylated liposomal emulgel was 0.47 ± 0.035 µg/cm^2·h, enhancing the permeability by 2.33 ± 0.174 folds, which was significantly higher than the flux from PEGylated liposome (0.321 ± 0.028 µg/cm^2·h) with ER value 1.603 ± 0.142 ($p < 0.05$). In fact, higher flux from PEGylated formulations could be attributed to the integration of DSPE-PEG, which would bind to water molecules and consequently increase the hydration of stratum corneum, resulting in augmented skin permeability [8]. Higher flux from liposomal emulgel than liposomal formulation itself could be related to the colloidal characteristics of surfactant as well as the presence of jojoba oil, where both work as a permeation enhancer that improved the permeability [43,44]. In addition, surfactant has the capability to interact with the lipids of rat skin, increasing its fluidity, which could seemingly improve the drug permeation. Another explanation for higher permeation detected in liposomal emulgel could be ascribed to the dual action of surfactant and emulsion that forms the emulgel preparation, which could probably diffuse across the narrow pores of the membrane [45]. Our results are in agreement with a study done by Shehata et al. that proved significant higher permeability of insulin from niosomal emulgel when compared to insulin solution and niosomal gel [46].

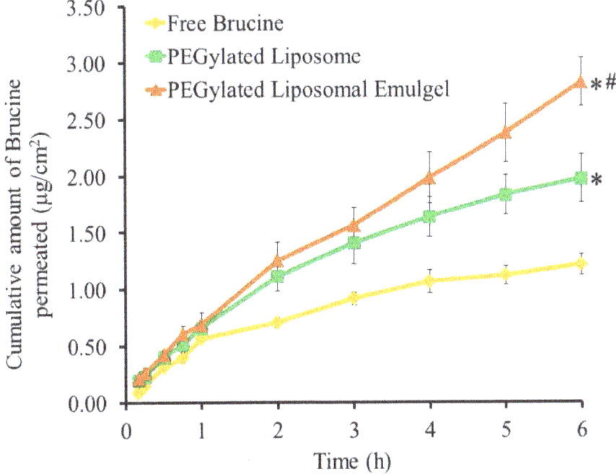

Figure 9. Outline of *ex vivo* permeability study of Brucine from diverse preparations through rat skin membrane. Results are identified in respect of mean ± SD (n = 3). * ($p < 0.05$) compared to free Brucine; # compared to PEGylated liposome formulation.

Table 4. Skin permeation parameters of different developed formulations.

Formula	SSTF µg/cm^2·h	ER
Free Brucine	0.202 ± 0.015	1
PEGylated liposome	0.321 ± 0.028 * #	1.603 ± 0.142 * #
Liposomal emulgel	0.47 ± 0.035 *	2.33 ± 0.174 *

Values are expressed as mean ± SD. * $p < 0.05$ compared to free Brucine suspension, # $p < 0.05$ compared to PEGylated liposomal emulgel formulation.

2.8. In Vivo Study

2.8.1. In Vivo Skin Irritation Test

Careful examination of animal back skin treated with investigated formulations was performed for checking any sensitivity reactions that might occur. No inflammation, irritation, erythema or edema was recognized on the inspected area during the whole 7 days of the investigation, which reflected safety of the formulations.

2.8.2. *In Vivo* Anti-Inflammatory Study: Carrageenan-Induced Rat Hind Paw Edema Method

The inspection of anti-inflammatory action on carrageenan-induced rat hind paw treated with formulations encapsulating Brucine was executed, and the profile of the *in vivo* study was displayed in Figure 10. Vigilant observation of the result showed that the maximum percentage of inflammation was reached following 4 h in the control group (99.7 ± 5.1%), which displayed a significant difference when compared to all other groups under examination ($p < 0.05$). In addition, after 12 h of the study, a significantly higher inflammation was still observed between the control group (84.5 ± 5.2%) and other groups in the study ($p < 0.05$). Likewise, maximum % of inflammation was detected following 2 h in groups treated with Brucine orally and the placebo-treated group, and no significant difference was detected between them during the whole experiment ($p < 0.05$). However, both groups exhibited significant reduction\compared to control group 3 h following the initiation of the experiment ($p < 0.05$). This finding confirmed the effect of the placebo-treated group in diminishing the inflammation and suggested the role of jojoba oil in reducing the inflammation. The result is in agreement with Habashy et al., who demonstrated the efficiency of jojoba in lowering inflammation in various examination models [47]. On the other hand, it was noted that % of inflammation reached in the Brucine orally treated group (64.7 ± 4.8%) and placebo-treated group (58.2 ± 4.6%) was significantly higher when compared with treated GP I (47.7 ± 4.8%) and treated GP II (34.2 ± 3.8%) ($p < 0.05$). Additionally, it was important to observe that at 6 and 12 h following the initiation of the experiment, treated GP II showed marked significant reduction of the inflammation, compared to all formulations under investigation ($p < 0.05$), which affirm the role of emulgel and its constituent in improving the anti-inflammatory effect of liposome and supports the synergistic action between Brucine and jojoba oil. This result is similar to that of Ibrahim and Shehata who proved that niosomal emulgel could considerably augment the anti-inflammatory effect of Ketorolac [45].

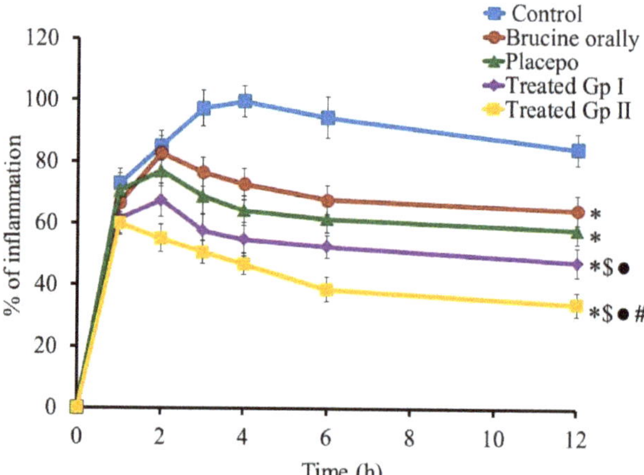

Figure 10. Representing the anti-inflammatory effect of various formulations on rat hind paw edema. Results are expressed as mean with the bar showing SD (n = 6). * $p < 0.05$ versus control-treated group; $ $p < 0.05$ versus Brucine orally treated group; • $p < 0.05$ versus placebo treated group and # $p < 0.05$ versus treated GP I.

3. Conclusions

In the present exploration, various PEGylated Brucine liposomal formulations were developed using thin film hydration method. The quality by design approach contributes to optimizing the formulation to be combined into jojoba-oil-based emulgel to form PEGylated Brucine liposomal emulgel. The formulation revealed respectable physical characterization, improved skin permeation performance and significant anti-inflammatory action. The result elucidated the incredible effect of jojoba oil toward the *in vivo* behavior of liposomal formulation that suggests the synergistic effect between Brucine and jojoba oil. Ultimately, liposomal emulgel could be considered as a prospective nanocarrier that successfully delivers the drug topically.

4. Materials and Methods

4.1. Material

Brucine was obtained from Alpha Chemika (Mumbai, India). Egg phosphatidyl choline (EPC) and cholesterol were purchased from Sigma Aldrich (St. Louis, MO, USA). Poly ethylene glycol-distearoylphosphatidyl ethanolamine (DSPE-PEG 2000) was procured from Lipoid LLC (Newark, NJ, USA). Jojoba oil was obtained from NOW® Essential Oils (NOW Foods, Bloomingdale, IL, USA). Ethanol, chloroform, polysorbate 80 (Tween 80) and Sodium carboxy methylcellulose (Na CMC) were acquired from Sigma-Aldrich Co. (St Louis, MO, USA). All other reagents were of the finest grade available.

4.2. Experimental Design

Optimizing the fabricated PEGylated Brucine loaded liposome was carried out using Central Composite Design (CCD), which is one tool of Response Surface Methodology (RSM). Fundamentally, two-factor, two-level (2^2) factorial design was constructed using two independent factors, EPC concentration (A) and cholesterol concentration (B), in which two levels were selected, low (-1) and high ($+1$), as illustrated in Table 5. The action of these independent variables on the investigated dependent variables was evaluated by operating Design-Expert software version 12.0 (Stat-Ease, Minneapolis, MN, USA). The examined dependent variables were particle size (R_1), encapsulation efficiency EE (R_2) and drug *in vitro* release after 6 h (R_3). The obtained data were analyzed via an analysis of variance (ANOVA) test followed by assembling model graphs and actual mathematical equations that clarify the influence of the independent variables on the explored dependent variables.

Table 5. CCD data showing independent variables and their level of variation.

Independent Variable	Character	Level of Variation	
		-1	$+1$
EPC concentration (mg)	A	50	100
Cholesterol concentration (mg)	B	5	15

4.3. Preparation of PEGylated Brucine-Loaded Liposome

Thin film hydration method that was previously described by Knudsen et al. was executed in order to develop different liposomal formulations [48]. Fifty milligrams of Brucine, ten milligrams of DSPE-PEG 2000 and a specified amount of EPC and cholesterol, as mentioned in Table 1, was added to a round bottom flask and mixed with a 6 mL ethanol: chloroform mixture (1:2). The flask was attached to a rotary evaporator (Heidolph, GmbH, Co. KG, Schwabach, Germany) for 2 h at 100 rpm maintained at 60 °C and operated till complete evaporation of the solvent and formation of thin lipid film on the inner wall of the round-bottom flask. The obtained lipid film was rehydrated with 5 mL phosphate buffer pH 7.4 for 30 min while vortexing using a classic advanced vortex mixer (VELP Scintifica, Usmate Velate, Italy). Next, the resultant dispersion was subjected to sonication for 30 s utilizing probe sonicator (XL-2000, Qsonica, Newtown, CT, USA) to obtain proper particle size. Eleven experimental formulations were fabricated using (CCD) alongside the values of their observed response as clarified in Table 1.

4.4. Characterization of PEGylated Brucine Loaded Liposome

4.4.1. Determination of Particle Size

Particle size of the fabricated liposomes was analyzed using Zetasizer apparatus (Malvern Instruments Ltd., Worcestershire, UK). Dynamic light scattering was used for evaluating the formulations keeping a scattering angle (90°) and temperature 25 °C [49].

4.4.2. Encapsulation Efficiency (EE)

A centrifugation method using centrifuge (Andreas Hettich GmbH, Co. KG, Tuttlingen, Germany) was applied to estimate the percentage of Brucine encapsulated into the PEGylated liposome. A sample of the preparation was added into Amicon® ultra-4 (Ultracel-10K, Merk Millipore Ltd., County Cork, Ireland) and allowed to be centrifuged at 6000 rpm maintained at 4 °C for 1 h. The filtrate was collected, diluted and measured spectrophotometrically at λ_{max} 262 nm using spectrophotometer (UV Spectrophotometer, JENWAY 6305, Bibby Scientific Ltd., Staffordshire, UK) for detecting the free drug [50]. EE% was calculated as follows:

$$\% \text{ EE} = ((\text{Total} - \text{Free})/\text{Total}) \times 100$$

4.5. In Vitro Drug Release from Different Liposomal Preparations

This investigation was designed to identify the percentage of Brucine released from the prepared PEGylated liposomal formulations. In this context, the ERWEKA dissolution system (ERWEKA, GmbH, Heusenstamm, Germany) functioned in order to perform the experiment. Briefly, liposomal samples were kept in glass tubes that closed from one side with a (Dialysis membrane Spctra/por® (MWCO 2000–15,000), New Brunswick, NJ, USA). The tubes were attached to the apparatus and immersed into the release media, which formed of phosphate buffer pH 7.4 maintained at 37 ± 0.5 °C and attuned at rotation speed 50 rpm. Two milliliters of the samples were taken at certain time intervals (0.25, 0.5, 1, 2, 4 and 6 h) and checked spectrophotometrically for the absorbance at λ_{max} 262 nm. The checked samples were substituted with the same volume of the vehicle [51].

4.6. Stability Studies of Optimized Liposomal Formulation

Stability of the optimized PEGylated Brucine liposomal formulation was certified relative to various parameters including the particle size, EE and *in vitro* drug release. The investigation was executed in agreement with the guiding principle of International Conference on Harmonization (ICH). The sample was stored at two environments, 4 ± 1 °C and at 25 ± 1 °C for 1 and 3 months.

4.7. Preparation of PEGylated Liposomal Emulgel Encapsulating Brucine

As mentioned previously, topical preparation should exhibit proper viscosity in order to spread properly and not be detached easily from the affected area of the skin. For that reason, the optimized liposomal formulation was integrated into a pre-prepared jojoba-oil-based emulgel. In short, 0.5 g Na CMC was added to 10 mL distilled water and allowed to stir till homogenous gel was attained. On the other side, emulsion was prepared using 1 g jojoba oil that mixed for 5 min with 1 g Tween 80. Afterward, aqueous phase was gently added over the oily phase with constant vortexing using classic advanced vortex mixer (VELP Scientifica, Usmate Velate, Italy) for 10 min till white emulsion was obtained. To attain the desired emulgel, the formulated emulsion was added to the pre-formulated gel and mixed with support of a mixer (Heidolph RZR1, Heidolph Instruments, Schwabach, Germany) till the development of homogenous jojoba-oil-based emulgel [52]. For developing PEGylated liposomal emulgel encapsulating Brucine, the pre-prepared liposomal formulation was mixed with the developed emulgel using a mixer (Heidolph

RZR1, Heidolph Instruments, Schwabach, Germany) till the desired formulation was obtained. Figure 11 displayed an illustrative scheme for the method of developing PEGylated liposomal emulgel loaded with Brucine.

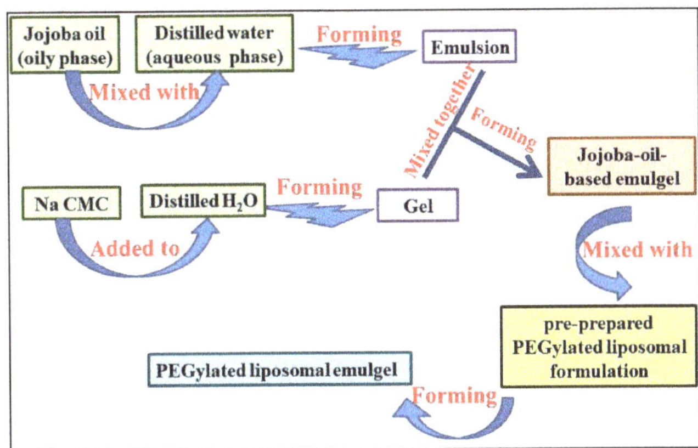

Figure 11. A schematic representation demonstrating the steps of manufacturing PEGylated liposomal emulgel incorporating Brucine.

4.8. Estimating the Features of Developed PEGylated Liposomal Emulgel Loaded with Brucine

4.8.1. Physical Inspection

This inspection was performed via visual assessment of the color and homogeneity that related to the developed PEGylated liposomal emulgel loaded with Brucine.

4.8.2. Validation of pH Value

The pH value of the topical formulation is very critical parameter to ensure the safety of the preparation. This evaluation was carried out using Standardized pH meter (MW802, Milwaukee Instruments, Szeged, Hungary) [53].

4.8.3. Spreadability Test

The current experiment is important for checking the competence of the formulation to spread evenly when smeared on the affected area of the skin. This is practically accomplished by measuring the spreading diameter where 1 g of the liposomal emulgel was added to be in between two glass slides (25 cm × 25 cm). A definite load of about (500 g) was put over the slides for 1 min. The spreading diameter of the formulation was assessed to provide an indication about the spreadability [54].

4.8.4. Viscosity

The measurement of the formulation viscosity was performed in order to determine its rheological behavior. This evaluation was employed using Brookfield viscometer (DV-II+ Pro., Middleboro, MA, USA) utilizing spindle R5 that rotate at 0.5 rpm at 25° [55].

4.9. In Vitro Drug Release from Liposomal Emulgel

The technique previously stated in Section 4.5 was followed in order to validate the percentage of Brucine released from the developed PEGylated liposomal emulgel formulation and compared to the free drug and the liposomal preparation [51].

4.10. Animal

All animal experiments were carried out using male Wistar rats of about 200–220 g. Housing condition of the animals was controlled to be switched every 12 h between light and dark cycles, and the temperature was kept at 25 ± 2 °C. The animals were obtained from the animal breeding center at the College of Science, King Faisal University. All experiments were conducted in agreement with the regulations and recommendations of Research Ethics Committee (REC) of Ha'il University (20455/5/42).

4.11. Ex Vivo Study

4.11.1. Animal Skin Preparation

Animal skin required for present study was obtained from male Wistar rats. The hair at the dorsal part of the animal was shaved carefully using an electric clipper followed by scarifying the rats and separating the skin. The adipose tissue was removed from the skin then it was preserved at 4 °C in phosphate buffer (pH 7.4) [45].

4.11.2. *Ex Vivo* Permeation Study

Permeation capability of the drug from Brucine suspension, optimized liposome and liposomal emulgel through animal skin was determined using amended Franz diffusion cells organized in our lab and used previously [46,56]. One hundred milliliters of phosphate buffer pH 7.4 containing 0.02% sodium azide was used, representing the release media and kept at 37 ± 0.5 °C. The rat skin was well attached to a glass tube that was held in the apparatus and suspended into the vehicle. The skin was used instead of cellophane membrane, where the upside stratum corneum was adhered to the sample, whereas the skin dermis was facing the media. All cells were shielded with Parafilm (Bemis, Oshkosh, WI, USA) in order to avoid evaporation of media. The system was operated and allowed to rotate constantly at 100 rpm [57]. Aliquots were taken to be analyzed spectrophotometrically at 0.25, 0.5, 0.75, 1, 2, 3, 4, 5 and 6 h. Certain parameters associated with the permeation across the skin were calculated including steady state transdermal flux (SSTF) and enhancement ratio (ER). SSTF characterizes the amount of permeated drug/(area × time); however, ER designates SSTF of test/SSTF of control. The experiment was assessed in triplicates.

4.12. In Vivo Study

4.12.1. *In Vivo* Skin Irritation Test

It is very important that topical preparations be safe and that no skin sensitivity is shown. To verify this, a skin irritation test was done using male Wistar rats that were prepared one day before beginning the study. The examined liposomal emulgel formulation was applied gently over the skin after being shaved using an electric clipper. Animals were kept under observation for 7 days to distinguish any sensitivity response including inflammation, irritation, erythema or edema. The detected response was clarified on the basis of a scale that ranged from 0, 1, 2 and 3 where it represents no reaction, minor reaction, moderate reaction or severe erythema that might be accompanied with edema, respectively [42].

4.12.2. *In Vivo* Anti-Inflammatory Study: Carrageenan-Induced Rat Hind Paw Edema Method

The anti-inflammatory influence of Brucine encapsulated into liposomal emulgel was appraised using male Wistar rats that went through a carrageenan-induced rat hind paw edema protocol as conducted previously by Shehata et al. [58]. Edema was initiated into the rat hind paw 30 min prior to the commencement of the study using subcutaneous injection of 0.5% *w/v* carrageenan in saline into the left hind paw [59]. Rats were arbitrarily categorized into 5 groups; each group carrying 6 animals as follows:

Group I was related to the control group, which is incited with inflammation only without treatment.

Group II treated orally with Brucine suspension (10 mg/kg) [60].

Group III was placebo that treated with PEGylated liposomal emulgel with no drug.

Group IV was treated with jojoba oil emulgel (Treated GP I).

Group V was treated with PEGylated liposomal emulgel formulation (Treated GP II).

The inflammatory response was assessed at diverse time (0, 1, 2, 3, 4, 6 and 12 h). Meanwhile, the distinctions in the thickness of rat hind paw following topical application of the examined formulations were measured by means of digital caliber. The % of inflammation was calculated from the following equation [61]:

$$\% \text{ of inflammation} = ((Th_t - Th_0)/Th_0) \times 100$$

where Th_t designates the thickness of carrageenan treated hind paw however Th_0 describes the hind paw at time zero.

4.13. Statistics

All investigations were performed at least three independent times, and the results were accompanied with mean $\pm$ SD. To distinguish the statistical differences between the groups, a Student's *t*-test was done. A one-way analysis of variance (ANOVA) followed by the least significant difference (LSD) as a post hoc test was employed to compare data and state the statistical significance. These assessments were designed using SPSS statistics software, version 9 (IBM Corporation, Armonk, NY, USA). If p-value < 0.05, it is considered to be statistically significant.

Author Contributions: M.H.A., Conceptualization, funding acquisition, writing—review and editing and supervision; H.S.E., A.S.A., K.A., methodology, software, data curation, validation, formal analysis, investigation, writing—review and editing; R.U., H.A.E., M.S.S., software and writing—original draft preparation. All authors have read and agreed to the published version of the manuscript.

Funding: This research was funded by Scientific Research Deanship at University of Ha'il, Saudi Arabia, through project number RG-20 126.

Institutional Review Board Statement: The study was conducted according to the guidelines of the Declaration of Helsinki, and approved by the Institutional Research Ethics Committee (IAEC), University of Ha'il, Saudi Arabia (approval no. 20455/5/42 at 27/11/2020).

Informed Consent Statement: Not applicable.

Data Availability Statement: Not applicable.

Acknowledgments: The authors thank Scientific Research Deanship at University of Ha'il, Saudi Arabia for funding this study through the project number RG-20 126.

Conflicts of Interest: The authors declare no conflict of interest.

References

1. Patra, J.K.; Das, G.; Fraceto, L.F.; Campos, E.V.R.; Rodriguez-Torres, M.D.P.; Acosta-Torres, L.S.; Diaz-Torres, L.A.; Grillo, R.; Swamy, M.K.; Sharma, S.; et al. Nano based drug delivery systems: Recent developments and future prospects. *J. Nanobiotechnol.* **2018**, *16*, 71. [CrossRef] [PubMed]
2. Abdallah, M.H.; Sabry, S.A.; Hasan, A.A. Enhancing Transdermal Delivery of Glimepiride Via Entrapment in Proniosomal Gel. *J. Young Pharm.* **2016**, *8*, 335–340. [CrossRef]
3. Deshpande, P.P.; Biswas, S.; Torchilin, V.P. Current trends in the use of liposomes for tumor targeting. *Nanomedicine* **2013**, *8*, 1509–1528. [CrossRef] [PubMed]
4. Din, F.U.; Aman, W.; Ullah, I.; Qureshi, O.S.; Mustapha, O.; Shafique, S.; Zeb, A. Effective use of nanocarriers as drug delivery systems for the treatment of selected tumors. *Int. J. Nanomed.* **2017**, *12*, 7291–7309. [CrossRef]
5. Matos, C.; Lobão, P. Non-Steroidal Anti-Inflammatory Drugs Loaded Liposomes for Topical Treatment of Inflammatory and Degenerative Conditions. *Curr. Med. Chem.* **2020**, *27*, 3809–3829. [CrossRef]
6. Faustino, C.; Pinheiro, L. Lipid Systems for the Delivery of Amphotericin B in Antifungal Therapy. *Pharmaceutics* **2020**, *12*, 29. [CrossRef] [PubMed]
7. Tiwari, G.; Tiwari, R.; Sriwastawa, B.; Bhati, L.; Pandey, S.; Pandey, P.; Bannerjee, S.K. Drug delivery systems: An updated review. *Int. J. Pharm. Investig.* **2012**, *2*, 2–11. [CrossRef]
8. Knudsen, N.; Rønholt, S.; Salte, R.D.; Jorgensen, L.; Thormann, T.; Basse, L.H.; Hansen, J.; Frokjaer, S.; Foged, C. Calcipotriol delivery into the skin with PEGylated liposomes. *Eur. J. Pharm. Biopharm.* **2012**, *81*, 532–539. [CrossRef]
9. Allen, C.; Dos Santos, N.; Gallagher, R.; Chiu, G.; Shu, Y.; Li, W.; Johnstone, S.; Janoff, A.; Mayer, L.; Webb, M. Controlling the physical behavior and biological performance of liposome formulations through use of surface grafted poly (ethylene glycol). *Biosci. Rep.* **2002**, *22*, 225–250. [CrossRef]

10. Teshima, M.; Kawakami, S.; Nishida, K.; Nakamura, J.; Sakaeda, T.; Terazono, H.; Kitahara, T.; Nakashima, M.; Sasaki, H. Prednisolone retention in integrated liposomes by chemical approach and pharmaceutical approach. *J. Control. Release* **2004**, *97*, 211–218. [CrossRef]

11. Albash, R.; El-Nabarawi, M.A.; Refai, H.; Abdelbary, A.A. Tailoring of PEGylated bilosomes for promoting the transdermal delivery of olmesartan medoxomil: In-vitro characterization, ex-vivo permeation and in-vivo assessment. *Int. J. Nanomed.* **2019**, *14*, 6555–6574. [CrossRef] [PubMed]

12. Jøraholmen, M.W.; Basnet, P.; Acharya, G.; Škalko-Basnet, N. PEGylated liposomes for topical vaginal therapy improve delivery of interferon alpha. *Eur. J. Pharm. Biopharm.* **2017**, *113*, 132–139. [CrossRef] [PubMed]

13. Alhakamy, N.A.; Aldawsari, H.M.; Ali, J.; Gupta, D.K.; Warsi, M.H.; Bilgrami, A.L.; Asfour, H.Z.; Noor, A.O.; Md, S. Brucine-loaded transliposomes nanogel for topical delivery in skin cancer: Statistical optimization, in vitro and dermatokinetic evaluation. *3 Biotech* **2021**, *11*, 288. [CrossRef] [PubMed]

14. Sah, S.; Badola, A.; Nayak, B. Emulgel: Magnifying the application of topical drug delivery. *Indian J. Pharm. Biol. Res.* **2017**, *5*, 25–33. [CrossRef]

15. Rao, M.; Sukre, G.; Aghav, S.; Kumar, M. Optimization of Metronidazole Emulgel. *J. Pharm.* **2013**, *2013*, 501082. [CrossRef] [PubMed]

16. Khullar, R.; Kumar, D.; Seth, N.; Saini, S. Formulation and evaluation of mefenamic acid emulgel for topical delivery. *Saudi Pharm. J.* **2012**, *20*, 63–67. [CrossRef]

17. Pagano, C.; Baiocchi, C.; Beccari, T.; Blasi, F.; Cossignani, L.; Ceccarini, M.R.; Orabona, C.; Orecchini, E.; Di Raimo, E.; Primavilla, S. Emulgel loaded with flaxseed extracts as new therapeutic approach in wound treatment. *Pharmaceutics* **2021**, *13*, 1107. [CrossRef]

18. Kim, J.H.; Kismali, G.; Gupta, S.C. Natural Products for the Prevention and Treatment of Chronic Inflammatory Diseases: Integrating Traditional Medicine into Modern Chronic Diseases Care. *Evid.-Based Complement. Altern. Med.* **2018**, *2018*, 9837863. [CrossRef]

19. Elsewedy, H.S.; Dhubiab, B.E.A.; Mahdy, M.A.; Elnahas, H.M. Development, optimization, and evaluation of PEGylated brucine-loaded PLGA nanoparticles. *Drug Deliv.* **2020**, *27*, 1134–1146. [CrossRef]

20. Lu, L.; Huang, R.; Wu, Y.; Jin, J.-M.; Chen, H.-Z.; Zhang, L.-J.; Luan, X. Brucine: A Review of Phytochemistry, Pharmacology, and Toxicology. *Front. Pharmacol.* **2020**, *11*, 377. [CrossRef]

21. Zhou, Y.; Zhao, W.; Lai, Y.; Zhang, B.; Zhang, D. Edible Plant Oil: Global Status, Health Issues, and Perspectives. *Front. Plant Sci.* **2020**, *11*, 1315. [CrossRef]

22. Sturtevant, D.; Lu, S.; Zhou, Z.-W.; Shen, Y.; Wang, S.; Song, J.-M.; Zhong, J.; Burks, D.J.; Yang, Z.-Q.; Yang, Q.-Y.; et al. The genome of jojoba (Simmondsia chinensis): A taxonomically isolated species that directs wax ester accumulation in its seeds. *Sci. Adv.* **2020**, *6*, eaay3240. [CrossRef] [PubMed]

23. Gad, H.A.; Roberts, A.; Hamzi, S.H.; Gad, H.A.; Touiss, I.; Altyar, A.E.; Kensara, O.A.; Ashour, M.L. Jojoba Oil: An Updated Comprehensive Review on Chemistry, Pharmaceutical Uses, and Toxicity. *Polymers* **2021**, *13*, 1711. [CrossRef]

24. Ranzato, E.; Martinotti, S.; Burlando, B. Wound healing properties of jojoba liquid wax: An in vitro study. *J. Ethnopharmacol.* **2011**, *134*, 443–449. [CrossRef] [PubMed]

25. Costa, I.; Rodrigues, R.; Almeida, F.; Favacho, H.; FalcÃO, D.; Ferreira, A.; Vilhena, J.; Florentino, A.; Carvalho, J.C.; Fernandes, C. Development of Jojoba Oil (Simmondsia chinensis (Link) C.K. Schneid.) Based Nanoemulsions. *Lat. Am. J. Pharm.* **2014**, *33*, 459–463.

26. Lin, T.-K.; Zhong, L.; Santiago, J.L. Anti-Inflammatory and Skin Barrier Repair Effects of Topical Application of Some Plant Oils. *Int. J. Mol. Sci.* **2017**, *19*, 70. [CrossRef] [PubMed]

27. Assaf, S.M.; Maaroof, K.T.; Altaani, B.M.; Ghareeb, M.M.; Alhayyal, A.A.A. Jojoba oil-based microemulsion for transdermal drug delivery. *Res. Pharm. Sci.* **2021**, *16*, 326. [CrossRef]

28. Ismail, T.A.; Shehata, T.M.; Mohamed, D.I.; Elsewedy, H.S.; Soliman, W.E. Quality by Design for Development, Optimization and Characterization of Brucine Ethosomal Gel for Skin Cancer Delivery. *Molecules* **2021**, *26*, 3454. [CrossRef]

29. Abdallah, M.H. Box-behnken design for development and optimization of acetazolamide microspheres. *Int. J. Pharm. Sci. Res.* **2014**, *5*, 1228–1239. [CrossRef]

30. Khalil, H.E.; Alqahtani, N.K.; Darrag, H.M.; Ibrahim, H.-I.M.; Emeka, P.M.; Badger-Emeka, L.I.; Matsunami, K.; Shehata, T.M.; Elsewedy, H.S. Date Palm Extract (Phoenix dactylifera) PEGylated Nanoemulsion: Development, Optimization and Cytotoxicity Evaluation. *Plants* **2021**, *10*, 735. [CrossRef]

31. Ibrahim, H.M.; Ahmed, T.A.; Hussain, M.D.; Rahman, Z.; Samy, A.M.; Kaseem, A.A.; Nutan, M.T. Development of meloxicam in situ implant formulation by quality by design principle. *Drug Dev. Ind. Pharm.* **2014**, *40*, 66–73. [CrossRef] [PubMed]

32. Shaker, S.; Gardouh, A.R.; Ghorab, M.M. Factors affecting liposomes particle size prepared by ethanol injection method. *Res. Pharm. Sci.* **2017**, *12*, 346–352. [CrossRef] [PubMed]

33. Wu, Y.; Xu, Y.; Sun, W. Preparation and particle size controlling of papain nano liposomes. *J. Shanghai Jiaotong Univ. (Agric. Sci.)* **2007**, *25*, 105–109.

34. Rahman, Z.; Zidan, A.S.; Habib, M.J.; Khan, M.A. Understanding the quality of protein loaded PLGA nanoparticles variability by Plackett–Burman design. *Int. J. Pharm.* **2010**, *389*, 186–194. [CrossRef]

35. Tefas, L.R.; Sylvester, B.; Tomuta, I.; Sesarman, A.; Licarete, E.; Banciu, M.; Porfire, A. Development of antiproliferative long-circulating liposomes co-encapsulating doxorubicin and curcumin, through the use of a quality-by-design approach. *Drug Des. Dev. Ther.* **2017**, *11*, 1605. [CrossRef]
36. Astarci, A.; Sade, A.; Severcan, F.; Keskin, D.; Tezcaner, A.; Banerjee, S. Celecoxib-loaded liposomes: Effect of cholesterol on encapsulation and in vitro release characteristics. *Biosci. Rep.* **2009**, *30*, 365–373. [CrossRef]
37. Wu, H.; Yu, M.; Miao, Y.; He, S.; Dai, Z.; Song, W.; Liu, Y.; Song, S.; Ahmad, E.; Wang, D.; et al. Cholesterol-tuned liposomal membrane rigidity directs tumor penetration and anti-tumor effect. *Acta Pharm. Sin. B* **2019**, *9*, 858–870. [CrossRef] [PubMed]
38. Maherani, B.; Arab-tehrany, E.; Kheirolomoom, A.; Reshetov, V.; Stebe, M.J.; Linder, M. Optimization and characterization of liposome formulation by mixture design. *Analyst* **2012**, *137*, 773–786. [CrossRef]
39. Laxmi, M.; Bhardwaj, A.; Mehta, S.; Mehta, A. Development and characterization of nanoemulsion as carrier for the enhancement of bioavailability of artemether. *Artif. Cells Nanomed. Biotechnol.* **2015**, *43*, 334–344. [CrossRef]
40. Buszello, K.; Harnisch, S.; Müller, R.H.; Müller, B.W. The influence of alkali fatty acids on the properties and the stability of parenteral O/W emulsions modified with solutol HS 15. *Eur. J. Pharm. Biopharm.* **2000**, *49*, 143–149. [CrossRef]
41. Mehmood, T.; Ahmed, A.; Ahmad, A.; Ahmad, M.S.; Sandhu, M.A. Optimization of mixed surfactants-based β-carotene nanoemulsions using response surface methodology: An ultrasonic homogenization approach. *Food Chem.* **2018**, *253*, 179–184. [CrossRef] [PubMed]
42. Soliman, W.E.; Shehata, T.M.; Mohamed, M.E.; Younis, N.S.; Elsewedy, H.S. Enhancement of Curcumin Anti-Inflammatory Effect via Formulation into Myrrh Oil-Based Nanoemulgel. *Polymers* **2021**, *13*, 577. [CrossRef]
43. Abdallah, M.H.; Abu Lila, A.S.; Unissa, R.; Elsewedy, H.S.; Elghamry, H.A.; Soliman, M.S. Preparation, characterization and evaluation of anti-inflammatory and anti-nociceptive effects of brucine-loaded nanoemulgel. *Colloids Surf. B Biointerfaces* **2021**, *205*, 111868. [CrossRef]
44. Matsumoto, Y.; Ma, S.; Tominaga, T.; Yokoyama, K.; Kitatani, K.; Horikawa, K.; Suzuki, K. Acute Effects of Transdermal Administration of Jojoba Oil on Lipid Metabolism in Mice. *Medicina* **2019**, *55*, 594. [CrossRef]
45. Ibrahim, M.M.; Shehata, T.M. The enhancement of transdermal permeability of water soluble drug by niosome-emulgel combination. *J. Drug Deliv. Sci. Technol.* **2012**, *22*, 353–359. [CrossRef]
46. Shehata, T.M.; Nair, A.B.; Al-Dhubiab, B.E.; Shah, J.; Jacob, S.; Alhaider, I.A.; Attimarad, M.; Elsewedy, H.S.; Ibrahim, M.M. Vesicular Emulgel Based System for Transdermal Delivery of Insulin: Factorial Design and in Vivo Evaluation. *Appl. Sci.* **2020**, *10*, 5341. [CrossRef]
47. Habashy, R.R.; Abdel-Naim, A.B.; Khalifa, A.E.; Al-Azizi, M.M. Anti-inflammatory effects of jojoba liquid wax in experimental models. *Pharmacol. Res.* **2005**, *51*, 95–105. [CrossRef]
48. Knudsen, N.; Jorgensen, L.; Hansen, J.; Vermehren, C.; Frokjaer, S.; Foged, C. Targeting of liposome-associated calcipotriol to the skin: Effect of liposomal membrane fluidity and skin barrier integrity. *Int. J. Pharm.* **2011**, *416*, 478–485. [CrossRef]
49. Ibrahim, T.; Abdallah, M.H.; Megrab, N.; Elnahas, H. Upgrading of dissolution and anti-hypertensive effect of Carvedilol via two combined approaches; self-emulsification and liquisolid techniques. *Drug Dev. Ind. Pharm.* **2018**, *44*, 873–885. [CrossRef]
50. Syed, M.A.; Veerabrahma, K. Biodegradable preparation, characterization and In vitro evaluation of stealth docetaxel lipid nanoemulsions for efficient cytotoxicity. *Int. J. Drug Deliv.* **2013**, *5*, 188–195.
51. Elsewedy, H.S.; Aldhubiab, B.E.; Mahdy, M.A.; Elnahas, H.M. Brucine PEGylated nanoemulsion: In vitro and in vivo evaluation. *Colloids Surf. A Physicochem. Eng. Asp.* **2021**, *608*, 125618. [CrossRef]
52. Shehata, T.M.; Khalil, H.E.; Elsewedy, H.S.; Soliman, W.E. Myrrh essential oil-based nanolipid formulation for enhancement of the antihyperlipidemic effect of atorvastatin. *J. Drug Deliv. Sci. Technol.* **2021**, *61*, 102277. [CrossRef]
53. Ibrahim, T.M.; Abdallah, M.H.; El-Megrab, N.A.; El-Nahas, H.M. Transdermal ethosomal gel nanocarriers; a promising strategy for enhancement of anti-hypertensive effect of carvedilol. *J. Liposome Res.* **2019**, *29*, 215–228. [CrossRef]
54. Abdallah, M.H.; Lila, A.S.A.; Anwer, M.K.; Khafagy, E.-S.; Mohammad, M.; Soliman, M.S. Formulation, development and evaluation of ibuprofen loaded nano-transferosomal gel for the treatment of psoriasis. *J. Pharm. Res. Int.* **2019**, 1–8. [CrossRef]
55. Ayoub, A.M.; Ibrahim, M.M.; Abdallah, M.H.; Mahdy, M.A. Sulpiride microemulsions as antipsychotic nasal drug delivery systems: In-vitro and pharmacodynamic study. *J. Drug Deliv. Sci. Technol.* **2016**, *36*, 10–22. [CrossRef]
56. Morsy, M.A.; Abdel-Latif, R.G.; Nair, A.B.; Venugopala, K.N.; Ahmed, A.F.; Elsewedy, H.S.; Shehata, T.M. Preparation and evaluation of atorvastatin-loaded nanoemulgel on wound-healing efficacy. *Pharmaceutics* **2019**, *11*, 609. [CrossRef]
57. Shah, J.; Nair, A.B.; Shah, H.; Jacob, S.; Shehata, T.M.; Morsy, M.A. Enhancement in antinociceptive and anti-inflammatory effects of tramadol by transdermal proniosome gel. *Asian J. Pharm. Sci.* **2019**, *15*, 786–796. [CrossRef]
58. Shehata, T.M.; Ibrahim, M.M.; Elsewedy, H.S. Curcumin Niosomes Prepared from Proniosomal Gels: In Vitro Skin Permeability, Kinetic and In Vivo Studies. *Polymers* **2021**, *13*, 791. [CrossRef] [PubMed]
59. Abdallah, M.H.; Abu Lila, A.S.; Unissa, R.; Elsewedy, H.S.; Elghamry, H.A.; Soliman, M.S. Brucine-Loaded Ethosomal Gel: Design, Optimization, and Anti-inflammatory Activity. *AAPS PharmSciTech* **2021**, *22*, 269. [CrossRef]
60. Chen, J.; Xiao, H.L.; Hu, R.R.; Hu, W.; Chen, Z.P.; Cai, H.; Liu, X.; Lu, T.L.; Fang, Y.; Cai, B.C. Pharmacokinetics of brucine after intravenous and oral administration to rats. *Fitoterapia* **2011**, *82*, 1302–1308. [CrossRef] [PubMed]
61. Sarigüllü Ozgüney, I.; Yeşim Karasulu, H.; Kantarci, G.; Sözer, S.; Güneri, T.; Ertan, G. Transdermal delivery of diclofenac sodium through rat skin from various formulations. *AAPS PharmSciTech* **2006**, *7*, 88. [CrossRef] [PubMed]

MDPI

St. Alban-Anlage 66

4052 Basel

Switzerland

www.mdpi.com

Gels Editorial Office

E-mail: gels@mdpi.com

www.mdpi.com/journal/gels